Manual of
MICROBIOLOGY
Tools & Techniques

Manual of
MICROBIOLOGY
Tools & Techniques

Kanika Sharma
Department of Botany
M. L. Sukhadia University, Udaipur,
Rajasthan, India

ANSHAN LTD

Ane Books India

First Published in 2007 by

ANSHAN LTD
6 Newlands Road
Tunbridge Wells
Kent.
TN4 9AT. UK
Tel: +44 (0) 1892 557767
Fax: +44 (0) 1892 530358
e-mail: info@anshan.co.uk
Website: www.anshan.co.uk

© Copyright 2007 Ane Books India

Published in arrangement with
Ane Books India

4821 Parwana Bhawan, 1st Floor
24 Ansari Road, Darya Ganj, New Delhi -110 002, India
Tel: 91 (011) 2327 6843-44, 2324 6385
Fax: 91 (011) 2327 6863
e-mail: anebooks@vsnl.com
Website: www.anebooks.com

ISBN-10: 1 904798 98 5
ISBN-13: 978 1 904798 98 9

British Library Cataloguing in Publication Data
A catalogue record for this book is available from the British Library

This edition not for sale in India, Pakistan, Nepal, Sri Lanka and Bangladesh

Printed at Gopsons Paper Ltd, Noida (U.P.), India

Dedicated to my parents for being a source of inspiration, to my in-laws for being a pillar of strength and to my husband Avinash and children Nidhi and Nikhil for their patience, moral support and unwavering help.

Foreword

This valuable book *Manual of Microbiology* is written in simple language and is easy to understand even for beginners. The author Dr. Kanika Sharma has fifteen years of experience teaching undergraduate, postgraduate and Ph.D. students. The author has covered all topics related to microbiology—general aspects like techniques and culture and identification of bacteria; bacterial genetics; water, soil and food microbiology; and study of viruses and fungi. She has ventured into the field of medical microbiology and briefly dealt with important aspects like sample collection and identification of common pathogenic bacteria.

This book is unique because a basic idea of the topic is given in brief followed by various laboratory methods presented in a lucid and systematic manner, keeping in mind problems faced by students and also stressing upon the do's and dont's while performing various experiments. Diagrams and flow charts help to make learning easier and more interesting. In the practical exercises provided in the last few chapters of this book, students are guided step by step to enable them to perform the exercises with confidence and ease.

I am sure this book will be of great help to both students, and teachers of microbiology. I wish the author all success.

Dr. (Mrs.) *Santha Dube* MBBS, MD.
Prof. & Head, Department of Microbiology
Darshan Dental College & Hospital, Udaipur, Rajasthan.

Former Prof. & Head, Department of Microbiology
R.N.T. Medical College, Udaipur, Rajasthan.

Preface

Microbiology has today come forth as one of the most sought-after subjects in the science stream of graduate courses. Its dynamism is perhaps linked to the prospects and career options that the subject has to offer. Microbes have a very significant role in this era of biotechnology, and hence it is essential that students of all branches of biological sciences develop a thorough understanding of this subject and learn the basics of handling microbes in the laboratory.

In this book, I have tried to simplify practical microbiology based on my experiences in the laboratory. From basic concepts to techniques and tricks, all relevant information has been made available in an organized and informative manner keeping the language as simple as possible. While attempting to put all the salient information in one document, I have also kept in mind that while consulting the book in the laboratory, students need not turn pages too often or consult the index as each chapter is a complete unit in itself.

A most frequently faced problem by students of microbiology is the absence of supporting theoretical information relevant to the practical exercises. This book attempts at providing precise and basic information about microorganisms, the historical landmarks in the field of microbiology, classical as well as modern grouping of microbes, and various tools and techniques used to study microbes. Most of the experiments are such that they can be easily performed in any laboratory having basic facilities. They have been kept under suitable chapter heads so as to make it easier for the student to understand their relevance.

I sincerely hope that this book will act as a step-by-step guide for students of microbiology.

Kanika Sharma

ACKNOWLEDGEMENTS

I wish to acknowledge the help and support given to me by my students Ms. Gunmala, Ms. Tripta, Ms. Rekha, Ms. Aarti and Mr. Yogesh Kumar in preparing this manuscript. I would also like to express my sincere gratitude to my family and friends for their constant encouragement and to my publishers for offering to bring out this book.

Kanika Sharma

CONTENTS

1

INTRODUCTION TO MICROBIOLOGY

Introduction

Fossils of single-celled creatures that date back to at least 3.5 billion years indicate that microbes were the pioneer life forms on earth (humans came into existence about 2 million years ago). In 1995, scientists revived a 30-million-year-old bacterium that had lived in the gut of an ancient bee. The bee with its microbe had got trapped in a drop of tree sap that later became amber and was thus preserved. The unique genetic pattern of this ancient microbe is similar to *Bacillus sphaericus*, a bacterium found in modern bees. *B. sphaericus* assists the digestive process of bees and also produces an antibiotic which protects them against diseases. The natural antibiotic produced by this ancient bacterium will be investigated for possible medical applications. In the year 2000, scientists announced that they had revived 250-million-year-old bacteria that had lain in suspended animation encased in salt crystals buried 1,850 feet deep in the earth in southwest United States. Based on the presence of fossils of *Bacillus* like tiny bacteria in a rock that formed on Mars about 4.5 billion years ago, it has been suggested that bacteria like organisms may have even inhabited mars. The rock crash-landed on Earth as a meteorite thousands of years ago.

Microbes are found everywhere. They can be found in air, water, soil and inside plants, animals and human beings. They can live in extreme conditions such as volcanic vents (350°C) to frigid waters of Antarctic lakes, atop high mountain summits and in polar snow. ***Thermus aquaticus*** is a thermophilic bacterium which can withstand high temperatures and from which **Taq polymerase** enzyme was isolated. ***Deinococcus radiodurans*** ("Strange berry that withstands radiation") can withstand radiations up to 1,500,000 rads (500 to 1,000 rads is lethal).

SIGNIFICANCE OF MICROBES

We derive great benefits from microbes especially bacteria. Microbial products include

Foods: Breads, cheeses, yogurt, mushrooms, wine, beer, soy sauce, sake etc.

Food additives: Amino acids, thickening agents, vitamins.

Solvents: Butanol, methane, hydrogen.

Biofuels: Ethanol, methane, hydrogen.

Pharmaceuticals: Antibiotics, vitamins, probiotics, biopesticides, etc.

Laboratory and Diagnostic reagents: Enzymes, biochemicals, proteins, etc.

Biogeochemical cycling of various elements is possible due to microbial activity. Bacteria are the main components of detritus food chain. Without microbes, earth would have been converted into a monstrously large garbage can. Anaerobes transform waste into biogas and fuels such as methane gas and ethanol. Nitrate is reduced to harmless nitrogen gas by enzymes produced by microbes. Nitrate-reducing enzymes can be electronically coupled to a gold electrode to act as a biosensor for this pollutant. Methane-munching bacteria, or methanotrophs, make an enzyme that can break down more than 250 nasty pollutants into harmless molecules. **Pentichloriphenol** (fungicidal wood preservative) linked to cancer can be successfully biodegraded for the first time with a new microbe called **Ahring's microbe**.

Highly specialised microorganisms capable of growth in extreme environments can be used to extract gold and base metals from low-grade sulphide ores **(Bioprospecting)**. High temperature Archaea extracted from geothermally heated source material are capable of mediating the release of copper from copper sulphide minerals during leaching. E.g. *Thiobacillus ferrooxidans* catalyses oxidation of metal sulphides, particularly copper sulphate so that they become soluble in water. *T. ferrooxidans* is also used in the organic leaching of gold and uranium with the help of other bacteria such as *T. thiooxidans* and *Leptospirillum ferrooxidans*.

Several **transformed bacteria** are now being used for various purposes. Genes that control bioluminescence have been taken from a marine bacterium and fused into *Pseudomonas fluorescens* to form a new bacterium *Pseudomonas fluorenscens* which can break down napthalene, anthracene, phenanthrene and other toxic chemicals. The bacterium glows as it breaks down the toxic hydrocarbons. This gives environmental cleanup workers a visual clue that the toxic waste treatment is effective.

Microbes make **enzymes** that are used to make soy sauce, soda, beer, wine, cheese, infant formula, chewing gum, leather goods, paper, laundry detergent, the stone-washed look on blue jeans, etc. **Human drugs** such as insulin for diabetics, growth hormone for individuals with pituitary dwarfism and tissue plasminogen activator for heart attack victims, as well as animal drugs like the growth **hormones**, bovine or porcine somatotropin, are being produced by the fermentation of transgenic bacteria in fermenters. *Listeria monocytogenes*, a bacterium found in spoiled cheese, has been genetically engineered to produce one of the HIV virus protein products. Researchers believe this may prove to be a safe and effective way to help build immunity to HIV without exposing patients to the HIV virus itself.

Microbes are harmful also. They are responsible for **food spoilage, diseases** of living organisms, etc. Due to this property microbes are also being used as weapons of **bioterrorism**. The most commonly used pathogens are causal organisms of anthrax, botulism, plague, smallpox, haemorrhagic fevers (Ebola), etc.

DEVELOPMENT OF MICROBIOLOGY

Birth of Microbiology

'Organisms which are not visible to the naked eye are known as microbes. The groups included in this category are Bacteria, Archaebacteria, Cyanobacteria, Fungi, Actinomycetes, Protozoa, Mycoplasma and Virus. These subgroups are not closely related but are linked to each other only due to their size. The branch dealing with the study of microorganisms is called as **Microbiology**. It has several sub-branches, the main branches being bacteriology and virology.

Fig.1.1 Anton Von Leewenhoek

Existence of microbes and their effects was known since ancient times. Roman philosopher **Lucretius** (98–55 BC) and physician **Girolamo Fracastoro** (1478–1553) were the first to suggest that invisible creatures caused disease. Between 1625 and 1630 **Francesco Stelluti** used a microscope to study bees and weevils, but the start of microbiology as a field of science can be attributed to **Anton Von Leeuwenhoek** (1632–1723), a Dutch draper from Delft, Holland (Fig. 1.1).

Leeuwenhoek is known as **Father of Microbiology** because he pioneered the start of this branch of science in 1676. His hobby was to grind lenses, build microscopes and examine animalcules from saliva, dental plaque, sewage, pond water, sperms, etc. Leeuwenhoek's microscope was an extremely simple 3-4 inches long device (Fig. 1.2). The body of the microscope was made up of a brass plate which had a single lens mounted in a tiny hole. The specimens were placed on the sample holder and adjusted so they lay in front of the tiny lens. A beam of sunlight passed through the specimen and the viewer observed the illuminated material through the tiny lens. The position and focus could be adjusted by turning two screws. In order to observe the sample, instrument needed to be held very close to the eye and adequate lighting as well as patience were essential prerequisites.He observed bacteria, protozoa, fungi, plant parts, minute seeds, etc., made accurate drawings of these and described them in his letters to Royal Society of London. His lenses gave magnifications up to 300 x and were free of distortion. Since his microscopes were difficult to focus, instead of changing the specimen he used to change the microscope. He made 419 microscopes.

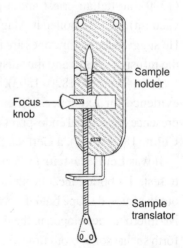

Fig.1.2 Leeuwenhoek's microscope

Spontaneous Generation v/s Germ Theory

Earlier people believed in the "**Theory of Spontaneous Generation**." It was **Francisco Redi** (1688) who first showed that maggots developed only in uncovered meat that flies could reach to lay eggs on and thus paved the way for disproving the theory of Spontaneous Generation (Fig. 1.3).

Although Redi had given evidence for disproving the theory of spontaneous generation, it still had its followers. One of them was **John Needham** (1745). He put boiled chicken broth in a flask and sealed it, still maggots appeared in the broth so he supported the theory of spontaneous generation (**Reason**: microbes grew because the flask was not properly sterilized).

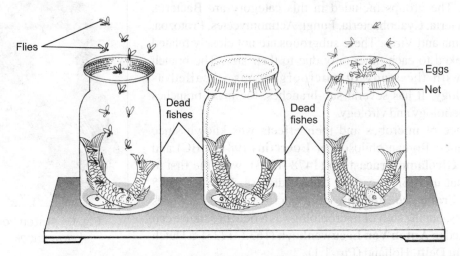

Flies

Dead fishes

Dead fishes

Eggs

Net

Fig. 1.3 Redi's experiment: A; maggots appear as jar is open. B & C; Maggots do not appear as jar is covered.

The first concrete evidence for rejecting this theory came from the work of **Lazarro Spallanzani** (1799), an Italian priest and a professor who first put broth in a flask, sealed it (thus creating a vacuum) and then boiled it. Maggots did not appear in the cooled broth till the flask remained sealed. He suggested that microbes are present in air and outside air is needed for the growth of microbes in the infusion (broth) and thus disproved spontaneous generation.

John Tyndall (1820–1893), an English physicist proved that soil carried germs. He also gave evidence for existence of heat resistant forms of bacteria. The presence of bacterial endospores in soil was also shown by **Ferdinand Cohn** (1828–1898), a German botanist.

It was **Louis Pasteur** (1859) (Fig. 1.4), who put this controversy to rest. He boiled meat broth in a flask, heated the neck and drew it out in a curved shape but left the end of the neck open to atmosphere. No microbes developed in the flask. Then he tilted the flask back and forth so that some broth flowed into the curved neck and then returned to the base of the flask. Once again maggots appeared in the broth bold (Fig. 1.5). (**Reason**: Due to gravity, microbes that had entered the flask with air and dust had settled in the neck. By tilting the flask they were washed into the broth, which explains their reappearance).

Fig. 1.4 Louis Pasteur

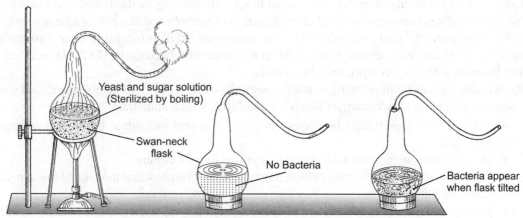

Fig.1.5 Swan neck Experiment of Pasteur

Labels in figure:
Yeast and sugar solution (Sterilized by boiling)
Swan-neck flask
No Bacteria
Bacteria appear when flask tilted

Landmarks of Microbiology

Pasteur's experiments disproved spontaneous generation and proved that microbes are present everywhere. He realized that growth of microbes causes decomposition of dead plant and animal tissue and is the reason for food spoilage. He developed the technique of **pasteurization** for preventing wine from spoiling. Pasteur also contributed to the development of vaccines. He developed vaccines for anthrax and rabies. He also gave the concept that **attenuated** bacteria can produce immunity. **Attenuated** means weakened virus or bacteria that is unable to cause the disease. Later it was discovered that killed microbes could also produce immunity.

The theory stating that microbes are responsible for disease got support from the work of **Agostino Bassi** (1835) and **M J Berkley** (1845), who respectively proved that silkworm disease and blight of potato which was responsiblie for the great Irish famine were caused by fungal infection.

Ignaz Semmelweiss (1857) was the first person to point out that the disease **Childbed fever** (*Puerperal sepsis*) in which healthy infants did not survive, was caused due to transmission of the causal agent *Streptococcus pyogenes* from the doctor to the patient. He realized that medical students first performed autopsy in the morgue and then went straight to the theatre for delivering children. He asked them to wash their hands with chlorine before entering the theatre. This reduced the mortality rate. Hence, we can say that he laid the foundations for germ theory of disease as well as the concept of asepsis.

Robert Koch (1876) (Fig. 1.6) finally gave "the Germ Theory of Disease" which states that, **"microbes (germs) cause disease and specific microbes cause specific diseases"**. Koch was the first scientist to identify *Bacillus anthracis* as the pathogen causing anthrax in cattle. He observed that the same microbes were present in all blood samples of infected animals. He isolated and cultivated these microbes. He then injected a healthy animal with the cultured bacteria. The animal became infected with anthrax and its blood sample showed the same microbes as the originally infected animals.

Fig. 1.6 Robert Koch

Koch utilised the ability of pathogens to react to specific staining methods and also made use of photomicrography, an objective method of examination of microbes unlike earlier assessment using only the human eye to study microbes. He also developed media containing meat extracts and protein digests to culture bacteria. Koch isolated the tubercular bacteria in 1882. He received the **Nobel Prize in 1905** for his work on tuberculosis.

Koch is also famous for developing scientific methods or rules (1884) for identifying causal agents of diseases, which are also known as **Koch's Postulates,** which state that:

- The causative agent must be present in every diseased individual but absent in healthy individual.
- The causative agent must be isolated and grown in pure culture.
- The isolated organism in pure culture must cause the disease when inoculated into a healthy experimental host.
- The same causative agent must be re-isolated from the experimental host and re-identified in pure culture.

Koch initially used cut surfaces of potatoes for culturing bacteria. He then devised a medium containing meat extract and protein digests. Next he used gelatin to solidify the liquid medium. But gelatin melted at room temperature and it could also be digested by bacteria. This problem was solved when **Fannie Eilshemius Hesse**, wife of Walter Hesse who was an assistant of Koch, **suggested use of agar instead of gelatin** to solidify the nutrient medium used to culture bacteria. Another assistant **Robert Petri** designed the containers for solid medium, now known as **petri dish**. Thus Koch and his assistants are responsible for laying the foundation of present day microbiology.

Fig. 1.7 Edward Jenner

The next giant step in this field was the discovery of vaccines and development of the concept of antisepsis. **Edward Jenner** (1796) (Fig. 1.7) is credited with the discovery of vaccines. Jenner observed that dairymaids seemed to become immune to smallpox, if they had a prior mild infection of cowpox. He inoculated a boy first with fluid from a cowpox blister and then with fluid from a smallpox blister. The boy contracted cowpox but did not contract smallpox. **The term vaccination thus came from *Vacca* for cow and was coined by Pasteur in honour of Edward Jenner**. He first used attenuated cultures of bacteria to introduce resistance in chicken against cholera. He also made vaccines for anthrax and rabies. The first person to be vaccinated against rabies was a thirteen-year-old boy, **Joseph Meister**.

Joseph Lister (1865) first developed antiseptic techniques. He used phenol (carbolic acid) soaked bandages to dress the wounds of a young boy named **James Greenless** and called this technique **antisepsis** meaning against infection.

Paul Ehrlich (1908) gave the principle of chemotherapy or selective toxicity (the drug must be toxic to the infecting microbe, but relatively harmless to the host's cells). He is known as the **Father of chemotherapy** for this reason. He also discovered a drug for treatment of syphilis.

I G Farben, a German company, made sulfa drugs in the 1930s, which were the first major class of antimicrobial drugs to come into widespread clinical use.

Alexander Flemming (1929) (Fig. 1.8) discovered penicillin, the first antibiotic (antibacterial compounds produced by fungi and bacteria). Later **Charles Chamberland** (1884) developed a porcelain filter for filtering bacteria, which also made the discovery of virus possible.

Griffith (1928) demonstrated the presence of genetic recombination in bacteria. He showed that exchange of genetic material takes place by transformation. **Avery, McCleod and McCarty** (1944) proved that DNA is responsible for transformation. **Lederburg and Tatum** (1946) discovered conjugation in bacteria. **Zinder and Lederburg** gave evidence for genetic recombination in bacteria by the process of transduction. **Hershey and Chase** (1952) showed that phage DNA is injected into host cells during transduction.

Fig. 1.8 Sir Alexander Fleming

TYPES OF CELLS

Living organisms are made up of cells. **Robert Hooke,** a curator at Royal Society of London was the first person to describe cells (*Cellulae*; Gk: small rooms)on the basis of his study of cork slices. His discovery led to the formulation of **Cell theory** by **Theodore Schwann** and **Mathias Schlieden** (1838-1839). This theory states that:

- All living organisms are made up of cells.
- Cell is the basic unit of structure for all living organisms.
- All cells arise from preexisting cells (*Omnis cellula e cellulae* : Rudolf Virchow 1855).

There are mainly two types of cells (viruses, prions and viroids are acellular – without a cell).

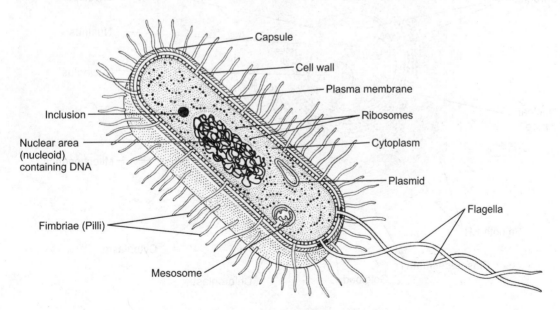

Fig. 1.9 Prokaryotic cell

1. **Prokaryotic** (**Before nucleus**): cells in which internal membrane bound structures are absent (no membrane-bound nucleus or membrane-bound organelles). Organisms of Kingdom Monera are prokaryotic. Prokaryotes belong to Kingdom Monera, which includes Archaea, Bacteria, Cyanobacteria, Mycoplasmas, Rickettsia, Chlamydia, Spiroplasma etc. All these organisms are microscopic, unicellular and have a prokaryotic cell structure. They resemble each other closely with minor differences, hence, usually bacteria are taken as a representative member of this group of organisms and by and large most characters exhibited by bacteria are seen in other prokaryotes also. Bacteria are cosmopolitan and ubiquitous. They do not have a true membrane bound, well defined nucleus and lack membrane bound cell organelles. Their DNA exists as a single circular double helix. The ribosomes are 70S type and they exhibit relatively very little morphological differentiation (Fig. 1.9).

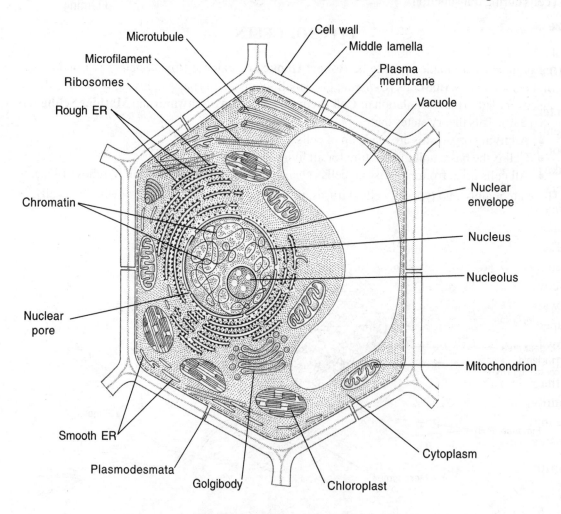

Fig. 1.10 Eukaryotic cell (Plant cell)

2. **Eukaryotic (True nucleus);** cells in which internal membrane bound structures (membrane bound nucleus and membrane-bound organelles) are present. Such a cell organization is found in organisms of the remaining four kingdoms. Cell structure of protists, algae, fungi, animals and plants is Eukaryotic. Most of these organisms are macroscopic and multicellular except protozoans and zoospores, which are motile and unicellular. Each organism has its own special habitat, nutrition and reproduction methods. They have unit membrane bound cell organelles as well as a double membrane bound true nucleus in which DNA is arranged in form of chromosomes (Fig 1.10).

GENERAL CHARACTERISTICS OF BACTERIA

Size of Bacteria

Prokaryotes are extremely small therefore the unit of measurement is microns (μ). Their size is 1/1000 the volume of a typical eukaryotic cell. Due to this they have high surface to volume ratio and a very high rate of multiplication. *Haemophilus influenzae* **was once thought to be the smallest bacteria** (0.2–0.3 by 0.5–2.0μ). In 1998 even smaller microbes have been discovered by Australian scientists in ancient sand stone dredged up from an oil drilling site about three miles below the ocean floor. These individual threads measure only about 20 to 150 nm in length. At the largest, that would make them no bigger than little viruses. They have been nicknamed "**Nanobes".** The most popular bacteriam E.coli is about 7 μm long and 1.8 μm in diameter. Some of the larger species are *Schuadinnum butschlii,* which is about 50 to 60 μm long and 0.75 μm thick. A typical very large rounded bacterium *Achromatium oxaliferum* is about 100 to 45 μm.

Earlier *Beggiatoa* was supposed to be the largest bacterium but now two new bacteria have been discovered which are large enough to be seen with naked eyes. *Epulopiscium fischelsoni* discovered in 1993 in gut of Surgeon fish which live in and around coral reefs in Australia, Hawaii and other island environments. It is a part of *Clostridium* group of anaerobes and is rod shaped. It can grow upto 600 micrometers in length and 80 micrometers in width. This bacterium shows vivipary. Two small bacteria are formed at one end of the mother cell and after the cell wall synthesis is complete for each cell, these are released through a slit that opens at the end of the mother cell. Another giant bacterium ***Thiomargarita namibiensis* (Sulphur Pearl of Namibia)** has been found in the ocean floor off the coast of S. Africa. It can grow upto almost one mm or 1/25th of an inch in diameter. It is closely related to sulphide eating nitrate respiring cousins *Thioplaca* and *Beggiatoa.*

Composition of a Bacterial Cell

70% of bacterial cell is made up of water. Macromolecules make up 26% which is 90% of dry wt. The most abundant molecules other than water are proteins, which constitute 50% of dry wt. Small molecules and ions make up 4% i.e. 10% of dry wt.

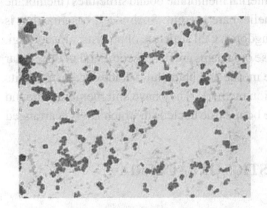

Staphylococcus

Spirochaetes

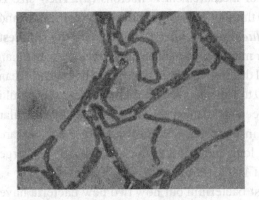

Streptobacilli

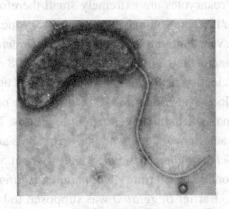

Monotrichous Bacillus

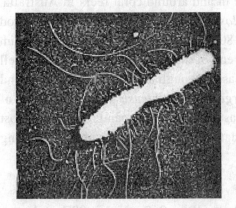

Peritrichous flagella

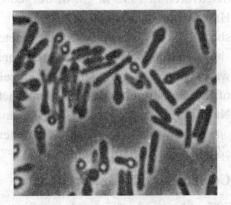

Bacilli with endospores

Photo plate I

Shape and arrangement of Bacterial Cells

Bacteria have characteristic shape. Their microscopic appearance is very important in classification and diagnosis. They commonly occur in three basic shapes (Fig 1.11). These shapes are:

- **Spiral:** Helically coiled cells e.g. *Treponema pallidum*. (cells with half a turn appear comma-shaped e .g *Vibrio cholerae* and cells with a complete turn are called **Spirillum.** Flexible spirilli are known as **Spirochaetes**.(Photo plate I).
- **Cocci:** Round or ellipsoidal cells e.g. *Staphylococcus* (Photo plate I), *Streptococcus*.
- **Bacillus:** Rod shaped cells e.g. *Clostridium*.
- **Actinomycetes** are made up of long, filamentous branched cells e.g. *Actinomyces, Streptomycs*. They are also called branching rods.

Along with these basic shapes some other unusual shapes also occur, such as *Pasteuria* which is pear shaped, *Sulfobolus* which is a lobed sphere, *Bacillus anthracis* is rectangular in shape and disc shaped cells of *Caryophanon* are arranged like a stack of coins. *Silebria* are rods with sculptured surface. Cells of club shaped rods of *Corynebacterium* are arranged in form of palisade tissue seen in plant leaf mesophyll.

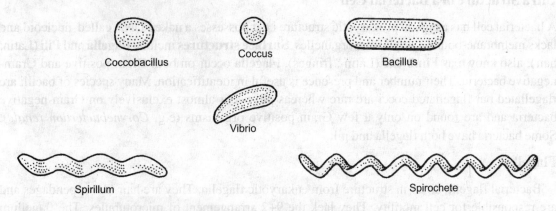

Fig. 1.11 Shapes of Bacteria

The **arrangement of cells** is also typical of various species or groups of bacteria. This arrangement is dependent on plane of division and separation of dividing cells (Fig. 1.12). When two dividing cell remain attached after division to form pairs the arrangement is termed **Diplo** with a suffix indicating the basic shape e.g *Diplococcus, Diplobacillus, Neisseria*. Some rods or cocci keep dividing in the same plane without separating. Such arrangement results in formation of characteristic chains which are called **Strepto** with a suffix e.g. *Streptococcus, Streptobacillus*. When cells divide in an irregular manner they form grapelike clusters designated as **Staphylo** with a suffix e.g. *Staphylococcus aureus*. Some round cocci form cubic packets e.g *Sarcina*. In trichome like arrangement rods get joined end to end just like in strepto arrangement but the area of contact between adjacent cells is much more.

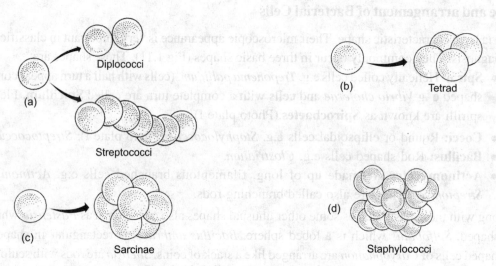

Fig. 1.12 Cell arrangements of Cocci

Ultra Structure of a Bacterial Cell

A bacterial cell has a typical prokaryotic structure i.e it possesses a naked DNA called nucleoid and lacks membrane-bound cytoplasmic organelles. **Surface structures** include **flagella** and **Pili** (Latin: hair), also known as **Fimbriae** (Latin : fringes). Flagella occur on both Gram-positive and Gram-negative bacteria. Their number and presence is useful in identification. Many species of bacilli are flagellated but flagellated cocci are rare whereas pili occur almost exclusively on Gram-negative bacteria and are found on only a few Gram-positive organisms (e.g. *Corynebacterium renale*). Some bacteria have both flagella and pili.

Flagella

Bacterial flagella differ in structure from eukaryotic flagella. They are hair like appendages and are responsible for cell motility. They lack the 9+2 arrangement of microtubules. The flagellum moves clockwise or anticlockwise about its long axis. Number and arrangement of flagella on the cell is diagnostically useful. They may be polar or lateral, 3 to 12 μm long, about 12 to 30 nm in diameter and made up of a protein called **flagellin**. Flagellins are immunogenic and constitute a group of protein antigens called the H antigens, which are characteristic of a given species, strain, or variant of an organism. The protein subunits of a flagellum are assembled to form a cylindrical structure with a hollow core.

A flagellum is made up of three parts:

1. **Long filament**, which lies external to the cell surface. It is a tapering hair like structure, 15mm long and 0.01 to 0.02mm wide. It is composed of eleven subunits arranged around a central hollow core

2. **Hook** connects the filament and basal body.

3. **Basal body** to which the hook is anchored imparts motion to the flagellum.

Basal body is a set of rings embedded in the cell envelope divided into two sets; **Proximalset** made up of two innermost rings, the **M** or **Membrane ring** and **S or Supermembrane ring** located in the plasma membrane and a second; **Distantset** made up of two outermost rings, the **Peptidoglycan** or **'P' ring** and **Lipopolysaccharide ring** or **'L-' ring**, which are located in the periplasm and the outer membrane respectively (Fig. 1.13).

Basal ring comprises the motor apparatus responsible for the rotary movement of flagellar filament. The rings function as bushes and support the rod where it is joined to the hook of the filament on the cell surface. As opposed to eukaryotic flagella, which are powered by hydrolysis of ATP, the energy required to move the flagellum comes from **proton motive force** or the **chemiosmotic potential** established across the bacterial membrane due to movement of protons.

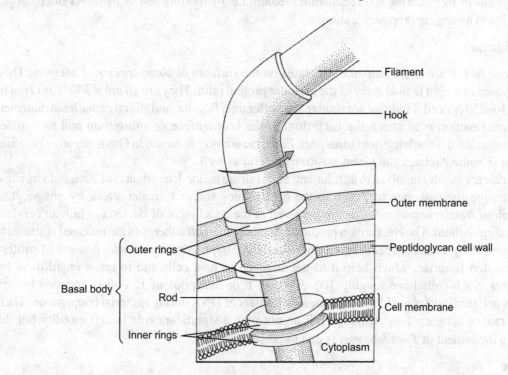

Fig. 1.13 Structure of flagellum showing basal body, hook and filament

Based on the arrangement of flagella bacterial cells are classified into following types:

Atrichous: Flagella absent e.g. *Diplococcus*.

Monotrichous: Single polar flagellum e.g. *Vibrio cholerae* (Fig. 1.6, photo plate I).

Cephalotrichous or Lophotrichous: Tuft of flagella at only one end e.g. *Bartonella bacilliformis* (Fig. 1.14).

Amphitrichous: A tuft of flagella is present on both ends .g. *Spirillum serpens* (Fig. 1.14).

Peritrichous: Flagella are present all over the boy surface e.g. *Proteus vulgaris* and *E. coli* (Fig. 1.14).

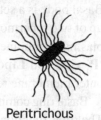

Monotrichous   Lophotrichous   Amphitrichous   Peritrichous

Fig. 1.14 Flagellar arrangements on bacterial cells

Detection of motility: Bacterial Motility can be detected by either staining the bacteria with specific stains or by culturing it on semisolid medium i.e. by motility test or by direct microscopic observation in a hanging drop preparation.

Pili (Fimbriae)

Fimbriae or Pili are short, hair-like structures on the surfaces of some species of bacteria. They are composed of a tight helical array of the globular protein pilin. They are 10 nm × 300–1,000 nm in size and 10–250 per cell. Fimbriae are shorter and stiffer than flagella, and slightly smaller in diameter. Their main function is to attach the bacterium to the host surface or substratum and to transfer genetic material from one bacterium to another. Fimbriae are very common in Gram-negative bacteria, but occur in some Archaea and Gram-positive bacteria as well.

Fimbriae are usually involved in attachment. Such pili are major determinants of bacterial virulence as they enable the pathogen to attach to the host tissues and/or to resist attack by phagocytes. Type I pili of *Neisseria gonorrhoeae* are involved in the attachment of bacteria to human cervical or urethral epithelium. The enterotoxin producing strains of *E. coli* adheres to the mucosal epithelium of the intestine by means of specific fimbriae. Similarly *Streptococcus pyogenes* possess M-protein and associated fimbriae, which help it to get attached to host cells and to resist engulfment by phagocytes. Such cells have roughly 100–300 pili. F or Sex pilus of *E. coli* is encoded by sex plasmids or F (fertility) plasmid and mediates the transfer of DNA during bacterial conjugation. Their number ranges between 1–4. Although pili do not play a significant role in cell motility but the twitching movement of *Pseudomonas* is due to presence of pili.

Capsule

Some bacteria are surrounded by a thick outer layer of high-molecular-weight, viscous polysaccharide gel called capsule. Capsule is mainly made up of polysaccharides. In some species capsule also contains amino sugars or peptides. The main function of capsule is adherence and protection. It also helps in attachment of cells to substratum, protects cells from engulfment by other cells or organisms such as protozoans and white blood cells (phagocytes) and antimicrobial agents. During adverse conditions capsule also protects cells from drying. Capsular materials (e.g. dextrans) also act as food reserves.

A true capsule is a discrete detectable layer of polysaccharides deposited outside the cell wall. A less discrete structure or matrix in which several cells are embedded is a called a **slime layer** or a **biofilm**. Some aquatic bacteria form chains or trichomes, which are enclosed in a hollow tube called

sheath. In addition to this some cells possess semi rigid expansions of cell wall or membrane called **prosthecae**.

Capsules can be **homopolymeric** e.g. *Streptococcus* which has a glucan capsule or **heteropolymeric** e.g. *Klebsiella pnuemoniae, Bacillus anthracis* in which capsule also contains a **polypeptide** along with polysaccharides.

Cell wall

All eubacterial cells are enclosed within a rigid cell wall of peptidoglycan. Peptidoglycan is a polymer of disaccharides (a glycan) cross-linked by short chains of amino acids (peptides). Peptidoglycan structure and arrangement in *E. coli* is representative of all *Enterobacteriaceae*, and many other Gram-negative bacteria, as well. The glycan backbone is made up of alternating molecules of **N-acetylglucosamine** (NAG) and **N-acetylmuramic acid** (NAM) connected by β**1, 4-glycosidic bond**. The 3-carbon of N-acetylmuramic acid is substituted with a lactyl ether group derived from pyruvate. The lactyl ether connects the glycan backbone to a peptide side chain that contains L-alanine, (L-ala), D-glutamate (D-glu), Diaminopimelic acid (DAP), and D-alanine (D-ala). All **Bacterial** peptidoglycans contain **N-acetylmuramic acid**, which is the definitive component of **murein**. D-glu, DAP and D-ala also are unique to bacterial cell walls (Fig. 1.15a, b 1.16).

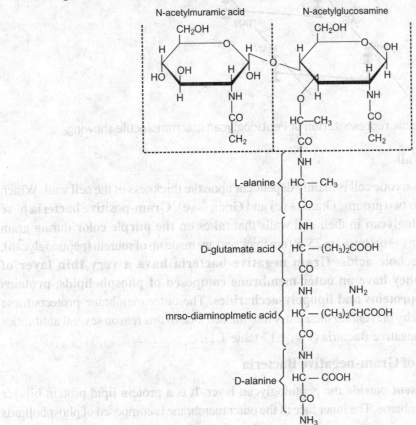

Fig. 1.15a Chemical structure of peptidoglycan

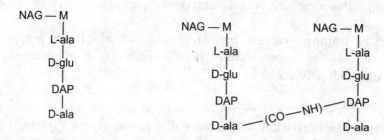

Fig. 1.15b Tetra peptide side chain and cross linkage

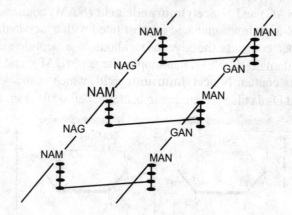

Fig. 1.16 Graphic representation of Peptidoglycan macromolecule showing

Gram Negatine Cell Wall

Gram staining of a prokaryotic cell is mainly dependent upon the thickness of the cell wall. Which distinguis has bacteria into two groups: Gram (+ve) and Gram (–ve). **Gram-positive bacteria** have a **thicker layer of peptidoglycan** in their cell walls that **takes on the purple color** during gram staining. Cell walls of Gram (+) bacteria are 25-50 nm thick and are made up of murein (peptidoglycan), techoic acids and lipotechoic acids. **Gram negative bacteria have a very thin layer of peptidoglycan, instead they have an outer membrane composed of phospholipids, proteins and contain special lipoproteins and lipopolysaccharides.** The outer membrane protects these bacteria against antibiotics by preventing their entry into the cell Due to this reason several antibiotics have no effect on Gram-negative Bacteria (Fig. 1.17 table 1.1).

The Outer Membrane of Gram-negative Bacteria

Outer membrane is present outside the peptidoglycan layer. It is a protein lipid protein bilayer resembling the plasma membrane. The inner face of the outer membrane is composed of phospholipids similar to the phosphoglycerides that compose the plasma membrane as well as special proteins called lipoproteins, which usually traverse the membrane and help in anchoring the outer membrane to the peptidoglycan layer. The outer face of the outer membrane contains a lipopolysaccharide (LPS) composed of a hydrophobic region, called **Lipid A**, that is attached to a hydrophilic linear polysaccharide region, consisting of the **core polysaccharide** and **the O-specific polysaccharide**. Bacterial lipopolysaccharides are toxic to animals and Lipid A is the toxic part.

The O-specific polysaccharide or O-antigen side chain is responsible for bacterial attachment, resistance to phagocytosis, antigenic properties of the cell and virulence of Gram-negative bacteria.

A group of trimeric proteins called **porins** form pores of a fixed diameter through the lipid bilayer of the membrane and allow passage of hydrophilic molecules of about 750 daltons. Porins allow passage nutrients but exclude passage of harmful substances from the environment (Fig 1.18).

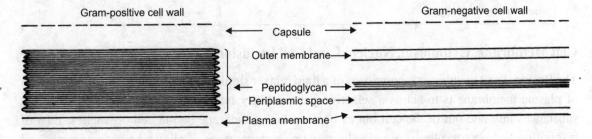

Fig. 1.17 Difference between Gram positive and negative cell wall

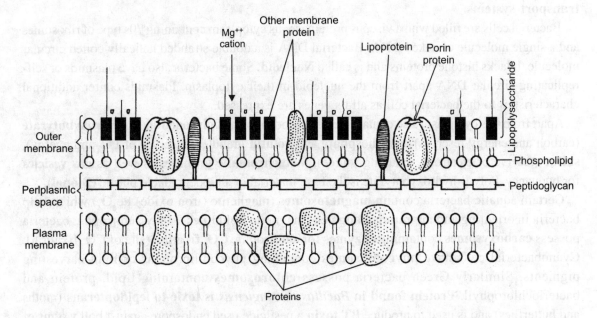

Fig. 1.18 Structure of Gram negative cell wall

Table. 1.1 Difference Betweeen Gram δ positive and Gram δ negative cell

Characteristic	Gram-positive cell	Gram-negative cell
Wall Thickness	thick (20-80 nm)	thin (10 nm)
Layers	1	2
Peptidoglycan (murein) content	>50%	10-20%
Teichoic acids	present	absent
Lipid and lipoprotein content	0-3%	58%
Protein content	0	9%
Lipopolysaccharide content	0	13%
Sensitivity to Penicillin G	yes	no
Sensitivity to lysozyme	yes	no

Cell Membrane, Cytoplasm, Nucleoid and Cell Inclusions

The Bacterial cell membrane is composed of a lipid bilayer similar to Eukaryotes. The main function of plasma membrane is to act as a **selective permeability barrier** that regulates the passage of substances into and out of the cell but as bacteria lack membrane bound cell organelles, ETS of respiration and photosynthesis are also located on the membrane. It allows passage of water and uncharged molecules of about 100 daltons, but does not allow passage of larger molecules or any charged substances. These substances enter the cell by special membrane **transport proteins** and **transport systems**.

Bacterial cells are filled with a viscous proteinaceous cytoplasm containing 70s type of ribosomes and a single molecule of naked DNA. Bacterial DNA is a double stranded helically coiled circular molecule. It lacks histone proteins and is called **Nucleoid.** Some bacteria also have plasmids or self-replicating circular DNA apart from the nucleoid in their cytoplasm. Plasmids confer additional characteristics to the bacterial cell as all its genes are expressed.

Apart from this cytoplasm also contains reserve food in form of **glycogen** and **β-hydroxybutyrate** (carbon and energy reserves) and **phosphate** and **volutin** granules (reserves of PO_4 and energy), **sulfur** (reserve of electrons, reducing source), **metachromatic granules** as well as **gas vesicles** for buoyancy. Poly-β-hydroxybutyrate is the base material of a biodegradable plastic (**Biopol**).

Certain aquatic bacteria contain **magnetosomes** (magnetite (iron oxide) Fe_3O_4) which help bacteria in orienting and migrating along geo- magnetic field lines. Many autotrophic bacteria possess **carboxysomes** containing enzymes for autotrophic CO_2 fixation (site of CO_2 fixation). Cyanobacterial cells have **phycobilisomes** containing phycobiliproteins or light-harvesting pigments. Similarly Green bacteria possess **chlorosomes** containing lipid, protein and bacteriochlorophyll. **Protein found in *Bacillus thuringiensis* is toxic to lepidopterans** (moths and butterflies) and is used to produce **BT toxin** a pesticide used endospore against boll worms of cotton.

Reproduction, genetic recombination and metabolism

Bacteria reproduce mainly by vegetative and asexual methods. Most bacteria reproduce vegetatively by a process called **binary fission**. Al though sexual reproduction is absent but genetic recombination takes place ane exchange of genetic material is brought about by either **conjugation** or **transformation** or **transduction.** Few bacteria produce endospores e.g. *Bacillus.*

Endospores

Endospores are highly heat-resistant, dehydrated resting cells formed intracellularly in members of the genera *Bacillus* and *Clostridium*. Endospores appear as round, highly refractile cells within the vegetative cell (Plate I). The spore is made up of spore protoplast, or core containing the DNA, ribosomes, and energy generating components surrounded by two layers of spore membrane. The space between the two membranes is filled with calcium dipicolinate. The outer membrane is enclosed by peptidoglycan spore wall followed by thick cortex that contains an unusual type of peptidoglycan. The entire structure is encased by a spore coat of keratin like protein (the exosporium). During maturation, the spore protoplast dehydrates and the spore becomes refractile and resistant to heat, radiation, pressure, desiccation, and chemicals. These properties can be attributed to calcium dipicolinate, cortical peptidoglycan and exosporium (Fig. 1.19).

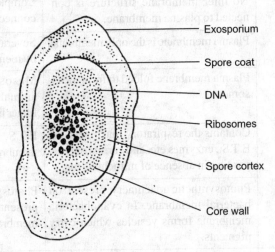

Exosporium
Spore coat
DNA
Ribosomes
Spore cortex
Core wall

Fig. 1.19 A Bacterial Endospore

Bacteria may be heterotrophic or autotrophic, producers or consumers, aerobic or anaerobic. **Photosynthetic autotrophs** use energy from the sun to produce their own carbohydrates for energy. **Chemosynthetic autotrophs** process inorganic molecules for energy (e.g. sulfur or iron). **Heterotrophs** depend on outside sources of organic molecules (e.g. carbohydrates or sugars) for energy. Temperature extremes range from –20ºC to 110ºC.

Differences Between Prokaryotes & Eukaryotes

Characters	Prokaryotes	Eukaryotes
Groups where found	Bacteria, Cyanobacteria, Archaea	Algae, Fungi, Protozoa, Plants and Animals
Size	1-2 by 1-4 μm or less	Greater than 5μm
Cell wall	Made up of peptidoglycan	Made of cellulose in algae and plants, of fungal cellulose or chitin in fungi, absent in protozoan and animals
Osmotic control	Wall possesses mechanical strength necessary to counter balance turgor pressure of cytoplasm	Maintained by contractile vacuole
Plasma membrane	Does not contain sterols except in Mycoplasma	Sterol present
	Fatty acids, exclusively saturated or monounsaturated, except in cyanobacteria which are able to synthesize polyunsaturated fatty acids	All lipids polyunsaturated i.e. contain more than one double bond
	No inter membrane structure is connected to plasma membrane.	Complex network of intermembrane connected with plasma membrane
	Plasma membrane is the only membrane.	Several cell organelles also covered by membrane (Unit mem-brane)
	Plasma membrane folded to form mesosomes.	Mesosomes are absent, but in animal cells, it is folded to form finger like microvilli.
	Contains the respiratory machinery i.e. E.T.S., enzymes etc. and carries out respiration in absence of mitochondria.	E.T.S. is present on mitochondrial membrane.
	Photosynthetic machiaery present on bacterial membrane. In cyanobacteria membrane forms vesicles which carry pigments.	Photosynthetic pigments and ETS present in the chloroplast membrane.
Cytoplasm	Granular due to presence of ribosomes, fibrous cytoskeleton absent	Fibrillar, made up of fibrous proteins which form fibrous cytoskeleton
Cytoplasmic streaming	Absent	Present
Pinocytosis, Exocytosis, Endocytosis, Phagocytosis	Absent, intracellular digestion, harbouring of endosymbionts absent	Present, intra cell was digestion, harbouring of endospymbionts present
Gas vacuoles	Present	Absent

Ribosomes	70S, freely distributed in cytoplasm	80S free as well bounded to ER. 70S in mitochondria and chloroplast.
Mitochondria	Absent	Present
Chloroplast	Absent, in cyanobacteria thylakoid vesicles present.	Present in plant cells and absent in fungi, protozoa and animal cell.
Golgi appratus	Absent	Present
Lysosomes	Absent	Present
Endoplasmic Reticulum	Absent	Present
Peroxisomes	Absent	Present
Centrioles	Absent	Present
Nucleolus	Absent	Present
True vacuoles	Absent	Present
Gas vesicles and carboxy-somes	Present	Absent
Flagella	Made up of flagellin fibrils. Flagellin sub-units arranged in a ring helically around a central core.	Made up of tubulin microtubules arranged in 9+2 formation.
Pili	Present	Absent
Cilia	Absent	Present in protozoans
Cell cycle	Is the life cycle and it is short	It is not necessarily the life cycle, it is long, made up of S, G_1, G_2 M and interphase.
Replication of DNA	Through out cell cycle	Only during 'S' phase
Synthesis of histone proteins	Absent	Only during 'S' phase
Cell growth, synthesis of RNA, membrane, macromolecules	Through out cell cycle	During Interphase
Cell division	Immediately after replication or simultaneously	During mitosis (M) phase
Chromosome distribution and number	Chromosomes attached to mesosomes to ensure each cell receives one copy.	Chromosomes distributed equally between daughter cells during telophase, chromosomes attached to spindle fibers which pull them towards poles.

Chromosome condensation and decondensation	Absent	Present: During interphase chromosomes decondense and are dispersed through out nucleoplasm. During mitosis or meiosis they condense and are clearly visible
Mitotic spindle Fibers (apparatus for cell division)	No special apparatus for cell division	Special spindle fibers present
Meosis or reduction division	Absent	Present
Chromosome number	Usually 1	Variable, 2- many
DNA	Circular, single chromosome, single or double stranded, helically coiled, usually haploid chromosome.	Single, linear, double stranded, helically coiled, chromosome diploid.
Amount of DNA	Less	More
Extranuclear DNA	Present in bacteria (plasmid 0.1-5% of bacterial chromosome)	Absent except in yeasts
Histone proteins	Absent	Present
DNA position, nuclear membrane	Usually central, not surrounded by nuclear membrane	Surrounded by nuclear membrane
Chromosomal attachment	Attached to plasma membrane.	Attached to nuclear membrane by nucleolus.
Chromosomal movement	Unidirectional from donor to recipient at times only exchange of plasmids.	Exchange of chromosome between two cells (gam etes)
Protein synthesis	Synthesis of m RNA and its translation simultaneously in cytoplasm.	Synthesis of mRNA is in nucleus. It's translation takes place in cytoplasm after some time, never simultaneous.
Stability of mRNA	Short half life,it starts degrading from 5' end just after its translation has started, and even before synthesis is completed.	Relatively stable, it degrades after translation is completed can be translated several times.
Structure of mRNA	Polycistronic	Monocistronic
Limitation of protein synthesis	Requires methylated base.	Methylated base not required.
Initiation complex	30S part 1st attaches to mRNA and than it gets attached to 50 S.	30S part 1st attaches to50 S and than it gets attached to mRNA.
Photosynthesis	Absent in most except cyanobacteria and autotrophic bacteria.	Present in plants. Absent in fungi, protozoa and animal cells.
N_2 fixation	Present in some bacteria and cyanobacteria.	Absent.

Respiration (generation of energy)	Performed by plasma membrane	Performed by mitochondria.
Anerobic respiration	Present in many bacteria	Shown by some e.g yeasts, few protozoans, muscle cells etc.
Aerobic respiration	Present.	Present.
Locomotion	Present in some bacteria, by flagella (swimming motility), no amoeboid movement but gliding present.	Absent in plants, algae, fungi, except in zoospores and few unicellular algae and fungi, present in animals and protozoans. (flagellated, ciliary, amieboid, gliding).
Pseudopodia	Absent.	Present in some protozoans.
Reproduction	Simple fragmentation or binary fission, no meosis, asexual spores produced in some.	Complex, meosis present, also by fragmentation, binary fission and asexual spores in some plants, algae and fungi and few protozoans.
Genetic recombination	Present in some but chromosomal cross over absent.	By chromosomal cross over during meiosis.
Nervous system	Absent.	Present in animals absent in plants, primitive in protozoans.
Vascular system	Absent.	Circulatory system present in animals and well developed vascular system in higher plants.
Extracellular digestion	Present	Present in animals.

CLASSIFICATION OF MICROORGANISMS

Earlier all organisms were classified into two groups: **Plants** and **animals** on the basis of mode of nutrition and presence or absence of cell wall. As more and more information about unicellular organisms became known, it became impossible to assign organisms like bacteria, slime molds, etc. to either of these groups. Thus, a third kingdom called **Protista** was formed by Haeckel (1866) consisting of unicellular organisms. As science progressed, differences in cell structure and cell wall came to be known and **Copeland** (1956) created a fourth kingdom **Monera**, containing unicellular prokaryotic bacteria. Gradually all unicellular prokaryotes having a peptidoglycan cell wall, irrespective of their mode of nutrition were kept in Monera and a fifth kingdom **Mycetae** was created by **Whittaker** (1969) to accommodate fungi.

Whittaker's classification is based on:

- *Life style:* Producer, consumer, and decomposer.

- *Phylogenetic relationship:* Based on morphological, cytological, genetical, biochemical, physiological and fossil records
- *Cell structure:* Prokaryotic or eukaryotic.
- *Body organisation:* Simple or complex
- *Modes of Nutrition:* Autotrophic or heterotrophic.

Later a super kingdom **Prokaryotae** was created having a single kingdom Monera consisting of all unicellular prokaryotic organisms. Microbiology, usually deals with organisms belonging to kingdom Monera now comprised of Eubacteria, Archaea, Cyanobacteria, Mycoplasma, Phytoplasma, Rickettsia, L-forms, Spiroplasma, Viruses, Viroids, Prions, etc. are also studied under this branch.

The artificial system of bacterial classification groups prokaryotes on the basis of similarities. This system is designed for use as a **Determinative key**, the best example of which is **Bergey's Manual of Systematic Bacteriology** (1984). It contains names, descriptions, literature citations, and determinative keys for classification of a new isolate. In Bergey's Manual prokaryotes are grouped on the basis of characteristics such as Gram stain, morphology (rods, cocci, etc.), motility, structural features (e.g. spores, filaments, sheaths, appendages, etc.), and on distinguishing physiological features (e.g. anoxygenic photosynthesis, methanogenesis, lithotrophy, etc.).

Nowadays organisms including prokaryotes are grouped on a genetic basis, i.e. by comparison of the nucleotide sequences of the small subunit **ribosomal** RNA (rRNA) that is contained in all cellular organisms. Bergey's manual, in its second edition in 1992 and continuing with its recent 3rd edition, has grouped prokaryotes on the basis of comparison of the nucleotide sequences in the small subunit rRNA, or in the gene that encodes for the small subunit rRNA.

The molecular analyses of organisms shows that life forms can be grouped into three domains: Bacteria, Archaea and Eukarya. C. Woese developed the **phylogenetic tree of prokaryotes** by analysis and comparison of highly conserved markers such as 16S rRNA sequence. 16S rRNA of 30S ribosomes yield highly conserved sequences, which indicate the similarities with almost all bacteria and dissimilarities with eukaryotes and some large groups of bacteria. Certain signature sequences have been identified which seem to be characteristic for a group of organisms.

These analyses produce **Dendrograms** or **phylogenetic trees**. The analysis of rRNA sequences has shown that Archaea are more similar to eukaryotes and that eubacteria such as *E.coli* and cyanobacteria, although morphologically dissimilar, are more closely related. Thus, the phylogenetic system of classification is very different from the artificial system, but it cannot replace it. The artificial system is still followed for identification and *de novo* description of bacteria. The phylogenetic system conveys information about evolutionary history.

Bergey's Scheme of Classification

The most widely used reference for bacterial classification is given in **"Bergey's Manual of Determinative Bacteriology"** It contains names and descriptions of organisms and diagnostic key for identification of bacteria. The manual was published by the **Society of American Bacteriologists** (now called the **American Society for Microbiology** and publishes the Journal of Bacteriology). They formed an **Editorial Board** consisting of Francis C. Harrison, Robert S. Breed, Bernard W. Hammer, and Frank M. Huntoon with **David H. Bergey as its Chairman**. This committee approved a system of classification, which was first published as **"Manual of Determinative Bacteriology"** with Bergey as its author. The first edition of the Manual came into print in 1923. The Board, with some changes in membership and Dr. David Bergey as Chairman, published a second edition of the Manual in 1925 and a third edition in 1930.

In 1934 during preparation of the fourth edition the Society transferred all of its rights, title, and interest in the Manual to Dr. Bergey who established a **Bergey's Manual® Trust** on 2nd January 1936 and transferred the ownership of the Manual, its copyrights, and the right to receive the income arising from its publication to the Trustees and their successors. **David H, Bergey, Robert S. Breed**, and **EGD Murray** were the initial trustees of this non profit organization whose sole purpose henceforth has been to prepare, edit and publish revisions and successive editions of the Manual and any supplementary publications, as well as provide for any research that may be necessary or desirable in such activities. Since then successively, the fourth, fifth, sixth, seventh, and eighth editions of the Manual in 1934, 1939, 1948, 1957, and 1974, respectively have been published.

Other publications of the trust include *Index Bergeyana* (1966), a *Supplement to Index Bergeyana* **(1981)**, and a planned future volume bringing the lists of published names up to date. Since its inception in 1934,The Bergey's Manual® trust has changed from a local body to an international organization. Along with its publication activities the trust also supports different facets of research in the field of bacterial taxonomy. It presents **The "Bergey Award"** to individuals who have made outstanding contributions to bacterial taxonomy. This award is cosponsored by funds from the trust as well as **Lippencott William & Wilkins** and **Springer-Verlag**.

The VIIth edition (1957) was prepared by Prof R S Breed, E G D Murray and Dr. Nathan and was titled as **"Bergey's Manual of Determinative Bacteriology"**. In this book bacteria were placed in class Schizomyctes with following orders: Pseudomonadales, Chlamydobacteriales, Eubacteriales, Actinomycetales, Spirochaetales. Mycoplasmatales, Rickettsiales and Virales.

In **1968 RG E Murray** proposed four divisions or phyla of bacteria namely Gram positive, Gram negative, Gram variable and organisms lacking a cell wall.

In **1977 the Shorter Bergey's Manual® of Determinative Bacteriology** containing the outline classification of bacteria, descriptions of all genera and higher taxa, all of the keys and tables for the diagnosis of species, all of the illustrations, and two of the introductory chapters was published. It was an **abbreviated version of the eighth edition**, however it did not contain the detailed species descriptions, most of the taxonomic comments, the etymology of names, and references to authors.

In **1982 Lynn Margulis** grouped bacteria as fermenting heterotrophs, respiratory heterotrophs and autotrophs.

In **1984 Edition Murray** proposed the breakdown of Kingdom Procaryotae into four divisions: Gracilicutes, Firmicutes, Tenericutes and Mendosicutes.

1st edition of **Bergey's Manual of Systematic Bacteriology** (Holt, J.G. editor-in-chief) was published in four volumes. It has been the standard authoritative guide to bacterial taxonomy and identification. These volumes are divided as follows:

Volume 1 (1984): Gram-negative Bacteria of general, medical or industrial importance

Volume 2 (1986): Gram-positive Bacteria other than Actinomycetes

Volume 3 (1989): Archaeobacteria, Cyanobacteria and remaining Gram-negative Bacteria

Volume 4 (1989): Actinomycetes

In **1994 a single volume** containing all the information was published and in **1995** it was made available in CD-ROM format.

Bergey's Manual of Systematic Bacteriology contains the phenotypic characteristics used to classify bacteria by conventional taxonomy and keys that can be used to identify unknown strains from their phenotypic characters. According to the Bergey's Manual of Systematic Bacteriology, all bacteria can be classified into four divisions or phyla according to the characteristics of cell walls. Each division is further subdivided into sections on the basis of:

- Grams stain reaction
- Cell shape
- Cell arrangements
- Oxygen requirement
- Motility
- Nutritional requirements
- Metabolic properties

Each section consists of a number of genera, which are grouped into families and orders. The four divisions are:

Gracilicutes: Gram-negative bacteria possessing an outer membrane containing lipids.

Firmicutes: Gram-positive bacteria with a single membrane and thick peptidoglycan wall.

Mollicutes: Gram-negative, wall less bacteria with no second membrane.

Mendosicutes: Archeabacteria

Brief outline of Bergey's Classification and characteristics of each division are as such:

Kingdom Prokaryotae: Prokaryotes, nucleoplasm lacks histone, nuclear membrane absent, cells single/filamentous/mycelial/colonial, immobile cytoplasm. Nutrients absorbed in molecular form, cells mostly enclosed in rigid peptidoglycan wall.

Division I; Gracilicutes: Phototropic or nonphototropic, nonsporulating, usually gram negative cells, variable shapes, thin cell wall, some sheathed or encapsulated, reproduction by binary fission or budding, anaerobic/aerobic/facultative anaerobes, some obligate intracellular parasites. If motile, motility by gliding or swimming.

Class I: Scotobacteria; Nonphotosynthetic, gram negative.

Class II: Anoxyphotobacteria: Light requiring bacteria that do not produce oxygen, anaerobic, contain bacteriochlorophyll.

Class III: Oxyphotobacteria: Light requiring cyanobacteria that produce oxygen, contain chlorophylls, gliding motility, aerobic, rigid multilayered wall with inner peptidoglycan layer.

Division II; Firmicutes: Usually gram positive, chemosynthetic heterotrophs, nonphotosynthetic, variable shape with occasional branching rods and filaments, thick cell wall, reproduction by binary fission, aerobic, anaerobic or facultative anaerobic, includes asporogenous and sporogenous bacteria, actinomycetes and their relatives.

Class I: Firmibacteria (Latin; *Firmus* = strong + Greek; *Bakterion* = small rod). Simple gram positive bacteria whose name means strong bacteria. They possess a thick peptidoglycan cell wall.

Class II: Thallobacteria (Greek *Thallos* = branch + *Bakterion* = small rod). Gram positive bacteria which show branching. Include actinomyces and related bacteria.

Division III: Tenericutes (Latin *Tener* = soft + *Cutis* = Skin). Prokaryotes which lack cell wall, pliable, soft nature, highly pleomorphic, enclosed by only a unit membrane, filamentous forms common with branching, wide size range, genome size smaller other prokaryotes, reproduction by budding, fragmentation, and /or binary fission, usually nonmotile, if motile- motility by gliding, complex media required for growth, form fried egg like colonies, saprophytic, parasitic or pathogens of animals, plants, humans and cultures.

Class I: Mollicutes: Prokaryotes with no cell wall, includes Mycoplasma.

Division IV: Mendosicutes: Ecologically and morphologically diverse prokaryotes, non spore forming, gram variable, variable shape, most strictly anaerobic, some aerobic, many motile with flagella, cell wall lacks muramic acid, contain protein macromolecules, heteropolysaccharides, live in extreme climate, known members are either methanogens, thermoacidophiles or strict halophiles.

Class I: Archaebacteria: diverse prokaryotes with unusual walls, membranes, lipids, ribosomes and RNA sequences.

In the revised edition of ***Bergey's Manual of Systematic Bacteriology*** these phyla are no longer believed to represent monophyletic groups. The Gracilicutes have been divided into many different phyla. Most gram-positive bacteria are placed in the phyla Firmicutes and Actinobacteria, which are closely related. However, the Firmicutes have been redefined to include the mycoplasmas (Mollicutes) and certain Gram-negative bacteria.

9th Edition of ***Bergey's Manual of Determinative Bacteriology*** (Holt, J.G. ed) published in 1994 by Williams & Wilkins, Baltimore has four volumes in which different bacteria are classified into 35 groups based on the phenotypic information. This manual does not attempt to offer a natural higher classification and is an abridged version of four volumes of *Bergey's Manual of Systematic Bacteriology*. The bacterial groups are comparable to the "Parts" in the eighth edition and the "Sections" in the Systematic volumes. These groups do not signify formal taxonomic ranks, but divide bacteria into easily recognized phenotypic compilations. This arrangement is practical and extremely useful for diagnostic purposes.

This manual is slightly different from the other editions as it serves as a reference to help in the identification of unknown bacteria. Whereas *Bergey's Manual of Systematic Bacteriology* provides Systematic information.

Bergey's Manual of Systematic Bacteriology

1st Edition, John G. Holt, (Editor-in-Chief), Williams & Wilkins, Baltimore

Volume 1 (1984): Gram-negative Bacteria of general, medical, or industrial importance
Volume 2 (1986): Gram-positive Bacteria other than Actinomycetes
Volume 3 (1989): Archaeobacteria, Cyanobacteria, and remaining Gram-negative Bacteria.
Volume 4 (1989): Actinomycetes

Phylogenetic analysis of the **Bacteria** has demonstrated the existence of at least eleven distinct groups of bacteria, but many groups consist of members that are phenotypically and physiologically unrelated.

The second edition of *Bergey's Manual of Systematic Bacteriology* (2001) has 5 volumes and recognizes 23 distinct phyla of bacteria (Phylum is the highest taxon in a Domain) where genetic similarities reflected in 16S and 23S ribosomal RNA sequences have been taken into consideration. Thus this edition gives a hierarchical approach to classification and taxonomy. It classifies **Prokaryotes** into **two domains Archaea and Bacteria**, which are further divided into Phylum, Class, Order, Family, Genus and Species. The Bacterial Domain contains 23 Phyla.

Bergey's Manual of Systematic Bacteriology

2nd Edition, George M. Garrity, (Editor in Chief), Springer-Verlag, New York

Volume 1 (2001): The Archaea and the deeply branching and phototrophic Bacteria
Volume 2 (2001): The Proteobacteria
Volume 3 (2002): The low G + C Gram-positive Bacteria
Volume 4 (2002): The high G + C Gram-positive Bacteria
Volume 5 (2003): The Planctomycetes, Spriochaetes, Fibrobacteres, Bacteriodetes and Fusobacteria

Salient features of some important bacterial groups of Bergey's Manual are:

1. **Spirochetes or coiled bacteria:** These bacteria were one of the first organisms to be described by Leeuwenhoek. They are usually found in contaminated water, sewage, soil, decaying organic matter as well as inside human and animal bodies. They are Gram-Negaitive, heterotrophic, flexible and resemble a metal spring. They are motile and possess two or more axial filaments found in the periplasmic space. Some Spirochetes are aerobic, and some are anaerobic, live freely, symbiotically, or parasitically this group includes several important pathogens e.g. *Treponema pallidum* (Syphilis), *Borrelia* (relapsing fever and Lyme's disease), *Leptospira* (Leptospirosis).

2. **Helical Bacteria or Vibroid motile Gram-negative Bacteria:** These are aerobic or microaerophilic bacteria. They lack axial filaments and usually have a single flagellum at one or both poles. Some possess tufts of flagella. This group includes mostly harmless aquatic bacteria such as *Azospirillum* (nitrogen fixing bacterium), and some pathogens like *Closteridium jejuni* (food borne intestinal diseases), *Helicobacter pylori* (gastric ulcers in humans).

3. **Gram-Negative Aerobic rods and Cocci:** This is one of the largest group of bacteria containing several harmful as well as economically important genera. Important bacteria of this group are:

   - Extremely resilient and tough **Psuedomonads** named after the genus *Pseudomonas*. These bacteria can grow even in soaps, antiseptics and are highly resistant to commonly used antibiotics. Most species of these rod shaped motile bacteria (polar flagella). **secrete water- soluble fluorescent as well as nonflourescent pigments** into medium giving it a colored appearance. This is an **important criterion for their identification**. *Pseudomonas aeruginosa* (infections of urinary tract, burns, and wounds, septicemia, abscesses, and meningitis) produces a blue-green pigment. Pseudomonads utilize nitrate as a source of oxygen and therefore are also responsible for denitrification of soil.

   - *Legionella* is frequently found in warm-water supply lines in hospitals and water cooling towers of air conditioner plants and cause Legionellosis, a form of pneumonia. Now it is possible to culture this bacterium in a special medium containing charcoal-yeast extract.

   - *Neisseria* is a non-endospore forming aerobic or facultatively anaerobic diplococci. It infects mucous membranes and its optimum growth temprature is near human body temperature. It causes gonorrhoea and meningococcal meningitis (*N. gonorrhoeae* and *N. meningitidis*).

   Other important species are ***Moraxella lacunata*** (obligate aerobic coccobacillus, causes conjunctivitis), ***Brucella*** (a small nonmotile coccobacillus, causes brucellosis), ***Bordetella pertusis*** (a nonmotile rod, causes whooping cough), pleomorphic ***Francisella tularensis*** (causes tularemia).

   Useful bacteria of this group include symbiotic Nitrogen fixing ***Rhizobium*** and ***Bradyrhizobium***, asymbiotic nitrogen fixing ***Azotobacter*** and ***Azomonas*** and ***Acetobacter*** and ***Gluconobacter,*** which convert ethanol to vinegar.

4. **Facultative Anaerobic Gram-Negative Rods:** This group contains pathogenic bacteria which mostly belong to family Enterobacteriaceae, Vibrionaceae and Pasteurellaceae. Important genera include:

   - Members of most important family Enterobacteriaceae consisting of motile as well as nonmotile **enteric bacteria** responsible for diseases of the gastro intestinal gast tract and other organs. Motile enteric bacteria have peritrichous flagella. Some possess fimbriae for attachment to mucous membranes and sex-pili used for exchange of genetic information. Most enterics are **fermenters of glucose and other carbohydrates** and also produce **bacteriocins,** which are proteins with antibiotic properties. e.g. *Escherichia;* a common habitant of human intestines, usually nonpathogenic but may cause urinary tract infections. Certain species produce enterotoxins, e.g. *Salmonella;* responsible for diseases like **Typhoid fever** and **Salmonellosis, Shigella**; causes **Bacillary dysentary** or **Shigellosis,** *Klebsiella* causes septicemia and pneumonia, *Serratia marcescens;* causal organism of hospital acquired infections, **Proteus**; which causes infection of urinary tract, *Yersinia pestis*; that causes plague and *Enterobacter*; causes urinary tract and nosocomial infections.

- Next important family is Vibrionaceae which is comprised of gram negative, facultative anaerobic, slightly curved rods. Important genera is *Vibrio cholerae*; causes cholera.

- Pasteurellaceae includes *Pasteurella* a pathogen of domestic animals and *Hemophilus* which infects mucous membrane of upper respiratory tract, mouth, vagina, and intestinal tract of humans.

5. **Anaerobic Gram-Negative Cocci:** This group is made up of nonmotile, non-endospore forming coccoid forms occurring in pairs or clusters singly or in chains, They constitute the normal flora of the mouth and are components of dental plaque.

6. **Rickettsias and Chlamydias:** These gram- negative, highly pleomorphic obligate intracellular parasites are cultured in yolk sac of chicken embryos as they can reproduce only within the host cell and are transmitted to humans by insects and ticks. Important genera are *Coxiella burnetti*, which causes Q fever and *Chlamydia trachomatis*, causal agent of trachoma as well as nongonococcol urethritis.

7. **Mycoplasmas:** These small (0.1–0.25um), aerobic or facultative anaerobic organisms are also highly pleomorphic as they a definite lack cell wall. The characteristic identification criteria is production of typical **"fried egg" like colonies** in artificial media, which are less than 1mm in diameter.

8. **Gram Positive Cocci:** The aerobic or facultative anaerobic Staphylococci occur in grape like clusters and are part of normal flora of skin but also produce toxins that contribute to bacterium's pathogenicity. e.g. *Staphylococcus aureus*. Other important genera belong to Streptococci which occur in chains except aerotolerant or facultatively anaerobic *Streptococcus pneumoniae* which occurs in pairs.

9. **Endospore forming Gram-Positive Rods and Cocci.** This group contains several useful as well as pathogenic gram positive bacteria. It includes some of the most dangerous rod shaped, nonmotile, facultative or obligate anaerobic bacteria such as:

- *Clostridium: Clostridium tetani, C.botulinum* and *C. perfringens* cause tetanus, botulism and common forms of food borne diseases respectively.

- *Bacillus anthracis*: causes anthrax, a disease of cattle, sheep and horses.
- *Lactobacillus*: This nonspore forming gram positive, aerotolerant, rod shaped bacterium is commonly found in vagina, intestinal tract and oral cavity. It has the ability to produce lactic acid from simple carbohydrates and has several applications in dairy industry.
- *Listeria monocytogenes* is a pathogenic species that can contaminate dairy products and meat. **Most important property of this genus is its ability to grow at refrigeration tempratures.**
- **Corynebacteria** are the other important genera which are highly pleomorphic gram positive, aerobic, anaerobic, or microaerophilic rods. e.g. *Corynebacterium diptheriae* which causes diptheria.
- Mycobacteria are also called **acid fast bacteria** as **they resist destaining by acid alcohol.** They are aerobic, nonspore forming, nonmotile, rod shaped. e.g. *Mycobacterium tuberculosis* (causal agent of tuberculosis) and *M.leprae* (causes leprosy).

Nomenclature of Prokaryotes

Description and naming of prokaryotes follows the **Bacteriologic code** established by the **International Committee of Systematic Bacteriology.** This code is different from the botanical code mainly in the rule that each new described strain must be deposited with a recognized culture collection centre to serve as reference or type culture.

Important: While describing prokaryotes especially bacteria, the following points should be mentioned:

- Morphological characters; shape, habit, presence or absence of capsule, flagella and their location, spore formation and reaction to gram stain
- Physiological and biochemical characters
- Serological properties
- GC content
- DNA-DNA hybridization
- Transformability by interspecies transformation
- 23S, 16S, 5S rRNA sequence
- Antibiotic sensitivities

The Binomial system of nomenclature is used for naming prokaryotes. Originally, generic names were given on morphological properties and specific name was based on biochemical or physiological characteristics. Lately several characters such as ecological, physiological and biochemical features (*Acetobacter, Nitrosomonas*), pigments (*Rhodomicrobium, Chromobacterium*), pathogenicity (*Pneumococcus, Phytomonas*), specific nutrient requirement (*Haemophilus, Amylobacter*), etc. are also being used for generic names. **Strain** is the basic unit. Strains are aggregated into species and several species form a genus. These collectively form a family which have names ending in –aceae.

CONSTITUENT ORGANISMS OF MAJOR GROUPS OF PROKARYOTES

Prokaryotes consist of millions of genetically distinct unicellular organisms. On the basis of shape, gram staining and oxygen relationships prokaryotes can be classified into the following groups:

a. **Gram-positive aerobic cocci:** *Micrococcus, Staphylococcus, Streptococcus, Leuconostoc, Pediococcus, Lactococcus.*

b. **Gram-positive anaerobic cocci:** *Peptococcus, Peptostreptococcus, Ruminococcus, Sarcina.*

c. **Gram-negative aerobic cocci:** *Neisseria, Moraxella, Acinetobacteria, Paracoccus, Lampropedia.*

d. **Gram-negative anaerobic cocci:** *Veillonella, Acidaminococcus, Megasphaera.*

e. **Gram-positive, aerobic, non spore forming rods:** *Lactobacillus, Listeria, Erysipelothrix, Caryophanon.*

f. **Coryneform aerobic bacteria and actinomycetes:** *Microbispora, Streptomyces, Corynebacterium, Arthrobacter, Brevibacterium, Cellulomonas, Propionibacteria, Eubacterium, Bifidobacterium, Mycobacterium, Nocardia, Actinomyces, Frankia, Actinoplanes, Dermatophilus, Micromonospora Streptosporangium.*

g. **Gram-positive, aerobic, endospore-forming rods and cocci:** *Bacillus, Sporolactobacillus, Sporosarcina, Thermoactinomyces.*

h. **Gram-positive, anaerobic, endospore-forming rods and cocci:** *Clostridium, Desulfotomaculum, Oscillospira.*

i. **Gram-negative, aerobic rods and cocci:** *Pseudomonas, Xanthomonas, Zoogloea, Gluconobacter, Acetobacter, Azotobacter, Azomonas, Beijerinkcia, Derxia, Rhizobium, Agrobacterium, Alcaligenes, Brucella, Legionella, Thermus.*

j. **Gram-negative, aerobic, chemolithotrophic bacteria:** *Nitrobacter, Nitrospina, Nitrococcus, Nitrosomonas, Nitrospira, Nitrosococcus, Nitrosolobus, Thiobacillus, Thiobacterium, Thiovulum.*

k. **Gram-negative, aerobic, sheathed bacteria:** *Sphaerotilus, Leptothrix, Streptothrix, Crenothrix.*

l. **Gram-negative, facultative anaerobic rods:** *Escherichia, Klebsiella, Enterobacter, Salmonella, Shigella, Proteus, Serratia, Erwinia, Yersinia, Vibrio, Aeromonas, Photobacterium.*

m. **Gram-negative, obligate anaerobic bacteria:** *Bacteroides, Fusobacterium, Leptotrichia, Fibrobacter.*

n. **Obligate anaerobic archaebacteria:** *Methanobacterium, Methanothermus, Methanosarcina, Methanothrix, Methanococcus.*

o. **Aerobic archaebacteria:** *Halobacterium, Haloferax, Halococcus, Sulfobolus, Thermoplasma, Thermus.*

p. **Anaerobic archaebacteria:** *Thermoproteus, Pyrodictum, Desulfurococcus, Pyrococcus, Thermococcus, Thermodiscus.*

q. **Gram-negative, aerobic, spiral and curved bacteria:** *Spirillum, Aquaspirillum, Azospirillum, Oceanospirillum, Campylobacter, Helicobacter, Bdellovibrio, Microcyclus.*

r. **Gram-negative, anaerobic, spiral and curved bacteria:** *Desulfovibrio, Succinivibrio, Butyrivibrio, Selenomonas.*

s. **Aerobic and anaerobic spirochaetes:** *Spirochaeta, Cristispira, Treponema, Borrelia, Leptospira.*

Large Special Groups

a. **Gliding Bacteria (are gram-negative only):** *Myxococcus, Archangium, Cystobacter, Melittangium, Stigmetella, Polyangium, Nannocyctis, Chondromyces, Cytophaga, Sporocytophaga, Flexibacter, Hepetosiphon, Saprospira, Beggiatoa, Thiothrix, Achromatium, Leucothrix, Vitreoscilla, Alysiella.*

b. **Bacteria with appendages, prosthecae and budding bacteria:** *Hyphomicrobium, Hyphomonas, Caulobacter, Asticcacaulis, Planctomyces, Ancalomicrobium, Prosthecomicrobium, Blastobacter, Selibria, Gallionella, Nevskia.*

c. **Rickettsia and Chlamydiae (Obligate parasitic forms):** *Rickettsia, Coxiella, Chlamydia.*

d. **Anaerobic, anoxygenic, phototrophic bacteria:** *Rhodospirillum, Rhodopseudomonas, Rhodobacter, Rhodomicrobium, Rhodocyclus, Rhodopilus, Chromatium, Thiocystis, Thiosarcina, Thiocasa, Thiospirillum, Thiopedia, Amoebacter, Ectothiorhodospira, Lamprocystis, Thiodictyon, Chlorobium, Prosthecochloris, Pelodictyon, Chloroherpeton, Chloroflexus, Chloronema, Oscillochloris.*

e. **Aerobic, oxygenic, phototrophic bacteria (Cyanobacteria):** *Synecococcus, Gloeocapsa, Gloeothecae, Gloeobacter, Pleurocapsa, Dermocarpa, Myxosarcina, Oscillatoria, Spirulina, Lyngbya, Phormidium, Plectonema, Anabaena, Nostoc, Calothrix, Fischerella.*

Groups Based on Genome Sequencing

Based on genome sequencing, especially of 16S ribosomal RNA (rRNA), bacteria have been classified into the following 11 groups:

1. **Proteobacteria or purple photosynthetic or nonphotosynthetic bacteria:** a clade of gram-negative rods, cocci and spiral bacteria having related rRNA sequences. They are further subdivided into 5 clades: alpha-, beta-, gamma-, delta-, and epsilon proteobacteria.

   a. **Alpha (α) Proteobacteria:** This subdivision includes *Hyphomicrobium, Methylococcus, Nitrobacter, Caulobacter,* purple nonsulphur bacteria, rhizobacteria and Rickettsias (small obligate intracellular parasites of arthropods and mammals). These cannot be clearly seen under the light microscope. Their genome is quite similar to that of mitochondria and they may be the closest relatives to the ancestors of mitochondria.

   b. **Beta (β) Proteobacteria:** This subdivision includes chemoautotrophic bacteria which oxidise reduced substance to obtain energy to manufacture food. e.g. Iron bacteria, the chemoautotrophs responsible for the brownish scale that forms inside the tanks of flush toilets. *Nitrosomonas* is a chemoautotroph which oxidises NH_3 (produced from proteins by decay bacteria) to nitrites (NO_2^-). Important human pathogenic β-proteobacteria are *Neisseria meningitidis, Neisseria gonorrhoeae* and *Bordetella pertussis.*

c. **Gamma (γ) Proteobacteria:** The largest and most diverse subgroup of the proteobacteria includes *Escherichia coli, Salmonella enterica var typhi* also known as *Salmonella typhi, Salmonella enterica var typhimurium* also known as *Salmonella typhimurium, Vibrio cholerae, Pseudomonas aeruginosa, Yersinia pestis, Francisella tularensis, Haemophilus influenzae*, purple sulfur bacteria which are photosynthetic bacteria that use solar energy to reduce electrons. They have bacteriochlorophylls. They lack photosystem II, which is why they are unable to use water as a source of electrons. Usually these bacteria are obligate anaerobes, therefore are found at the bottom of shallow ponds and estuaries. The absorption spectrum of their bacteriochlorophylls lies mostly in the infrared region of the spectrum so they can trap energy missed by the green plants above them.

d. **Delta (δ) Proteobacteria:** Myxobacteria or group of gliding bacteria that aggregate together to form a multicellular fruiting body in which development and spore formation takes place. They are found in soil and include some important sulfur and sulfate reducers such as *Bdellovibrio, Desulfovibrio,* etc.

e. **Epsilon (ε) Proteobacteria:** Includes *Helicobacter pylori,* the bacteria responsible for stomach ulcers and *Campylobacter jejuni* which causes gastrointestinal upsets.

2. **Gram-Positive Bacteria including Mycoplasma:** These bacteria are divided into two groups:

- **A high G+C group:** contain Actinomycetes or filamentous bacteria, which are found in soil. They play a major role in the decay of dead organic matter. Many of them produce antibiotics like streptomycin, erythromycin and the tetracyclines. It also contains Mycobacteria and Corynebacteria; Gram-positive organisms closely related to the actinomycetes which include *Mycobacterium tuberculosis* causes tuberculosis (TB), *Mycobacterium leprae* causes leprosy, *Corynebacterium diphtheriae* causes diphtheria.

- **A low G+C group:** This contains aerobic and anaerobic gram-positive rods such as *Clostridium tetani* responsible for tetanus, *Clostridium botulinum* responsible for food poisoning, Gram-positive cocci like *Staphylococcus albus,* a common inhabitant of skin, *Staphylococcus aureus* responsible for food poisoning and toxic shock syndrome, *Streptococcus mutans* a common inhabitant of the mouth. Streptococci cause strep throat, impetigo, middle ear infections, scarlet fever, and rheumatic fever. *Streptococcus pneumoniae* causes bacterial pneumonia. *Bacillus anthracis/cereus/thuringensis* are also included in this group. Difference lies in their plasmids. *Lactobacillus*, several species of which are used to convert milk into cheese, butter, and yogurt is also present in this group.

- **Mycoplasmas:** The smallest living organisms. Can be seen only under the electron microscope E.g., *Mycoplasma genitalium, Mycoplasma urealyticum, Mycoplasma pneumoniae.*

3. **Spirochetes:** Thin, corkscrew-shaped, flexible organisms, e.g., ***Treponema pallidum;*** causes syphilis, ***Borrelia burgdorferi*** causes Lyme disease. It is transmitted to humans through the bite of a deer tick.

4. **Chlamydiae:** Obligate intracellular parasites. They have a unique life cycle and lack peptidoglycan. *Chlamydia trachomatis* responsible for **sexually-transmitted disease (STD)** also causes pelvic inflammatory disease. Trachoma or eye infection is caused by a strain of *C. trachomatis. Chlamydia psittaci* causes psittacosis (a.k.a. ornithosis). It usually infects birds, but can also infect humans.

5. **Cyanobacteria (blue-green algae):** These are photosynthetic prokaryotes, possess **chlorophyll a** and like plants use water as the source of electrons to reduce CO_2 to carbohydrate. These **contain both photosystems** and blue pigment **phycocyanin** ("blue-green") and red pigment **phycoerythrin**. Possibly these organisms are ancestors of chloroplasts. Periodic blooms of red-coloured cyanobacteria are responsible for name of Red Sea.

6. **Planctomycetes:** Includes walled, budding bacteria, which lack peptidoglycan. E.g., *Planctomyces* and *Isosphaera.*

7. **Bacteroides, Flavobacteria and relatives:** includes genera like *Bacteroides, Cytophaga, Flavobacterium,* etc.

8. **Green Sulphur bacteria:** This group includes members of Chlorobiaceae which are anaerobic,photosynthetic and can oxidise sulphur and sulphide. E.g., *Chlorobium, Pelodictyon.*

9. **Green Non Sulphur bacteria and relatives:** Example included in this group are *Chloroflexus, Thermomicrobium,* etc.

10. **Radioresistant micrococci and relatives:** These are gram positive cocci, capable of resisting very high radiations and belong to family Deinococcaceae.

11. **Thermotoga and Thermosipho:** Thermophilic bacteria isolated from geothermally heated deep sea vents. Have not been studied as yet.

DOMAIN ARCHAEA

Archaea differ from bacteria in the absence of murein in cell wall, having ether-linked membrane lipids, tRNAs without thymidine in the T or TΨC arm, distinct RNA polymerases and ribosomes of different composition and shape. They prefer extreme habitats.

On the basis of ssrRNA analysis Archaea are divided into three phylogenetically-distinct groups:

- **Crenarchaeota:** consists mainly of hyperthermophilic sulfur-dependent prokaryotes
- **Euryarchaeota:** contains the methanogens and extreme halophiles

- **Korarchaeota:** only their ssrRNAs have been obtained from hyperthermophilic environments, similar to those inhabitated by Crenarchaeota. No organisms have been isolated or cultured so far.

Based on their physiology Archaea are of three types:

- **Methanogens:** These organisms are obligate anaerobes that will not tolerate even brief exposure to air (O_2). They are found in marine and fresh-water sediments, bogs and deep soils, intestinal tracts of animals, and sewage treatment facilities. Methanogens can use H_2 as an energy source and CO_2 as a carbon source for growth and produce methane (CH_4) in a unique energy-generating process. The end product (methane gas) accumulates in their environment. This group has three orders; Methanococcales containing several genera (*Methanococcus*), Methanobacteriales (*Methanobacterium*) and Methanomicrobiales (*Methanosarcina*). Methanogens represent a microbial system that can be exploited to produce energy from waste materials.

- **Extreme halophiles:** These archaea live at very high concentrations of salt (NaCl) and will not grow at low salt concentrations. Their cell walls, ribosomes and enzymes are stabilised by Na^+. This group contains a single order Halobacteriales with six genera. They are found in natural environments such as the Dead Sea, the Great Salt Lake, or evaporating ponds of seawater where the salt concentration is very high (as high as 5 molar or 25 percent NaCl). E.g. *Halobacterium halobium,* which possesses "purple membrane", containing patches of light-harvesting pigment **bacteriorhodopsin** that reacts with light to form a proton gradient on the membrane allowing the synthesis of ATP. This is the only example in nature of **non-photosynthetic photophosphorylation**. These organisms are aerobic heterotrophs.

- **Sulphur dependent extreme (hyper) thermophiles:** These are archaea that live at very high temperatures (80 degrees to 105 degrees). Their membranes and enzymes are unusually stable at high temperatures. They are found in hot, sulfur-rich environments such as volcanoes, hot springs, geysers and fumaroles in Yellowstone National Park, thermal vents ("smokers") and cracks in the ocean floor. Most of these Archaea require elemental sulfur for growth. This group contains anaerobes that use sulfur as an electron acceptor for respiration in place of oxygen and lithotrophs that oxidise sulfur as an energy source. Sulfur-oxidizers grow at low pH (less than pH 2) because they acidify their own environment by oxidizing sulfur to sulfuric acid. Discovery of *Sulfolobus* and *Thermus aquaticus* in Yellowstone National Park, launched the field of hyperthermophile biology. (*Thermus aquaticus* is the source of the enzyme **taq polymerase** used in the polymerase chain reaction (PCR). ***Thermoplasma* a unique thermophile** is the only genus of a distinct phylogenetic line of Archaea which **resembles the bacterial mycoplasmas** in that it **lacks a cell wall**. *Thermoplasma* grows optimally at 55 degrees and pH 2 and has only been found in self-heating coal refuse piles, which are a man-made waste.

2

TOOLS OF MICROBIOLOGY

MICROSCOPES AND MICROSCOPY

The word microscope was first coined by **Giovanni in 1625**. Microscope is an optical instrument, comprised of one or more lenses and is used to enlarge and/or magnify images of microscopic objects. The earliest microscope was just a tube with a plate for the object at one end and a lens at the other, which gave a magnification ten times the actual size. These were called **"flea glasses"** since they were used to view fleas or tiny creeping things. **Hans Lippershey** (developed the first real telescope), **Hans Janssen** and his son **Zacharias** together have been given credit for the invention of microscope.

Janssen first used a second lens to enlarge the image formed by primary lens to the order of 50–100X. **Galileo** invented the first compound microscope in 1610. Robert Hooke later improved it in 1665 and published a book **"Micrographia"**. The highest magnification of this microscope was 200X, but the honour of giving birth to the science of microbiology goes to **Antony Van Leeuwenhoek**– (1632–1723) who designed a simple compound microscope (Fig. 2.1). His crude microscope consisted of a biconcave lens enclosed in two metal plates. It had a single lens, therefore it is now referred to as a single-lens microscope. Its convex glass lens was attached to a metal holder and was focused using screws. Present day microscopes are capable of magnifications greater than 250,000 times.

Time Line of Development of Microscopes

1500 B.C. Magnification: The ancient Egyptians
1000 A.D. Lenses
1200 A.D. Spectacles
1590 A.D. Optical microscope
1667 A.D. Compound microscope
1679 A.D. Reflecting microscope
1934 A.D. EM: Electron Microscope
1940 A.D. TEM: Transmission Electron Microscope
1953 A.D. SEM: Scanning Electron Microscope,
             STEM: Scanning Transmission Electron Microscope
1956 A.D. FIM: Field-Ion Microscope
             XM: X-ray Microscope
             PXM: Projection X-ray Microscope
             RXM: Reflection X-ray Microscope
             STXM: Scanning Transmission X-ray Microscope
1981 A.D. SAM: Scanning Acoustic Microscope
1981 A.D. STM: Scanning Tunneling Microscope
1987 A.D. AFM: Atomic Force Microscope

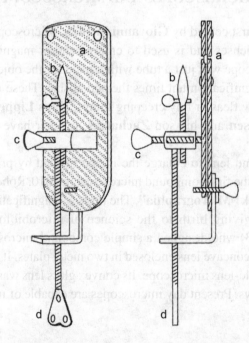

Fig.2.1 Leeuwenhoek's microscope showing (a) lens, (b) mounting pin, (c & d) focusing screws.

Types of Microscopy

a. **Light microscopy:** Uses visible or UV light to illuminate the objective. Light passes through a series of lenses, which alter the path of light and produce a magnified image of the object.

b. **Electron microscopy:** Uses a beam of electrons to illuminate the objective. The electron beam is moved around by magnets, which act like the lenses in an ordinary microscope. Electron microscopes can magnify objects over 200,000 times.

c. **Reflected Light Microscopy:** Microscopy using oblique or epi-illumination for the study of specimens that are opaque, including semiconductors, ceramics, metals, polymers, and many others.

d. **Stereomicroscopy:** This class of microscopes is extremely useful in a multitude of applications. There is a wide choice of objectives and eyepieces, enhanced with attachment lenses and coaxial illuminators that are fitted to the microscope as an intermediate tube. Working distances can range from 3–5 cm to as much as 20 cm in some models, allowing for a considerable amount of working room between the objective and specimen.

Principles of Microscopy

Image Magnification and Magnifying Power

The main purpose of microscope is magnification of image. Image magnification depends on:

- Magnifying power of objective and eyepiece
- Numerical aperture and resolving power of the instrument
- Wavelength of the light (λ)

Image is magnified due to interaction between visible light waves and the curvature of the lens. When a beam of light transmitted through air strikes and passes through the convex surface of glass it experiences **refraction**. The degree of bending depends on the refractive index of the media. Shape of the lens is an important feature affecting the degree of magnification. In a light microscope a series of ground glass lenses bend the visible light to magnify the image of the object. Use of two convex-convex lenses i.e. ocular and objective enlarges the image. Objective magnifying power is 4X to 100X and ocular magnifying power is 10X to 20X. Total magnification is equal to product of magnifying powers of both lenses.

Power of objective	×	Power of ocular	Total magnification
10X low power	×	20X	200X
40X high dry objective	×	10X	400X
100X oil immersion	×	10X	1000X

The 10X, 20X and 40X objectives are dry, meaning that there is air between the lens and the slide. 100X objective is also called oil immersion objective as it used by immersing the lens in oil. The advantage is that the working distance between the object and objective is reduced, and as the refractive index of the medium i.e. oil is better than air or water, RP is improved and the image is highly magnified, therefore, 100X is used to observe bacteria (Fig. 2.2).

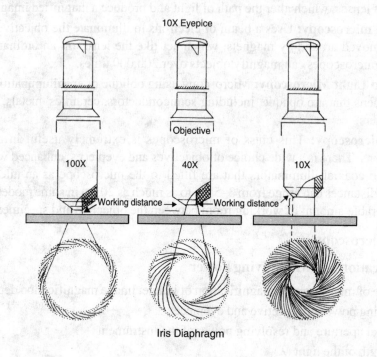

Fig. 2.2 Relation between objective lens and working distance.

Contrast and Resolution

Contrast is the difference between the brightness of various details in the object, and the difference as compared with the background. Resolution is the finest detail actually visible in the image and the most important thing in microscopy is to find a good balance between contrast and resolution. In "**brightfield**" microscopes objects are seen against a bright background therefore each is observed at the expense of the other. This is the reason phase-contrast and interference-contrast microscopes were developed.

Numerical Aperture (NA)

Numerical aperture is the property of a lens that describes the amount of light that can enter it and determines the resolving power of an objective, but the total resolution of a microscope system is also dependent upon the numerical aperture of the substage condenser. The higher the numerical aperture of the total system, the better the resolution. Numerical aperture is also dependent on the refractive index of the medium filling the space between specimen and objective and on the angle (θ) at which light rays enter the objective.

Medium having refractive index (n), same as glass, will prevent distortion of light rays and therefore the θ value will be greater and hence greater NA. Since the refractive index (n) of air is one i.e., less than that of glass, light rays bend as they pass through the glass slide and thus, objective used in air will never have NA greater than one. This is the reason microbes, especially bacteria are viewed with oil immersion lens (Fig. 2.3). Refractive index of substances used as mounting medium are given in table 2.1.

Resolving Power (RP)

Resloution is the degree to which the detail in a magnified image is retained. RP is the ability to differentiate between two closely placed objects when light passes through the space between them. RP depends on wavelength of light (λ) and numerical aperture of objective. Theoretically RP of a light microscope is 200 nm, therefore bacterial cells can be seen by light microscope. As the wavelength of visible light is 550 nm, it cannot resolve distances less than 220 nm. The wavelength of UV light lies between 100–400 nm, therefore, it can resolve distances less than 110 nm. As electrons behave both as particles and waves with a wavelength of 0.005 nm, hence EM can resolve distances upto 0.2 nm (Fig. 2.2). Use of colored filter improves RP (Blue filter: 400–500 nm). Shorter the wavelength more is the resolution.

Calculation of RP

$RP = \dfrac{0.5\,\lambda}{NA}$ where, λ = Wavelength of light used and NA= $n \sin \theta$ (NA is the function of effective diameter of objective in relation to its focal length and refractive index of medium between the specimen and object). $n \sin \theta$ is the refractive index of the medium between the object and the objective.

$$n = \frac{\text{velocity of light in vacuum (c)}}{\text{velocity of light in medium (v)}}$$

$n \sin \theta$ = angle of light or the measure of aperture. Magnitude of this is always expressed as sin value. Sin value of half aperture angle multiplied by refractive index of medium gives NA. Value of NA differs with magnification and is engraved on the sides of the objective lens.

For a compound microscope, $RP = \dfrac{\lambda\ (\text{in nm})}{2\ NA}$ or $\dfrac{0.5\,\lambda}{NA}$ or $\dfrac{\lambda}{2\,n \sin \theta}$

Smaller the value of λ, greater is the value of NA. If $\lambda = 564$ nm, NA = 1.25 (oil immersion lens) then RP = $564/2 \times 1.25$. Therefore, RP or the distance that can be resolved in this case = 225.6 nm. Higher the NA, better is the resolution. Each objective has a fixed NA. The NA of commonly used objectives are as follows:

10X————0.25, 40X————0.65, 100X————1.25

Refractive Index

The degree of deviation of light rays when they pass from one medium to the other depends on the defracting power of the medium which is called the refractive index and it effects the image magnification. Different substances have different refractive index (Table 2-1).

Table 2.1 Refractive indices of different substances used as mounting medium

Substance	Refractive index (n)
Glass	1.56
Distilled water	1.33
Cedar wood oil (immersion oil)	1.51
CCl_4 (carbon tetrachloride)	1.46
Sandal wood oil	1.51
Olive oil	1.47
Glycerol	1.47
Canada balsam	1.54
Euparal	1.48
Polysterene	1.59
Air	1.00
Naphrax	1.74

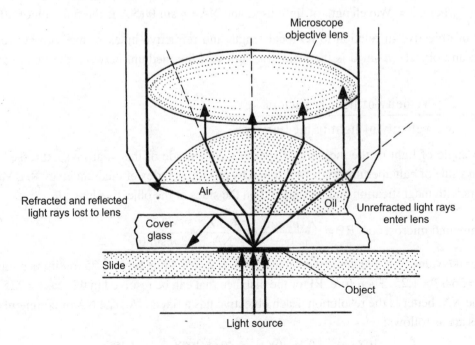

Fig. 2.3 Working of oil immersion lens

Image Illumination

"Illuminating train" is important to ensure the best image quality lenses can yield. The principle is that the light-source is focused in the object itself. So when you see the object in focus, the light source should also be in focus. It is of two types:

1. **Critical illumination:** Here the filament of the light bulb itself is the light-source. Because it's in focus in the field of view, the image of the filament would be very objectionable and therefore the light bulb is lightly frosted.

2. **Köhler illumination:** Here the light source consists of an iris diaphragm in the lamp. This lamp iris is not to be confused with the condenser iris. As the lamp iris is in focus in the field of view, its size can be so adjusted so that only the field of view is illuminated. This reduces glare.

Types of Microscopes

1. Based on the number of lenses, microscopes can be classified into two types. They are:

   a. **Simple microscope**: Consists of only a single optical lens.

   b. **Compound microscope**: Consists of at least two lenses; Objective and Eyepiece.

2. Based on the principle involved in magnification, there are two types of microscopes. They are:

   a. **Light microscope**: A beam of visible light is focused through an optical lens to achieve magnification. Various types include: Bright field, Dark field, Fluorescence, Phase contrast, Confocal Scanning Microscope, and Nomarsky Differential Interference Contrast Microscope (NDIC).

   b. **Electron microscope**: Uses a beam of electrons instead of light for magnification. There are three types of Electron microscope. They are; Transmission Electron microscope (TEM), Scanning Electron microscope (SEM), and Scanning Tunneling Microscope.

LIGHT MICROSCOPE

Components of Light Microscope

The optical and mechanical components of the microscope, including the mounted specimen on a glass micro slide and cover slip, form an optical train with a central axis that traverses the microscope base and stand (Fig. 2.4).

Mechanical Components

These are required for operating the instrument and include:

- **Stand** or frame for all parts.

- **Base or foot** is horseshoe-shaped and supports the stand.
- **Pillar** arises from the base and carries an arm or limb and an inclination joint at its top end.
- **Stage** is a flat plate on which the specimen (usually mounted onto a glass slide) is placed for observation. It is attached to the top of the pillar. It has a circular opening in the center called stage aperture. It is also equipped with two stage or spring clips, which hold the slide. Stages are often equipped with a mechanical device that holds the specimen slide in place and can smoothly move the slide back and forth as well as from side to side. Other stages are designed to allow rotation of the specimen through 360 degrees or to provide anchors for auxiliary light sources, specimen manipulation tools, and other accessories.

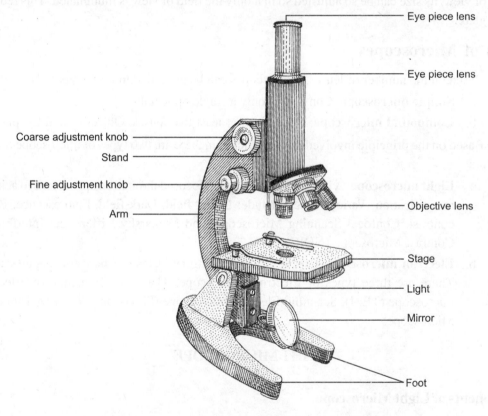

Fig.2.4 Compound microscope

- **Sub stage** is positioned below the stage and holds the condenser.
- **Body tube** is hollow, cylindrical, usually having a standard diameter. It is attached to the arm in such a manner that it is in line with the stage aperture. It carries the ocular on the top. It can be moved up and down by two screws or focusing knobs. The mechanical tube length of an optical microscope is defined as the distance from the nosepiece opening where the

objective is mounted to the top edge of the observation tubes where the eyepieces (oculars) are inserted. Earlier microscopes had a fixed tube length ranging from 160 mm to 210 mm, depending upon the manufacturer and application. Modern microscopes are equipped with infinity-corrected objectives that utilize a tube lens in the microscope body to form a parallel region of light waves into which optical accessories can be inserted without seriously affecting image quality.

- **Focusing knobs** are attached to the body tube. They help focus the image of the specimen through the eye piece and are of two types:
    1. Coarse adjustment knob with a pinion head.
    2. Fine adjustment knob with a micrometer head. It is used to sharpen the image.
- **Nosepiece** is a saucer shaped structure and is attached to the base of the body tube. It has fixed places where the objectives can be attached. The whole assembly i.e. nosepiece with the objectives can revolve in such a manner that objectives can be changed according to the required magnification.

Optical Components

These are required for image formation and include:

- **Mirror** is mounted on a frame and attached to the pillar just below the condenser. It has two surfaces, one plane and the other concave. It can rotate freely and can be focused in three different directions. It collects and directs the beam of light used to illuminate the specimen. Modern microscopes have a fixed illumination system.
- **Objectives** determine the quality of images that the microscope is capable of producing. Usually, they are of three magnifications; 10X, 40X and 100X (oil immersion), but objectives of other magnifications can also be used. The objective lens is usually a compound lens, a combination of two lenses made from different kinds of glass. **Water Immersion Objectives** use water in the place of oil as the immersion medium.

Objectives are classified into the following classes based on the degree of correction:

- a. **Nonchromatic:** without correction.
- b. **Achromatic:** with improved correction. These compound objectives are used for routine microscopy. They correct both distortion and aberration.
- c. **Apochromatic:** with greatly improved correction. It reduces chromatic aberration as well as spherical aberration. Chromatic aberration is more finely corrected which produces very high quality image revealing the true colours of the specimen without distorting the image. (Used mainly for photomicrography).
- **Ocular lens or eyepiece** in combination with objectives further magnifies the image. Eyepieces are usually of standard dimensions and are made up of two or more lenses; upper component or eye lens is the **magnifier** while the lower is called **field lens**. Lenses of variable power can be exchanged with each other. Eyepieces are also designed to work

together with objectives to eliminate chromatic aberration. Commonly used eyepieces are; Huygens (–) and Ramsden (+) compensating and high eye point available in 1X, 2X, 5X, 10X, 15X and 20X magnification.

- **Objective correction collars** allow adjustment for fluctuations in cover glass thickness. When objective numerical aperture is 0.4 or less, cover glass of a standard thickness of 0.17 mm and a refractive index of 1.515 is satisfactory. Cover glass thickness variations of only a few micrometers result in dramatic image degradation due to aberration especially when using high numerical aperture dry objectives (numerical aperture of 0.8 or greater), and it becomes worse with increasing cover glass thickness.

- **Condenser** fills the objective evenly with a cone of light. This cone of light is narrow for an objective of low NA (e.g. 0.25) but needs to be very wide for one of high NA (e.g. 1.25). The cone of light exiting from the condenser must be adjusted with the condenser iris diaphragm. The condenser can be moved with the condenser focus control but the position of the condenser is rigidly defined and is not arbitrary. Most common mistake is to rack it up or down to adjust the brightness of the image. For a normal condenser with NA between 0.9 and 1.4, the top lens of the condenser almost touches the lower surface of the slide and brightness of the image is controlled with a light dimmer fitted to the illuminator.

  For low magnifications (e.g. 3X objective), it is difficult or impossible to illuminate the large field of view entirely with a condenser of NA 0.9 or more. For such conditions special low-power condenser can be used or top lens of a normal condenser is screwed off or swung aside in some models. For correct illumination, the position of the condenser will then be much lower than with the top lens on.

Condensers are of following types:

a. **Abbe-type:** Cheap but have a poor image-forming quality.

b. **Aplanatic:** Less expensive, yield good illumination.

c. **Achromatic-aplanatic:** Expensive but give the best illumination.

d. **Darkfield condensers:** illuminate the objects "from the side" instead of "straight through" as in brightfield.

e. **Phase-contrast condensers:** are actually often of the simplest uncorrected type and only differ from the ordinary condenser in having a set of ring-shaped "annular stops", one for each phase-contrast objective, centerable and mounted on a rotating disc.

f. **Interference-contrast condensers:** are of complex optical design and are used in conjunction with additional special components in the microscope.

   - **Filter holder:** Holds the color filters used to change the wavelength of light.

   - **Iris diaphragm** is attached to the condenser and regulates the light reflected from the mirror into the condenser and on to the object.

Distortion and Optical Aberrations

When only one lens is used, we often get image distortion. Distortion of images is known as optical aberration. When we use a compound lens, the second lens corrects any distortion from the first lens. Optical aberrations are mainly of two types:

   a. **Spherical aberration:** distortion based on the shape of lens.
   b. **Chromatic aberration:** distortion based on the color of light, as the same amount of all the colors making up light are not refracted the same amount when passing through a glass lens.

   Correction of aberration: Distortion can be corrected by using compound lenses i.e. lenses of different shapes and composition.

Useful tips for Viewing Microbes

The following steps are useful in visualizing microbes with a microscope :

   • Reduce light by adjusting the iris diaphragm. This will make the viewing field darker, but the thin, unstained cells will have much greater contrast and will be clearly visible.

   • First focus with 10X objective, once you have found the cells, which will appear as very little dots with the 10X objective, then switch to the 40X objective.

   • For better resolution and higher magnification use oil immersion objective.

Types of Light Microscopes

Bright field

It is the most commonly used compound light microscope with a resolution of 200 nm. It uses visible light. **Light source is placed below the specimen and light is transmitted through the specimen.** Requires specimen fixing and staining, therefore it is very rarely used for observing live specimens. The technique is best suited for utilization with fixed, stained specimens or other kinds of samples that naturally absorb significant amounts of visible light. Images produced with bright field illumination appear dark and/or highly colored against a bright, often light gray or white background.

Parfocal

This microsope also uses visible light. Various lenses are so adjusted that after the specimen has been focused with one lens, it remains in focus even when another objective is switched in its place.

Phase contrast

Fritz Zernike received **Nobel Prize** for physics in **1953** for this invention. Principle involved is that light rays move at different speeds through materials having different refractive indices. Greater the refractive index of cell or cell structure, greater is the retardation of light. When these out of phase rays are brought together, contrast is generated due to interference (Fig. 2.5). In a phase contrast microscope **light is placed below the specimen** and is split into two types:

- **Direct or transmitted light**, which passes through the specimen and is focused on the phase shifting plate. This plate either delays or advances the ray.
- **Scattered or diffracted light**, which passes through the margin of specimen and does not strike the phase shifting plate, therefore, its phase is not disturbed.

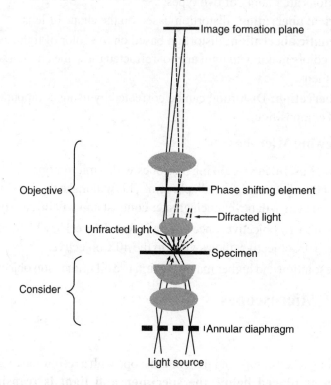

Fig. 2.5 Path of light and image formation in phase contrast microscope

When these two unite again, they are not in phase. This **microscope converts the differences in phases of light waves into visible difference in contrast.** Depending on the phase shifting element used, specimen may appear light against dark background (dark contrast) or vice versa (bright contrast).

Application: It has a limited use in microbiology but is useful for examining growth, cell division, flagellar motility, spore and capsule formation and cytopathic effects of virus.

Merits: Fixing and staining is not required, therefore, live specimens can be visualized.

Demerits: Image may be surrounded by a halo of light.

Dark field microscope

In this type of microscope, the condenser does not permit the transmission of light directly through the specimen and into the object. Instead, it focuses light on the specimen at an oblique angle in such a manner that light, which does not reflect off an object, does not enter the objective (Fig. 2.6).

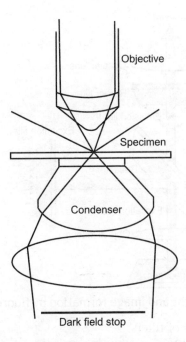

Fig. 2.6 Path of light and image formation in dark field microscope

Light passing through the slide is not seen, therefore, the field appears dark, but light reflecting off the specimen is seen and, therefore, the specimen appears bright on a dark background. Arrangement of condenser and the objective is such that none of the zero order rays enter the objective. Specimen scatters the thin cone of light incident on it and acts as a secondary illuminator. Hence, it appears as a glowing object against a dark background. Specimens, which do not require higher magnification, are screened with the use of Abbe condensers with dry objectives. To observe specimens with oil immersion objective, special condensers with high NA such as cardoid and paraboloid are used.

Merits: Fixing and staining is not required, therefore, live specimens can be visualized.

Precautions: Equipment, slides etc. should be dirt and dust-free.

Fluorescence Microscope

In this type of microscope specimen is stained and illuminated by one wavelength (excitation wavelength), and observed due to light emitted at a different wavelength (emission wavelength). The excitation wavelength is usually in the UV range and the emitted light is in the visible range. Use of dye makes the organism appear larger, therefore, it is easy to use low power objective and study a comparatively larger area. It is not used with oil immersion lens because oils become fluorescent after exposure to air and especially to UV rays (Fig. 2.7). Analar grade glycerol can be used instead but it is also not the solution to this problem.

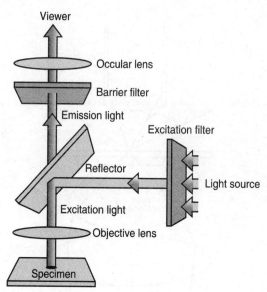

Fig. 2.7 Path of light and image formation by fluorescence microscope

The microscope has two types of filters:

    a. **Excitation or primary filter:** permits the passage of excitation wavelength. It is placed between illumination source and object. It is usually blue and allows UV to pass.

    b. **Barrier or secondary filter:** prevents the passage of all wavelengths except the emission wavelength i.e. light from flourescent specimen. It is placed in the eyepiece or anywhere between the eye and the object.

Types of fluorescence microscopes

Fluorescence microscopes are of two types:

    a. **Transmitted fluorescence**: excitation light is transmitted from below the specimen.

    b. **Epifluorescence** *(Gk., epi: upon)*: excitation light is transmitted to the specimen through objective lens.

    Applications: useful for immuno-identification of pathogenic microbes as dyes can be linked to antibodies and observed by immuno-fluorescence. It is also used to study structural details.

    Precautions: When using UV light, always use barrier filters to prevent UV light from reaching the eye as it can cause blindness.

    Flourochromes or fluorescent dyes frequently used: Acridine orange, auramine-o, thioflavin S, thiazoyellow-G, morin and flourescein isothiocyanate.

Confocal Scanning Microscope

This is actually a type of light microscope, which does not form the conventional 2D image. A beam of light emitted from a laser is focused to a point by an objective lens and scanned through the sample with the help of two scanning mirrors. One scans while the other objective magnifies the image.

Light detector measures the interaction with each point in the object as it is scanned. Light detected is the light diffracted from a specific point of focus and light diffracted from other points is not detected at all. This eliminates the diffracting light which blurrs the image in a light microscope. Image is placed on a digital computer screen for analysing. Medium for mounting specimen is air. Specimen is stained and mounted on glass slides. Focusing is entirely computerized and magnification is digitally enhanced.

Nomarsky Differential Interference Contrast Microscope (NDIC)

Georges Y Nomarsky laid the principles for this in the 1950s. Light waves, which are out of phase are combined to produce interference, which alters the intensity, produces high contrast and shows transparent specimens in 3D. It is made up of polarizing filter, interference contrast condenser and prism analyzer plate.

Light source is below the specimen. It emits unpolarized light, which passes through field diaphragm and gets polarized by the polarizing filter. This polarized light has a defined pattern of the aligned light waves. It is further split into two beams, which lie at right angles to each other. These beams travel parallel to each other through the specimen, combine and finally enter the analyzer. In this way, when two rays pass through the specimen they are differentially diffracted. Recombination of these produces interference. A false 3D picture is thus produced due to the stereoscopic effect.

Degree of the 3D image is dependent on the refractive index differences at the boundary of the specimen. At the edges or borders, due to large refractive index, a much better contrast is produced. Different structures appear in different colors and are related to phase changes in light as it passes through each structure. Hence, images produced are brilliantly colored.

Difference from phase contrast: Although like phase contrast it uses difference in refractive index and produces contrast by interference, NDIC uses prisms in the place of objective, while phase contrast uses complementary pair of opaque rings in the same position.

Working with a Light Microscope

When using a light microsope concave side of the mirror is adjusted so that a beam of light is focused up on the body tube. Light is reflected from it on to the object and then passes on into the objective and the ocular. This light forms an inverted image of the object called the **primary image**, which is once again inverted and magnified by the eyepiece to form a virtual image. The distance between the eye and the **virtual image** is called the **projection distance**.

Condenser position is adjusted by removing the eyepiece and observing the back of the objective. Condenser diaphragm is closed, till its edge is just visible and then with the help of screw adjustments, condenser is positioned in the center till its iris aperture is just in the center of the objective lens aperture. Iris diaphragm is adjusted so as to fill 2/3rd of the back aperture of the objective with light.

Both the eyepiece and the objective are aligned in a straight line also coinciding with the stage aperture. The slide carrying the object is kept on the stage with the specimen positioned directly above the stage aperture. With the help of the coarse adjustment screw, body tube is lowered with 10X objective in position till the object is visible and the image is sharpened with the help of fine adjustment knob.

Tips: If the light is very bright, adjust the condenser and iris diaphragm. For making observations on higher magnification, change the objective by rotating the nosepiece. For using oil immersion, place a drop of cedar wood oil on the specimen, lower the body tube till the 100X objective dips in the oil but is not touching the specimen.

Merits: Easy to use. Does not require special light source. Can work in sunlight also.

Demerits: RP not very good. Requires specimen fixation and staining, therefore, not suitable for observing living specimens.

Precautions: The objective should not touch the specimen. Sufficient light should be provided which is neither too bright nor too diffused. Always clean the objectives especially the oil immersion with lens cleaning paper after use. Never use cloth or tissue as it scratches the lens or leaves lint on it.

ELECTRON MICROSCOPE (EM)

Electron microscopes can magnify objects over 200,000 times. **Max Knoll** and **Ernest Ruska** invented it in **1931** at the Technical University in Berlin. An electron gun generates electrons, which are concentrated by other components of the microscope into a fast moving, narrow beam, which illuminates the specimen. Electromagnetic lenses made up of wire and encased in soft iron casing focus the beam. When an electric current is passed through the coil, an electromagnetic field is generated which focuses the electrons.

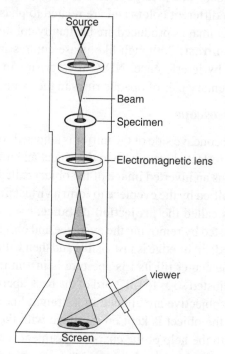

Fig. 2.8 Image formation in TEM

In all, three lenses are used. One is placed between the source of illumination and the specimen. It focuses the electron beam on to the specimen and can be compared to the condenser of a light microscope. The second and third lenses are placed on the other side of the specimen. They act like the objective and the eyepiece, and magnify the image. A phosphorus-coated screen, which fluoresces upon irradiation by electrons, receives the beam and converts it into the final image. Permanent record of the image can be made on photo plates.

Merits: It has a greater resolution, as the wavelength of the electron beam is much shorter than the visible light. This beam is generated at an extremely high voltage—60,000 V, therefore, the wavelength is 0.005 nm with an RP of 0.2 nm approximately. This is 1000 times better than the light microscope. The magnification is also much higher. It is used to study ultra structural details.

Demerits: The microscope is very expensive and requires technical skills, as the specimen needs to be specially prepared. Specimens are dried, therefore, cannot be used to study live organisms. There is a distortion of structures and appearance of artifacts due to exposure to electrons. These microscopes can only view dry specimens in vacuum.

Precautions: The insides of the microscope should be kept under high vacuum (10^{-4}–10^{-6} mm) in order to prevent entry of air in it as air interferes with electron movement. Specimen should be ultra thin as the electron beam has a poor penetration power.

Types of EM

Electron microscopes are of two types,

- Transmission electron microscope which provides a 2D internal view.
- Scanning electron microscope, which provides a 3D surface view.

Transmission electron microscope (TEM)

TEM is **electron illuminated**. It has very high magnification and resolution. TEM uses hot tungsten filament at 30–150 KV in an electron gun. Heat draws electron from the filament and causes them to accelerate, forming a fine electron beam that passes by an anode via high voltage that is present between filament and anode. Electron beam is focused on the specimen with an electromagnetic condenser lens by varying the current to the lens (Fig. 2.8). Thin slice of specimen is obtained and embedded in collodion films or other supporting material on copper grids. The electron beam passes through the section to reveal the internal structure. Electron beam travels through vacuum where the specimen is placed. Lenses used are of two types : one electrostatic and few electromagnetic. Focusing and magnification produced by electrical contrast is generated by electron scattering.

Scanning electron microscope (SEM)

SEM is an excellent equipment for studying the surface topography. It was invented by **Knoll, Von Ardenne, Zworytein** *et al.* in the **1960s**. It is a **combination of an EM and television**. Electrons are bombarded on the object from above unlike light microscope where it is transmitted from below and do not pass through the specimen. Specimens are freeze-dried and then coated with a thin film of metal such as gold or platinum and surface scanned by a 20,000 V beam of electrons generated by hot tungsten or lanthanum hexoboride cathode gun. On striking the object, the electron beam emits secondary electrons, which are transmitted to a collector. Some primary electrons are reflected or back scattered from the specimen but their number is very small or negligible when compared to the secondary ones emitted from the specimen.

Thus, the signal generated from the secondary electrons is much more stronger. From the collector, electrons are transmitted to the detector, which consists of a substance that emits light when struck by an electron. Emitted light is converted into an electrical current and is used to control the brightness of an image on cathode ray tube (CRT) screen like that of a TV. The final image is produced on a cathode tube screen. The secondary electrons are characteristic of the surface, therefore, the intensity of the image reflects components and topography of the specimen. Medium for mounting is vacuum. Focusing and magnification are electrically controlled. Specimen contrast is provided by electron scattering.

Precautions: Complete coating is essential as beam produces charge on uncoated specimens which causes image distortion.

Scanning Tunneling Microscope (STM)

Gerd Binnig and Heinrich Rohrer invented STM in 1981. They shared the **Nobel Prize** for Physics in **1986** for this work with the German scientist Ernst Ruska, who designed the first electron microscope. It provided the first images of individual atoms on the surfaces of materials.

STM have atomic resolution. Electrons are emitted from a sharp point of a conductor which is computerized. The computerized STM moves above a material. Electrons jump from the tip of the STM to the surface of the material being scanned. This creates an electric charge. The electric current changes as the tip is moved across the material. By moving the tip back and forth across the material, the computer can map the electron clouds (and the atoms) (Fig. 2.9). Because the tunneling current is so sensitive to distance, STM's have a vertical resolution of 1/100th of an atomic diameter.

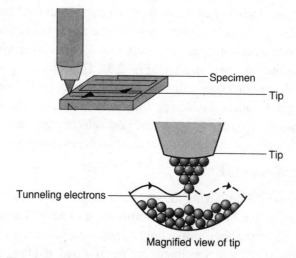

Fig. 2.9 Path of light and image formation in STM

SOME NEW GENERATION MICROSCOPES

Field-Ion Microscope (FIM): Whereas an EM forms images by bombarding the specimen with electrons, the FIM forms an image of the electron source itself. Electrons are emitted from the very sharp point of a conductor, which can then be imaged with resolution down to the atomic level.

Scanning Acoustic Microscope (SAM): Uses high frequency sounds as a correspondingly short wavelength. Resolution of 50 nm has been achieved with 8 gigahertz sound waves.

Scanning Ion-Conductance Microscope (SICM): This microscope is used for surface mapping. An SICM head has a very narrow pipette end, clogging of which impedes an electric current between the medium and the pipette interior. Clogging occurs when it is near the cell's ion channel and results in its surface mapping. SICM can also detect chemical differences.

Near-field Scanning Optical Microscope (NSOM): This uses visible light with a detector smaller than the wavelength of the light and works like an optical stethoscope.

Atomic Force Microscope (AFM): Works on the same principle as a record player, except it's tip is much sharper and the tracking force is about one millionth greater.

Friction Force Microscope (FFM): This is a modified AFM which measures how much the tip drags when it moves across the surface.

Electrostatic Force Microscope (EFM): Measures electric charge on the surface. It is used to study contact electrification i.e. the static electric charge created when two surfaces touch or rub against one another.

Scanning Chemical-Potential Microscope (SCPM): Measures chemical reactivity or the chemical makeup of the surface atoms. It can therefore tell which atoms are where.

Table 2.2 Comparison of working range of different microscopes

Types of microscopes	Specimen size	Organisms viewed
Light microscope	1 mm–10^{-3} mm	Algae, Fungi, Protosoa, RBC, Bacteria, Rickettsia, PPLO.
UV microscope	1 mm–10^{-4} mm	Algae, Fungi, Protosoa, RBC, Bacteria, Rickettsia, PPLO, Viruses.
SEM	1 mm–10^{-3} mm	Algae, Fungi, Protozoa, RBC, Bacteria, Rickettsia, PPLO.
TEM	10^{-2}–10^{-6} mm	RBC, Bacteria, Rickettsia, PPLO, Viruses, Macromolecules.

CARE AND USE OF THE MICROSCOPE

The following points should be noted while handling and using a Microscope:

- Never blow on the lens as it deposits saliva on it.
- When using pressurized air dusting, always use an inline filter to trap oil and other contaminants.
- Never clean the lens with a dry lens cleaning paper.
- Never use a facial tissue paper or bibulous paper to clean lenses as they may contain glass filaments, which can scratch lenses.
- Linen or chamois may be used for cleaning but may not be as convenient as lens tissue.
- Stubborn stains can be removed by xylene but lens mounting glues are soluble in alcohol, therefore use xylene sparingly.
- For dirty lenses, use lens cleaner solution available at any camera supply store.
- Always keep the microscope tubes, eyepiece and objective closed with dust plug or as appropriate.
- Do not touch lenses since fingerprints degrade image quality.
- Internal optics should always be professionally serviced if needed.
- Use proper immersion liquids on immersion objectives.
- Never use immersion liquid on non-immersion objectives as it can damage the lens mounting glue.
- Always keep the microscope covered when not in use.
- Avoid extreme temperature and high humidity in work areas.
- Store the microscopes in circulating air.
- In very high humidity, store optical parts in tightly covered containers with desiccant and keep them very clean to prevent mold growth on the optic coating.
- Never use sub-stage diaphragm to control brightness.
- Wipe body of the microscope with a soft cloth dipped in alcohol or soap water.
- Keep sliding parts lubricated with a petroleum jelly or manufacturer-recommended lubricant.

How to clean a lens or any other Optical Instrument

While cleaning a lens (or any other optical instrument), the following points should be noted:

- First blow away dust and grit by a rubber bulb.
- Breathe on lens to fog it, and then use lens tissue for light cleaning. Fogging is basically water and is not harmful.
- Crumple lens tissue to create several folds.
- Do not touch that part of the tissue that will come into direct contact with the lens. Touching transfers natural oils from the fingers to the lens tissue.
- Apply very little lens cleaning solution to the lens tissue and blot the tissue against absorbent material to prevent fluid from entering the lens mount.
- Wipe the lens very lightly to remove gross dirt that was not blown away by the rubber bulb.

- If required, repeat the cleaning process with a new piece of lens tissue and with more pressure to remove oily or greasy residue.
- Once again breathe on lens to fog it, and then use lens tissue to clean it.

Adjustment, set-up and illumination of a Binocular Compound Microscope

The following points should be followed during adjustment, set-up and illumination of binocular compound microscope:

- Both the eyepieces (oculars) must be adjusted according to the user's eyes.
- Adjust interpupillary distance by lengthening or shortening the distance between centers of oculars to match the distance between centers of pupils of the eyes.
- One eyepiece, or the tube into which it fits, is usually adjustable. Adjust the microscope for each eye.
- First focus the specimen slide at a low magnification.
- Next cover the eyepiece, which has the focusing eye tube with a card and then with both eyes open, bring the specimen into focus for the other eye with the fine focus knob.
- Keep relaxing vision by blinking or looking up frequently at distant objects or by staring into infinity to prevent eyestrain.
- When a sharp focus is obtained, cover the other eyepiece and now use the focusing ring on the open eyepiece to bring the same point on the slide back into focus.
- Distance between the eye and the eyepiece is very important. Too close or too far reduces the field of view and the specimen appears blurred. To find the right distance, move in slowly until the field appears the widest and sharpest. This distance, from the pupil of the eye to the lens, is the eye point or exit pupil of the microscope.
- Use of eyeglasses during microscopy is not essential as focusing of the microscope and adjustment for each eye corrects most conditions of near or farsightedness.
- Moderate astigmatism does not hamper usage of microscope but for serious astigmatism and other eye problems, prescription eyeglasses should be worn.
- Koehler illumination technique should be used to achieve optimum resolution and illumination.
- For this, close the lamp (field) iris diaphragm, if present, at the base of the microscope, and bring the image into focus by vertical adjustment of the condenser.
- Now bring the focused circle in the center field by using centering screws or knobs on the condenser, present on the condenser or on the base near the lamp diaphragm. Open the lamp diaphragm until it is just past the field of view. In the case of low power objectives, this may not be possible until the auxiliary lens in the condenser is correctly adjusted.
- If the microscope has an auxiliary swing-in (or swing-out) lens in the condenser, follow manufacturer's recommendations for its correct use with various objectives.
- Use the coarse and fine adjustments to focus the specimen.
- If the microscope has no lamp diaphragm, place a piece of paper with a small hole cut in it over the lamp opening and make the same adjustments to focus on its inner edge.
- Next adjust the substage (aperture) diaphragm. This adjustment affects the resolution and contrast of the image.

- The substage diaphragm adjustment requires that it should be set at 2/3 the size of the microscopic field in an open or closed position when seen by removing the eyepiece and looking down the tube.
- To achieve this setting without removing the eyepiece, open the substage diaphragm fully and gradually close it while looking through the microscope until the image gains a sudden increase in sharpness and detail.
- This can be checked by removing the eyepiece and looking down the tube.
- Turn down light if it is too bright by using the rheostat or by adding neutral density or other filters.
- Never use substage diaphragm to control brightness as resolution will be affected if it is closed too far or opened too much.
- Closing gives more contrast, but impairs resolution and false images are formed due to diffraction lines or fringes.
- Repeat the procedures for Koehler illumination with each objective.
- For phase-contrast microscopy, all basic steps are same but the substage diaphragm is not used to make adjustments in order to bring the phase annulus and annular diaphragm into coincidence. Use of a green filter is usually recommended.

SPECTROPHOTOMETER

Spectrophotometer (Fig. 2.10 a) measures the transmission or absorption of light in liquids or solids as a function of wavelength. It consists of two parts, viz. **Spectrometer** for producing light of any selected color (wavelength) and **Photometer** for measuring the intensity of light.

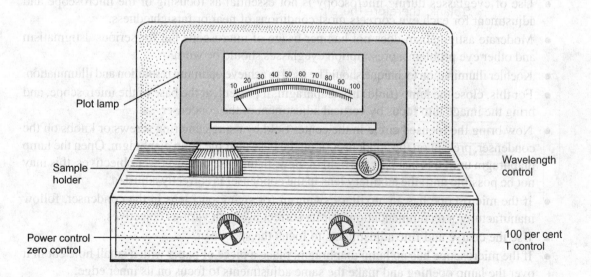

Fig. 2.10 (a) Graphic representation of a Spectrophotometer.

Principle

The concentration of solid or particulate matter such as solute, cells etc. in a solution can be measured by a spectrophotometer as a function of amount of light transmitted (**% Transmittance**) or absorbed (**Absorbance**). Absorbance is represented as **Optical Density (O.D) of the solution.**

The assay solution is filled in a special container called cuvette, which is then placed between the spectrometer beam and the photometer. Spectrometer shines light of a definite wavelength through the sample. Amount of light passing through and absorbed by the sample in the cuvette is measured by the photometer, which delivers a voltage signal to a display device. The signal changes as the amount of light absorbed by the liquid changes (Fig. 2.10 b, c).

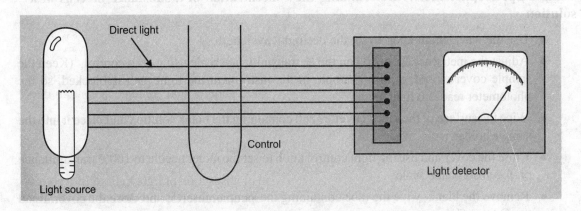

Fig. 2.10 (b) Placement of sample between spectrometer and photometer. The cuvette contains blank, therefore there is 100% transmission of light.

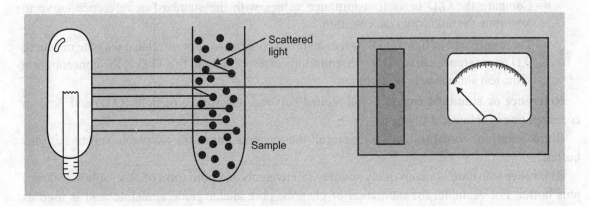

Fig. 2.10 (c) Placement of sample between spectrometer and photometer. The cuvette contains assay solution, light gets scattered due to solute molecules therefore transmission of light is less than 100% and O.D of the solution is more.

Concentration of the substance can be measured by determining the extent of absorption of light at the appropriate wavelength. By comparing this amount with a graph of known concentrations **(standard or reference curve/graph),** the concentration of an unknown solution can be determined. Degree of absorbance of light is proportional to the concentration of substance/cells present in the sample. Particulate matter turns the solution turbid. Denser the liquid, lesser is the light transmitted i.e % transmittance is less and optical density is more. Thus % transmittance and O.D are reciprocal of each other.

The spectrophotometer uses either the visible range of light or UV light. In the visible range the light source emits yellow light which can be changed to blue, green or blue by adding colored filters.

Using a Spectrophotometer: Determining the concentration of a substance or cells in a solution

- Use the wavelength knob to set the desired wavelength

- Adjust the meter needle to "0" on the % transmittance scale using zero control. (Keep the sample cover closed, with no sample in the instrument the light path is blocked, so the photometer reads no light at all).

- Wipe the surface of the tube / cuvette / cell containing the blank solution and place it into the sample holder.

- Close the cover and use the light control knob to set the meter needle to 100% transmittance on the absorbance scale.

- Remove the blank, wipe the tube containing the sample, insert it and close the cover. Read the O.D or % transmittance.

- Remove the sample tube, readjust the zero, and recalibrate if necessary before checking the next sample.

- Compare the O.D or % transmittance values with the standard or reference curve to determine the unknown concentration.

- The absorbance of light (O.D) by the unknown concentration is correlated with the matching O.D on reference curve. The concentration corresponding to this O.D is the concentration of the test substance in the solution.

Reference or standard curve: graph plotted between absorbance of light (O.D) and known concentration of solute in a solution.

Blank solution: contains assay solution (all the reagents minus the assayable substance) plus buffer.

Reference solution: contains assay solution (all reagents and pure form of assayable substance) plus buffer. For example, for estimation of phenols, pure analar grade cinnamic acid is used as sample in the assay solution.

Sample or test solution: a solution that contains unknown amount of assayable substance plus all reagents plus buffer.

Assayable substance: substance whose concentration in the solution has to be assayed.

Assay solution: solution containing the reagents needed for estimation of concentration of assayable substance.

Difference between the blank solution and test (sample) solution

The concentration of the test substance in the blank solution is zero whereas test solution contains an unknown amount of assayable substance. Blank solution transmits as much light as is possible with the assay solution minus the test substance used whereas, test solution absorbs more light than the blank solution therefore, transmits less light to the photometer.

Water cannot not be used as reference solution. Test solution contains; assayable substance + assay solution (reagents + buffer). Therefore when its O.D is measured, it will be a sum of light absorbed by the assay solution and the amount of the test substance present in it. Since water does not contain any of these substances, it transmits 100% light. Thus, O.D of water will be nil and the results of O.D values of test solution will not be accurate. Therefore, the assay solution is always used as blank solution unless the assay solution contains only water + assayable substance.

Preparation of a Reference Graph or Standard Curve

A standard curve or calibration graph of absorbance v/s concentration is made by measuring the O.D of known concentrations of pure chemical and then using it to find out the concentration of same chemical in an unknown solution. For example in order to find out the concentration of protein in a test solution, standard curve is first generated with pure albumin or any other protein using the same assay method as used for assay of protein in test solution. A graph between concentration and O.D. values is then plotted with O.D on x axis and concentration on y axis. This graph can now be used to find out the concentration of protein in unknown solution by simply correlating the concentration corresponding to O.D of unknown substance.

Method

- Make a stock solution of the chemical for which reference curve is to be plotted. For example analar grade of cinnamic acid is used to plot a standard curve for phenols.
- Dilute the stock to known concentrations such as 1 mg, 2 mg, — 10 mg, etc.
- Switch on the spectrophotometer half an hour before starting the experiment.
- Set the required wavelength.
- Set 100% transmittance with blank.
- Perform the quantitative test for assay of each known concentration of cinnamic. Measure the O.D of each concentration.
- Plot a graph between absorbance and concentration.
- This graph serves as a reference curve for finding the concentration of phenolics in an unknown solution.

The absorbance of light (O.D) by the unknown concentration is correlated with the corresponding O.D on reference curve. The concentration corresponding to this O.D is the concentration of the test substance in the solution.

Things to remember when using the spectrophotometer

- Always switch on the instrument 30 minutes before use.
- Set the required wavelength.
- First set the % transmittance to 0.
- Clean the cuvette or test tube provided to hold the liquid. Any marks or fingerprints on it will hinder transmittance.
- Now set 100% transmittance or zero O.D with the blank (distilled water or the liquid/broth used for experiment).
- Use a separate cuvette or test tube for measuring O.D of test solution/suspension or wash and dry the cuvette thoroughly before using it again.
- Fill the cuvette with test solution/suspension and note the readings.
- Always reset the instrument before using it again for the next sample.

pH METER

When an acid or a base is added to water, it dissociates into positive and negative ions, for e.g. Hydrochloric acid (HCl) dissociates into H^+ and Cl^-. Sodium hydroxide (NaOH) dissociates into Na^+ and OH^-. Stronger acids give more protons and stronger bases give more OH^- ions. Neutral substances have an even balance of protons and OH^- ions. The negative log of hydrogen ion concentration of any solution is known as its pH and indicates the amount of hydrogen ions available for reaction. pH of some common substances are given in Table 2.3.

Table 2.3 Some common substances and their pH

Bases	Neutral	Acids
Sodium Hydroxide: 14.0	Pure Water: 7.0	Milk: 6.6
Lime (Calcium hydroxide): 12.4		Tomato: 4.5
Ammonia: 11.0		Wine and Beer: 4.0
Milk of Magnesia: 10.5		Apple: 3.0
Baking Soda: 8.3		Vinegar: 2.2
Human Blood: 7.4		Lemon Juice: 2.0
		Battery Acid: 1.0
		Hydrochloric Acid: 0

pH meter (Fig. 2.11a, b) is an instrument, which uses an ion-selective electrode (ISE) that ideally responds to only one specific ion, such as the H^+ concentration of the solution, to measure the pH of a solution. The electrode produces a current, proportional to the concentration of the H^+ concentration, which is converted into a pH reading and displayed on the meter. Based on the type of display, pH meters are of two basic types: Digital and Analog.

Fig. 2.11a. pH meter

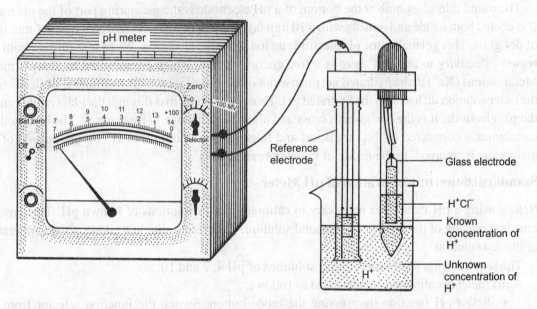

Fig. 2.11 b. Graphic representation of a pH meter showing reference and glass electrodes.

A pH meter consists of:

- H^+-selective membrane
- Internal reference electrode
- External reference electrode
- Meter with control electronics and display

Functions or modes of a pH meter are: Standby (0), pH, mV, and ATC (automatic temperature control is used when measuring pH of solutions whose temperatures are changing. It needs a special temperature probe).

Working Design and Principle

Earlier pH meters used to have two separate electrodes; glass and reference. Nowadays single or combination electrode is being used.

Combination electrode combines both the glass and reference electrodes into one body. The internal reference electrode is filled with a saturated KCl and a 0.1 M HCl solution. The cathode terminus of the reference probe is also present within this electrode. The inner tube is the reference end. It does not come in contact with the test solution, thus KCl (reference solution) remains unchanged. The anode terminus is wrapped around the outside of the inner tube. Reference solution is present in both the inner as well as the outer tube. The outer tube is in contact with the test solution by means of a porous plug, which acts as a salt bridge. It contains solution of KCl, which mixes with the outer solution. Since due to ion loss and evaporation, KCl solution gets depleted, hence it needs to be replenished regularly.

The round thin glass bulb at the bottom of a pH electrode is the measuring part of the pH meter. It is coated both inside and outside with a 10 mm layer of a hydrated gel separated by a 0.1 mm layer of dry glass. This gel makes the pH electrode an ion selective electrode. When ions diffuse from one region of activity to another, there is a free energy change, which is measured by the pH meter. Metal cations (Na^+) in the hydrated gel diffuse out of the glass into the outer solution, while H^+ from the outer solution diffuse into the hydrated gel. Thus, H^+ do not cross through the glass membrane of the pH electrode, it is the Na^+, which cross and allow for a change in free energy. The hydrated gel membrane is connected by Na^+ transport and thus the concentration of H^+ on the outside of the membrane is 'relayed' to the inside of the membrane by Na^+.

Standardization or Calibration of pH Meter

Before using a pH meter it is necessary to calibrate it with solutions of known pH. This ensures errorless reading of the test or experimental solution. Calibration should be done at the temperature of the test solution.

The requirements for this are: buffer solutions of pH 4, 7 and 10.

A pH meter is calibrated/standardised as follows:

- Select pH function by pressing the mode button. Switch the function selector from the Standby position to pH position so that the electrodes are connected to the meter and the pH can be read.

- Set the temperature according to the temperature of test solution with temperature knob.

- Clear the existing standardisation.

- To begin a new calibration, press the STD button.

- Remove electrode from the storage solution, rinse well with distilled water and blot dry. Immerse the electrode in pH 4 buffer, and measure its pH.

- Adjust display to read 4 with calibration 1 knob.

- Once the reading becomes stable, press the STD button again to calibrate with a second buffer solution.

- Remove the electrode from pH 4 buffer, rinse well and immerse in pH 10 buffer.
- Calibrate with this buffer. Adjust meter to read 10 with calibration 2 knob.
- Remove electrode from pH 10 buffer, rinse and blot dry.
- Measure pH of 1st buffer once again. pH should read 4. If not, readjust calibration 1 knob.
- Lastly measure pH of buffer with pH value 7. Meter should read 7. If not repeat the entire procedure once again till meter reads exactly 7.
- The meter is now calibrated and is ready to use for measuring the pH of any solution.

Measurement of pH

pH of any solution is measured as follows:
- Remove electrodes from the standardisation solution.
- Rinse electrodes (and blot dry).
- Place electrodes in an unknown solution.
- Turn function selector from the standby to pH.
- Read pH of the solution.
- Return function selector to standby.

pH scale is logarithmic which means that a change of one unit of pH is equal to a ten fold change in hydrogen ion concentration of the solution. Thus a solution of pH 6 is ten times more acidic than pH 7 (neutral pH) and similarly a solution of pH 8 is ten times more alkaline than that of pH 7. For example pH of 1mol/l HCl is approximately 0. pH of 0.1 mol/l HCl is approximately 1 and pH of 0.01 mol/l HCl is approximately 2.

Maintaining the pH of a Solution

Change in pH of a solution can be prevented by adding **buffer** to the solution. Buffers tend to resist pH change. They are formed by mixing weak acid with its salt. Weak alkali with its salt is also used. The principle behind it is that weak acid is weakly dissociated whereas its salt with an alkaline metal is strongly dissociated. Increase in hydrogen ions will result in their reaction with high concentration of salt anions to form non-ionized weak acid that does not ionize appreciably and its ionization is opposed by high concentrations of anions present. Therefore addition of hydrogen ions is balanced by their removal with no or minor change in pH.

An equimolar solution of a weak acid and its salt has the greatest buffering power. Buffers in the range of 1 pH unit on each side of the pH of the solution can be prepared from such mixtures.the buffering power falls off rapidly outside this range.

Good Buffers are buffers described by **Good et al (1966)**, which can be used at the biological range of pH i.e. 6-8. Usual buffers such as phosphate, tris etc are not suitable for this pH range especially in cell free system. Phosphate buffer is a poor buffer above pH 7.5. It precipitates polyvalent ions and is also a metabolite and an inhibitor of biological systems. Tris is often inhibitory and a poor buffer below pH 7.5. Most Good buffers are **zwitterionic N-substituted amino acids** such as N-hydrooxyethylpiperazine-N-ethanesulphonic acid (**HEPES**). Adjustment of pH is done by HCl or NaOH of equal concentrations.

Points to Remember

The following points should be noted while using a pH meter:

- Calibration should preferably be done everyday.
- Calibration should be done at the same temperature at which the pH of the test solution is to be measured.
- Electrodes should be stored in distilled water and should be connected to the meter. If left dry, it takes at least 24 hours of soaking before they can be used again. Bottle containing the storage solution should always remain full.
- If the pH meter is to be standardized with only a single standard solution, then only the Cal 1 knob is used; and the Cal 2 knob is set to 100%.
- In the Two Point Method, the meter is standardized using two solutions of different pH which spans the pH range to be measured; and both the Cal 1 and Cal 2 knobs are used to adjust the pH values.
- Remove electrodes from storage beaker before use.
- Always rinse the electrodes with distilled water from a wash bottle into an empty beaker.
- Never rub the electrodes, always blot dry with a soft tissue.
- This procedure should be carried out whenever the electrodes are transferred from one solution to another to minimize the chance of contamination.
- Lower the electrodes into the standardization solution carefully so they do not strike the bottom of the beaker and break.
- Swirl the solution to fully saturate the electrode with buffer.
- If an experiment requires the use of a magnetic stirring bar in the solution whose pH is being measured, be careful that the bar does not hit the electrodes.
- A buffer solution is used as the standardization solution because its pH is known and it will maintain its pH in case of contamination as long as it is not excessive. A buffer solution of pH 7 is commonly used although the instrument should be standardized in the pH region of the unknown solutions.
- Never view the meter at an angle as it gives incorrect reading due to parallax error. To eliminate error, view the meter straight on.
- If the pH of several solutions is to be measured, the pH meter should be periodically restandardized.
- Before switching off the pH meter, remove electrodes from the last solution. Rinse, blot dry and place electrodes in storage beaker. Check that function selector is on standby, then turn power switch off and/or disconnect line cord and clean the work space.

MICROMETER AND MICROMETRY

Micrometer (Gr.μ; small, meter; a measure) is an instrument attached to microscopes for measuring dimensions of small objects. **William Gascoigne** invented the micrometer. Robert Hooke improved Gascoigne's micrometer by substituting parallel hairs for the parallel edges.

Micrometry or morphometrics is the practice of measuring linear, area, and volume dimensions of a specimen with the microscope. Microscopic measurements were first performed in the late 1600s by Anton von Leeuwenhoek, who used fine grains of sand as a gauge to determine the size of human erythrocytes. Microscopic measurements are in the range of the average field diameter of wide field eyepieces i.e. 0.2 μ –25 μ.

Measurements below 0.2 μ are beyond the resolving power of the microscope. Lengths larger than the field of view of a wide field eyepiece are usually measured with a stereomicroscope. The measurements are expressed in mm, μm (light microscope) or nm (EM). The parameters that can be measured are: length (linear), thickness (volume), area and angles.

Morphometric Methods

1. Comparison with micrometer scale in the *x–y* plane of the microscope by using calibrated mechanical stages, which have a vernier scale that allows reading of the stage displacement with an accuracy of 0.1 mm.

2. Comparing projected real images and those made by means of a traditional or digital camera system combined with a stage micrometer.

3. Comparing the size with a scale or objects of known size, included in the field of view. For e.g. polystyrene or glass beads included with specimens.

4. Use of measuring scale fitted in the eyepiece and calibrated with a stage micrometer having a scale of known dimensions etched on it.

5. Use of calibrated slides and counting chambers.

6. Comparing the view field size i.e. fixed dimensions of the microscope. This gives a rough estimate of relative linear dimensions of the specimen.

7. Utilization of a calibrated fine focus adjustment on the microscope for measuring vertical distances along the microscope optical axis (*z*-direction).

8. Use of transfer scale that can be placed in contact with the object or superimposed on the object, allowing the length (or width) of the object to be directly compared. The absolute dimension of the object is later determined by comparison to a calibrated scale (or ruler). The scale can be placed on a transparent material, photographed or engraved onto a glass element and placed in the optical path at one of the image-forming planes also.

9. Measurement by comparison with the image of a stage micrometer taken at the same magnification as the specimen.

Stage Micrometers

Stage Micrometers are microscopic scales of defined length etched directly on the surface of a standard microscope slide (1 × 3 inches) or sandwiched beneath a cover glass of known thickness (0.17 mm). Micrometer scale is one or two mm in length, subdivided into units that are one-tenth mm in length (100 µm units). Each 100-µm unit is further subdivided into ten equal sections, resulting in the smallest graduation representing 10 µm (Fig. 2.12a).

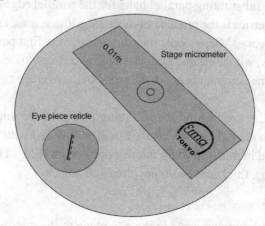

Fig. 2.12a Eye piece reticle and stage micrometer

Reticle and it's Calibration

Graduations of eyepiece scale are referred to as **reticle.** The most common technique used to measure the size of microscopic organisms with an optical microscope is to compare the size with the graduations of reticle. The reticle is inserted in the eyepiece by unscrewing the retaining ring from the bottom of the eyepiece and after the reticle is properly seated at the fixed diaphragm, the retaining ring is reinserted and tightened. Entire assembly is inserted into the eyepiece barrel and moved towards the eye lens until proper focus is achieved (Fig. 2.12b).

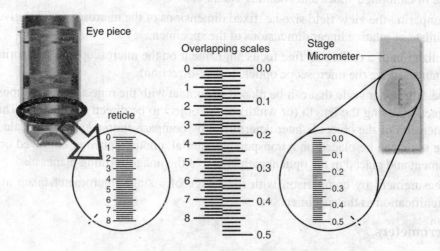

Fig. 2.12b Calibration of micrometer

Calibration of the reticle involves determining **micrometer value**, or **calibration factor**, which is the distance on the stage micrometer (imaged in place of a specimen), corresponding to one division of the scale in the eyepiece reticle. The size of any specimen or its feature can be then calculated by multiplying the number of eyepiece reticle divisions spanned by the specimen with the calibration factor for the objective in use. Calibration is done as follows:

- Align and configure the microscope for Köhler illumination.
- Insert the reticle into the microscope eyepiece and adjust the eye lens so that the engraved scale on the surface of the glass reticle disk appears sharply focused.
- Check the orientation of the reticle to verify that the numbers positioned above or below the engraved lines are not reversed.
- If the microscope is equipped with compensating adjustments on both eyepieces, the reticle calibration values will be correct for any interpupillary spacing. Otherwise adjust the microscope binocular interpupillary spacing and record this value for subsequent measurements.
- Place a stage micrometer on the microscope stage and bring the micrometer scale into focus using the microscope coarse and fine focus control knobs.
- Rotate the desired objective into position.
- Both the stage micrometer and the eyepiece reticle should be visible at the same focus.
- Bring the two scales into parallel alignment, using the *x–y* movement control knobs or handles, and/or rotating the eyepiece (and its reticle).
- If the mechanical stage is provided with rotational movement around the microscope optical axis, loosen the thumbscrew (usually located at the front of the stage, beneath the specimen platform) and rotate the stage until the micrometer and the eyepiece reticle are parallel.
- Position the eyepiece reticle directly over the stage micrometer and align the left-hand rule in the reticle with one of the longer, numbered (100 μm) division lines on the stage micrometer.
- Determine two points at which the reticle and micrometer scales exactly match. For accurate measurements, utilize the largest possible range of divisions on both scales.
- With reticles manufactured for specific eyepieces, reticle and stage micrometer graduations coincide over the entire length visible in the eyepieces.
- Determine the apparent length of the eyepiece scale in reference to the divisions on the stage micrometer (Fig.2.12b).
- Divide the known length of the selected region of stage micrometer by the corresponding number of divisions of the eyepiece scale to calculate micrometer value for the objective.
- The obtained value is the **calibration constant**.

Once the eyepiece reticle has been calibrated, linear dimensions of the specimen can be measured by positioning the reticle scale in concision with the specimen. Move the specimen till the left edge coincides with a numbered line on the eyepiece reticle, and count the number of scale divisions spanned by the target region and multiply it with calibration factor to get the actual size. Minimum resolvable distance in an optical microscope is approximately 0.2 μm (under optimal circumstances), a linear measurement below this value cannot be accurately determined.

Calibration factor differs at different magnifications, therefore it should be calculated separately for each magnification and measurements taken at a particular magnification should be multiplied by the calibration factor calculated at that magnification. For example if spore size has been measured at low power i.e. 100 × magnification (10 × 10), it should ne multiplied with calibration factor estimated at low power only.

Calculation of calibration factor for ocular micrometer:

X divisions of ocular micrometer = Y divisions of stage micrometer

Since one division of stage micrometer = 0.01 mm

Therefore Y divisions of stage micrometer = Y × 0.01 mm

Hence X divisions of ocular micrometer = Y × 0.01 mm

Therefore one division of ocular micrometer = 0.01 Y/X mm

Since one mm is equal to 10^{-3} microns, the obtained value is divided by 10^{-3} to convert it into microns.

Example

3 divisions of ocular micrometer = 20 divisions of stage micrometer

One division of stage micrometer = 0 .01mm

3 divisions of stage micrometer = 3 × 0.01mm

20 divisions of ocular micrometer = 0.03 mm

One division of ocular micrometer = 0.03/20 mm or 0.001.5 mm

Since 1 mm = 10^{-3} micron, 0.0015mm = 0.0015/1000 microns or 1.5 microns.

Points to remember when measuring microscopic objects

When measuring microscopic objects, the followings points are to be noted:

- Take several measurements to increase accuracy.
- When circular or oval specimens are being measured (such as blood cells, yeast, bacteria, etc.,) take at least 20 observations from different fields.
- Each objective should be independently measured as magnification varies even for similar objectives from the same manufacturer and inscribed with the same magnification factor.
- Highest magnification objective should be chosen for all measurements. It enables the entire feature of interest to fall within the span of the reticle scale.
- Microscope objectives are usually corrected for use with a cover slip of standard thickness.
- Micrometers without a cover slip should be used with reflected light objectives that are corrected for zero-thickness or the absence of a cover slip.
- Some stage micrometers utilize a metal plate, as a carrier for a small circular glass insert that is imprinted or engraved with the graduated scale.

- Unprotected scales are vulnerable to damage and must be treated with great care to avoid scratching and contamination with dust, dirt, fingerprints, or other debris.
- The calibration information for each combination should be recorded and stored to avoid repeating the procedure.
- Ragged edges of the lines of photographically produced micrometer scales, coupled with the occurrence of randomly distributed isolated silver grains between the lines, make some micrometers unsuitable for precise measurements.
- Errors can occur if an objective having significant curvature of field or image distortion is used, hence it is advisable to use flat field or plan objectives.
- If an objective having some field distortion must be utilized, restricting measurements to the central portion of the view field will minimize measurement errors.
- Difficulty in precisely aligning eyepiece reticle lines with those of the stage micrometer used for the calibration arises very frequently. A precision, graduated mechanical stage can be utilized to make this procedure much easier to accomplish.
- The quality of the graduations on a stage micrometer has a significant effect on the accuracy with which a calibration can be conducted, and this is especially true at high magnifications.
- Calibration factor will differ for low power, high power and oil immersion objectives.
- Calibration factor used to determine the size of object should be calculated at that magnification at which the object is finally observed.
- For measuring the size of bacteria, calibration factor determined at oil immersion is used as bacteria are best observed at this magnification.
- Use low power objective to first locate the circle surrounding the scale, and then the scale itself.
- The ring encircling the micrometer scale is visible with the naked eye and should be used to position the stage micrometer in the center of the microscope optical path (stage aperture).
- Several stage micrometer designs have a line engraved from the ring to the edge of the scale, which is also helpful in locating the scale when using high magnification objectives.
- The value of ocular division depends on the length of body tube and objective used.

HEMOCYTOMETER

Hemocytometer (also known as hemacytometer) is a glass slide with two counting chambers etched in a surface area of 9 mm^2 (Fig.2.13a). Each chamber (Fig.2.13b) is divided into nine 1.0 mm squares. Each 1 mm square is divided into groups of 25 medium sized squares of 0.2 mm × 0.2 mm each. These squares are further divided into 16 small squares of 0.5 mm × 0.5 mm each. In this manner the counting chamber is divided into total 400 squares. Each medium sized square is separated by triple lines, the middle one acts as the boundary.

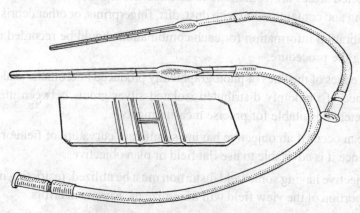

Fig.2.13a Hemocytometer

It has raised sides which keep the cover slip 0.1 mm above the chamber floor so that the total volume of each square becomes 0.0001 ml (1.0 mm × 1 mm × 0.1 mm) or 10^{-4} cm^3 (length × width × height). Staining of cells identifies viable cells. Stains generally used are: Trypan Blue, Erythrosin B and Nigrosin. Nuclei of damaged or dead cells do not take up the stain. If more than 20% of the nuclei are stained, the result is probably significant.

Accuracy of results depends on:
- Proper mixing of the sample.
- Number of chambers counted.
- Number of cells counted.

Faulty results may occur due to following errors:
- Unequal cell distribution in the sample.
- Improper filling of chambers (too much or too little).
- Inconsistent counting.
- Statistical errors.

Things to remember:
- Mix the cell suspension thoroughly.
- Dilute the cell suspension so that each such square has between 20 and 50 cells.
- Count at least 300–400 cells, as counting error is approximated by the square root of the total count.
- Count cells that touch the middle lines (of the triple lines) to the left and top of the square.
- Do not count cells similarly located to the right and bottom.
- Cells settle due to gravity within a few seconds so work quickly.
- If over 10% of the cells represent clumps, repeat entire sequence.
- If fewer than 200 or more than 500 cells are present, repeat with a more suitable dilution factor.
- Repeat count to check accuracy and reproducibility of result (± 15%).
- Cell suspension should not overflow during filling of the chambers.

Requirements: Hemocytometer and cover glass, or cover slips, pasteur pipettes or transfer pipettes, balanced salt solution (BBS or PBS), Trypan blue (0.4% in BBS or PBS), microscope, test tubes and cell suspension.

Method:

- Dilute 0.2 ml of Trypan blue with 0.8 ml of BBS.
- Place cover glass over hemacytometer chamber.
- Transfer 0.5 ml of agitated cell suspension to a 15 ml tube and add 0.5 ml of diluted trypan blue.
- With a Pasteur or transfer pipette, fill both chambers of the hemocytometer by capillary action
- ml of cell suspension is put in each chamber.
- Place under the microscope at a low power magnification.
- Count the cells by either method A or method B.
- Calculate the number of cells per ml, and the total number of cells in original culture.
- Cells/ml = average count per square $\times 10^4$
- Total cells = cells per ml × any dilution factor × total volume of cell preparation from which the sample was taken.

Cell Counting Methods

Two methods are available for cell counting. They are:

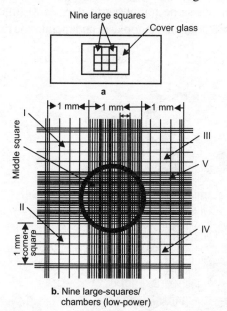

b. Nine large-squares/
chambers (low-power)

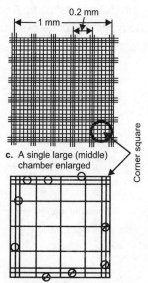

c. A single large (middle)
chamber enlarged

d. Corner square enlarged : indicates
procedure for counting for cells/spores

Fig. 2.13b Haemocytometer used to count spores of microorganisms and total count of cells. **(a)** Haemocytometer, **(b)** Standard haemocytometer chamber. The circle indicates the approximate area covered at 100 × microscope magnification (10 × ocular and 10 × objective). **(c)** A single large (middle) square enlarged. **(d)** Corner square enlarged. Count cells on top and left touching middle ling (O) and do not count cells touching middle line at bottom and right θ.

Method 1

Count the number of cells in the four outer squares. This is usually done for large cells, fungal spores etc. The cell concentration is calculated as follows:

Cell concentration per milliliter = Total cell count in 4 squares × 2,500 × dilution factor

Method 2

Estimate cell concentration by counting five squares in the large middle square. This is usually done of cells are small. The cell concentration is calculated as follows:

Cell concentration per milliliter = Total cell count in 5 squares × 50,000 × dilution factor.

AUTOCLAVE

Autoclave (Gk: *auto-*, + Latin *clavis*, key (self-locking from pressurization)) It is a double walled vessel made up of thick stainless steel or copper. It is fitted with a separate gauges which measure temperature and pressure, an exhaust valve (steam cock) for release of steam from the chamber, a safety valve to avoid accidents due to development of excess pressure, controls for adjusting temperature and pressure. A rubber seal inside the lid maintains airtight condition. Screw fasteners or fly nuts are used to secure the lid. used to sterilize all those substances that cannot be sterilized by dry heat such as, nutrient media, water, reagents, metabolites, instruments etc. It uses the principle of steam sterilization. Steam sterilization is a process of killing microorganisms through the application of moist heat (saturated steam) under pressure. Saturated steam is the water vapour at the temperature at which it is produced. As compared to hot air at the same temperature it is 2500 times more efficient. Heat damages the cell's essential structures including the cytoplasm and membrane rendering the cell inviable (Fig. 2.14).

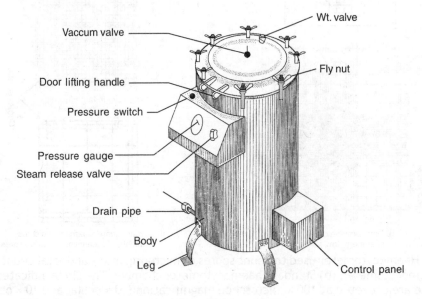

Fig.2.14 Autoclave

The rate at which bacterial cells are thermally inactivated depends on the temperature and the time of heat exposure to which they are exposed. In practical terms, this means that it would take a longer amount of time at lower temperatures to sterilize a population than at a high temperature. Additionally, the higher the concentration of organisms that need to be killed, the longer it will take to kill all of the cells in that population at the same temperature (Table 2.4.).

Table 2.4. Temperature and Pressure Relationship

Steam pressure (Psi)	Temperatue (°C)				
	Air discharge (%)				
	100	66	50	33	0
0	100.0				
1	101.9				
2	103.6				
3	105.3				
4	106.9				
5	108.4	100	94	90	72
6	109.8				
7	111.3				
8	112.6				
9	113.9				
10	115.2	109	105	100	90
11	116.4				
12	117.6				
13	118.6				
14	119.9				
15	121.0	115	112	109	100
16	122.0				
17	123.0				
18	124.1				
19	125.0				
20	126.0	121	118	115	109
21	126.9				
22	127.8				
23	128.7				
24	129.6				
25	130.4	126	124	121	115
26	131.3				
27	132.1				
28	132.9				
29	133.7				
30	134.5	130	128	126	121

Around the working range of 15 psi, each psi raises the autoclave temperature by approximately 1°C.

Difference between boiling and autoclaving

When we heat water in an open pan the boiling temperature rises up to 100°C, water ceases to warm any more.Thus heating water above the boiling point in an open vessel is impossible. In this system evaporation takes all the heat and prevents further heating of water. Any further heating simply leads to evaporation of water till there is no water left in the pan.

In a sealed vessel as temperature reaches approximately 90°C, extensive evaporation takes place but the vapor cannot escape therfore excessive pressure is produced within the vessel and a time comes when vapor pressure in the vessel reaches the value that corresponds to the temperature. At this point evaporation ceases and in this manner not all water turns into vapor. The higher the temperature, the higher is the pressure of the vapor and it becomes possible to increase the boiling point. This latent heat of vaporization is used in the autoclave to increase the temperature above boiling point. Autoclaves work by allowing steam to enter, and maintaining pressure at 103 kPa (15 psi) causing the temperature of steam to reach 121°C which is further maintained for at least 15 minutes. There are physical, chemical and biological indicators that can be used to ensure an autoclave reached the correct temperature for the correct amount of time

Substances which should not be autoclaved

- Heat susceptible products such as plastics cannot be sterilised this way as they will melt.
- Paper, chemicals such as sugars, vitamins, enzymes and other products may be damaged by steam.
- Inoculum should never be autoclaved as it will be killed and will be of no use.

Items should be autoclaved in autoclave bags and a rigid secondary container (typically polypropylene or stainless steel).

Autoclave resistant organisms

A single-cell **organism Strain 121**, survives autoclave temperatures. Prions also may not be destroyed by autoclaving.

Autoclave operating procedures

The operating procedures of an autoclave are as follows:
- Fill the autoclave and heat it till steam strats building up.
- Place items to be sterilized in the chamber.
- Close and lock the autoclave door.
- When temperature reaches 100°C, close the operating valve.
- Check the jacket pressure.
- Measure the sterilization time from the moment temperature reaches 121°C rather than 15 lbs pressure.

- Sterilize for 20 minutes.
- After 20 minutes, close the steam supply valve and wait till the pressure falls to zero.
- Unlock and open the autoclave door.
- Remove autoclaved items from the chamber.
- Clean up.

Precautions

The following points should be noted while using an autoclave:
- Do not open the door unless the chamber pressure comes down.
- Do not open the lid immediately as pressurized steam is hotter than boiling water and causes severe burns.
- Sudden release of pressure causes the liquids to boil up, which will wet the plugs or blow them out.
- Let the pressure fall gradually when liquids are being sterilized.
- If air has not been completely displaced from the autoclave, even though the pressure gauge may indicate 15 lbs pressure the temperature will not reach 121°C.
- Heat of vaporization released upon the condensation of steam causes much more severe damage than does the same quantity of boiling water.
- Wear insulated gloves when handling hot autoclaved items.
- Clean and disinfect the autoclave after every use.
- Large loads require increased sterilization time. Therefore it is advised to sterilize in small convenient loads.
- Divide 5-litre medium in five one-litre flasks rather than autoclaving it in one 5-litre flask.
- Cotton plugs should be tightly fitted. Cotton fibers absorb moisture and after drying stick to each other, therefore the plug becomes loose later on.
- Plastic caps, rubber screws and stoppers should be set slightly loose so as to allow air to escape, otherwise the flasks or tubes will burst due to pressure of steam generated within.
- In case of burn immerse burn area in cold water and get medical aid.

Reasons for Autoclave Failure

The following may be the reasons for autoclave failure:
- Not enough water in the autoclave.
- Autoclave container/bag is too large.
- Over-filling the autoclave bag/container.
- Use of containers that block access of steam to the load.
- Faulty sealing.
- Lid not fitted properly.

Testing of Autoclave

The following is the procedure for the testing of an autoclave:
- There are physical, biological and chemical indicators for testing whether the autoclave reached the correct temperature for the correct amount of time or not.
- Physical indicators include placing of an alloy designed to melt only after being subjected to 121°C for 15 minutes.
- In biological methods accuracy is tested by autoclaving attest devices which are vials containing spores of a heat resistant bacterium. These vials are autoclaved along with a test load of approximately same weight as the material that needs to be sterilized routinely.
- Usually autoclave efficiency is tested by by autoclaving test indicator kit containing spores of *Bacillus stearothermophilus* along with a non-autoclaved control vial.
- It is necessary to incubate the autoclaved test vial at 56 to 60°C. Incubation causes surviving spores to grow.
- Efficient dispersal of steam throughout the autoclave is tested by placing the spore vials on the top, bottom, and sides and in the center of a test load prior to autoclaving in new autoclaves.
- Chemical indicators can be found on medical packaging and autoclave tape, and these change colour once the correct conditions have been met. This indicates that the object inside the package, or under the tape, has been autoclaved sufficiently.
- Autoclaves already in use should be tested periodically after every 40 hours of use if used to inactivate human or non-human primate blood, tissues, clinical samples, or human pathogens, and every 6 months if used to inactivate other material.
- For periodic testing, place a spore vial in the center of a test load prior to autoclaving.

CENTRIFUGE

A centrifuge separates a heterogeneous mixture of solid and liquid by spinning it. It is a device used to separate two or more substances of different density, e.g. two liquids or a liquid and a solid by using centrifugal force (Fig. 2.15). After a successful centrifugation, the solid precipitate settles to the bottom of the test tube and the clear solution, called the centrifugate is obtained.

Centrifuge consists of a fixed base or frame and a rotating part in which the mixture is placed and then spun at high speed. Disposable plastic or glass tubes are used to place the specimen/sample. Tubes are placed in the rotating part in holders, so arranged that when the rotary motion begins, the test tubes swing into a slanted or a horizontal position with the open ends toward the axis of rotation; the heavier, solid part of the solution settles in the bottom of the tube and the lighter liquid part forms a layer on the top.

Types of Centrifuges

The various types of centrifuges are as follows:
   a. **Low speed:** go up to about 5000 rpm.
   b. **High speed:** go up to about 25,000 rpm.

c. **Airfuge:** in which rotor is suspended and driven by a stream of air. These can reach over 100,000 rpm in a very short time but are limited to very small sample sizes.

d. **Microcentrifuge or microfuge:** These are simple machines used with 0.5 or 1.5 ml disposable plastic tubes. Most of these machines are single speed and operate between 10,000 and 13,000 rpm.

Fig. 2.15 Centrifuge

e. **Ultracentrifuge:** operate at speeds of more than about 20,000 rpm. It was devised in the 1920s by the Swedish chemist Theodor Svedberg. Svedberg determined accurately the molecular weights of substances including proteins and viruses by using an optical system with the centrifuge to observe sedimentation rates. The unit used for sedimentation rate is therefore called as **Svedberg unit** (S).

f. **Vacuum-type ultracentrifuges:** centrifuge rotor is located inside the vacuum chamber. Centrifugal field of these machines is more than 300,000 g. The rotor of a typical vacuum-type ultracentrifuge is 18 cm (7 inches) in diameter and carries 300 ml (10 ounces) of liquid.

g. **Centrifuge General Purpose (Table Top):** Table top centrifuge is of robust construction and is provided with stepless speed regulator, transparent acrylic lid and is suitable for accommodating different types and capacities of rotor heads.

h. **High Speed Refrigerated Research Centrifuge:** This range of centrifuges are designed for high speed centrifuging applications. It includes a temperature control-cum-indicator with a range of $-20°C - +40°C$, digital timer with a range of 0–99 minutes and alarm at automatic switching off, dynamic brake, zero start interlock, safety cut-off in case of imbalance and lid locking switch etc.

i. **Programmable High Speed Research Centrifuge (Microprocessor Based):** This type of centrifuge is provided with a brush-less indication motor, programmable microprocessor control for pre-setting rpm with a range of 100–20000, RCF temperature ranging from 10°C to 40°C, timer with a range of 1–120 minutes and acceleration/deceleration time. Functional displays include rpm indication by default, RCF by selection, remaining time, working chamber temperature, lid lock status, rotor imbalance etc.

Using a Centrifuge

Place test tube in the centrifuge holder. Balance with another test tube filled to the same level in the opposite holder set the spm and time. Close cover and turn knob. Note that you must turn off the centrifuge with the switch and wait for it to stop spinning, to effectively separate the precipitate and the solution.

Precautions

Mechanical stress can result in rotor failure. Improper loading and balancing of rotors can cause the rotors to break loose while spinning.

Prior to centrifuge operation, check:

- The total mass and balance.
- Distance/position of all counterweights.
- Secure all elements.
- The centrifuge chamber should be clear of obstructions.

BALANCE

A **balance** is a device used to measure the mass of an object by comparing it with known mass. The **beam balance** or **laboratory balance** consists of a beam with a fulcrum at its center. The substance to be weighed is placed in a weighing pan and a standard mass or combination of standard masses is placed in the scale pan. Standard weights are added to the scale pan until the beam is in almost at equilibrium. To achieve fine balance a slider weight is placed on the beam and moved along a scale till fine balance is achieved. The slider position gives a fine correction to the mass value (Fig. 2.16).

Accurate measurements are ensured by attaching a pointer to the beam which amplifies any deviation from a balance position, by keeping the fulcrum of the beam friction-free and also by using the lever principle, which allows fractional weights to be applied by movement of a small weight along the measuring arm of the beam, as described above. Buoyancy in air and densities of the weights and the sample effect the total accuracy.

Mass unlike weight is independent of the force of gravity. It is measured in grams, kilograms, pounds, ounces, or slugs. In a balance moments of force on either side of balance and the acceleration of gravity on each side cancels out, so a change in the strength of the local gravitational field will not change the measured weight.

For high degree of precision measurement an **Analytical balance** is used. The weighing pan or pans are enclosed in a closed chamber with glass doors. This prevents collection of dust and interference by air currents (Fig. 2.17a). Chainomatic® attachment combined with a notched beam, speeds the weighing process. All weight adjustments from 0.1 mg. to 100 mg. can be made from outside the case, by increasing or decreasing the length of the chain. The rider, the rider lifter, and the lifter are also controlled by knobs outside the case so that weight adjustments can be made without opening the balance door. The notched beam permits the adding of additional weight up to one gram, in 100 mg. amounts, without opening the balance door. The magnetic damper brings the indicator to rest quickly by stopping the swinging or oscillation of the beam. A magnifying optical system provides a direct reading right from the screen equipped with a vernier scale or similar device for fractions of divisions on the scale. Modern balances are monopan balances with a digital liquid crystal display (Fig. 2.17b). They have built-in calibration weights to maintain accuracy.

Top Loading Balance

Top loading balance is used to weigh solid material when a precision of 0.1 g is adequate. For more accurate mass measurements or small amounts, use an analytical balance.

Using a Top-loading Balance

Press the on/off button and wait until the display reads 0.0 g. Place the butter paper on the balance pan. Record the weight of the paper. Set the required weight. Carefully add substance to the container or paper up to the desired mass. Record mass. Use the brush provided to clean any spills.

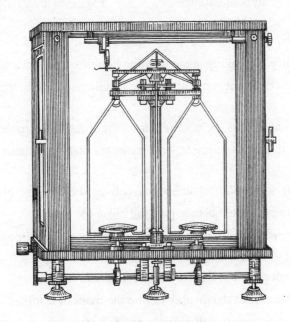

Fig. 2.16 Physical balance

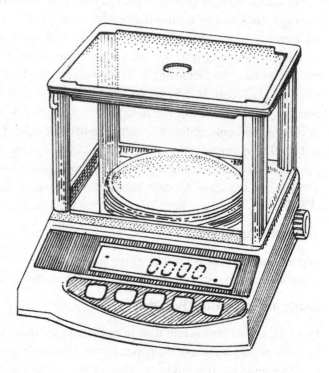

Fig. 2.17 a. Analytical balance   **Fig. 2.17 b.** Analytical balance

Things to Remember

- Check the balance by looking at the leveling bubble on the floor of the weighing chamber.
- If it is not centered, center it by turning the leveling screws on the bottom toward the back of the balance.
- Close the doors and press the control bar on the front of the balance.
- Wait for a row of zeros to appear. This indicates that the balance is zeroed and ready for use.
- Liquid, powder, or granular substance must always be weighed in a weighing container.
- Place the container on the balance pan and close the doors.
- Tare the container by briefly pressing the control bar so that the display reads zero with the container sitting on the pan.
- Place the substance to be weighed in the container.
- Close the doors and read the display to find the mass of sample.
- Solid objects can be weighed directly on the pan after setting the balance to zero.

- Be careful not to spill chemicals on the balance.
- Clean up any chemicals that may have spilled on the balance.
- Turned off the balance after finishing the work.
- Balance should be placed upon a solidly constructed bench or table.
- The level of the support should be permanent. This elimi-nates repeated relevelling of the balance case and the readjust-ment of the zero point.
- The balance support should be free from any vibration that produces a visible effect upon the operation of the balance itself.
- Older balances should be calibrated periodically with a standard weight.

Precautions

The following precautions are to be followed:
- Don't pick up paper or containers with bare hands since your fingerprints add mass.
- Don't lean on the bench while weighing.
- Record the mass of paper or container.
- Check the level indicator bubble before weighing.
- Use a brush to clean spills in the weighing chamber.

PIPETTES

Pipettes are used for volumetric measurements and transfer of known quantity of solution from one container to another vessel. Pipettes may be made up of glass or plastic. They may be graduated or digital.

Types of Pipettes

The different types of pipetted include:
1. **Positive displacement type**: extremely accurate and used for transferring high viscosity and volatile liquids. The piston is in direct contact with the liquid.
2. **Air displacement type**: is highly accurate, used for standard pipetting applications. Piston is separated from the liquid by a specified volume of air. Performance is affected by atmospheric pressure, specific gravity and viscosity of the solution.
3. **Pasteur style type:** small glass tubes with a bulb at the end used for dispensing small amounts of fluids.

Micropipettes

1. Micropipettes are precision instruments. They are of two types: single and multichannel. They are provided with plastic disposable tips for dispensing liquid.

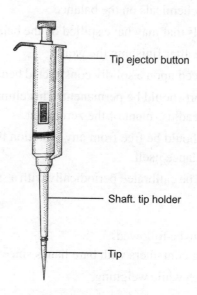

Tip ejector button

Shaft. tip holder

Tip

Fig. 2.18 Micropipette

These tips are usually packed into autoclavable plastic boxes. Micropipettes have three positions: rest, first stop and second stop (Fig. 2.18).

Using a micropipette

The following points are to be followed while using a micropipette:
- Attach the tip to the end of the micropipette shaft. Adjust the setting. For e.g. to 100 ml.
- Press and twist slightly to make sure it is airtight.
- Hold the pipette vertically and press the plunger to the first stop. This displaces air equal to the volume of the setting.
- Now immerse the tip into the liquid.
- Release the plunger back to the rest position.
- Wait a second for the liquid to be sucked up into the tip. The volume of the liquid in the tip is equal to the volume of the setting of the micropipette.
- Place the tip at an angle (10°–45°) against the wall of the tube or flask or a micro-well plate and once again press the plunger to the first stop.
- After one second, press the plunger to the second stop to expel all the liquid.
- Remove the tip away and bring the plunger to the rest position.

Graduated Pipettes

These pipettes are made of glass and are marked into graduations to show the amount of liquid present in it. The volume of liquid is measured by reading the value indicated on the graduated scale.

For clear liquids, the bottom of the meniscus is read and for colored liquids, the surface of the meniscus is read. Plastic pipettes are also available in the market these days. Disposable pipettes are useful for pipetting toxic or viscous substances. These pipettes are emptied either by keeping the tip in contact with the vessel or by blowing the liquid out with a pipette filler.

Pipette fillers

This is a device attached to the end of the graduated pipette and used to draw up the liquid. For e.g. simple rubber bulb, a triple valve rubber bulb, a hand operated pump or an electronic pipette filler.

Precautions

Following precautions need to be taken:
- The tips should fit tightly onto the end of the pipette.
- Pipette should always be kept in an upright position to prevent liquids from running inside the shaft of the pipette.
- Always leave the pipettes on the pipette stand.
- Most pipettes have a digital display of volume. Adjust the volume before use.
- The liquid drawn up should reach the expected level in the tip and there should be no air bubbles in the tip.
- In multichannel pipettes, the volume of the liquid should be same in each tip.
- Never draw liquid by mouth. Use pipette fillers for drawing up the liquids.
- Drawing up the liquid by mouth can result in infection if the liquid/pathogenic suspension is sucked up into the oral cavity.
- Even when using safety pipetting aids, always use cotton-plugged pipettes when pipetting biohazardous or toxic materials.
- Perform pipetting operations in a biological safety cabinet while handling biohazardous or toxic fluid.
- Biohazardous materials should not be bubbled through a liquid with a pipette. Do not blow into the pipette to expel biohazardous material out.
- To prevent accidental spillage or release of infectious droplets, place a disinfectant-soaked towel on the work surface and autoclave the towel after use.
- Do not discharge material through a pipette from a height.
- Hold the pipette at an angle to the wall and allow the liquid to run down the container wall for complete discharge.
- Immerse used/contaminated, reusable pipettes in a disinfectant. Place them horizontally for complete immersion. Autoclave the pan and pipettes before reuse.
- Discard contaminated/blocked/broken pipettes.
- Graduated pipettes should be plugged with cotton wool on the top and sterilized before use to minimize the contamination of fluids being measured.

- Before using graduated pipettes, check the volume scale and measuring accuracy by checking that the pipette empties or fills from full volume to zero or from zero to full volume.

Repipet

A repipet is a hand operated pump that dispenses solution. Its volumes are accurate to within about 2%. For a more precise volume measurement, use a more precise piece of glassware, like a graduated cylinder, pipette, or burette.

Using a Repipet: Place a clean, empty container at the outlet. Pull up the top of the repipet as far as it will go. Push down gently and completely to deliver the full amount of solution.

VACUUM FILTRATION

Vacuum filtration is a technique for separating a solid product from a solvent or a liquid reaction mixture. The mixture of solid and liquid is poured through a filter paper in a Buchner funnel. The filter traps the solid and the liquid is drawn through the funnel into the flask below, by a vacuum pump (Fig. 2.19).

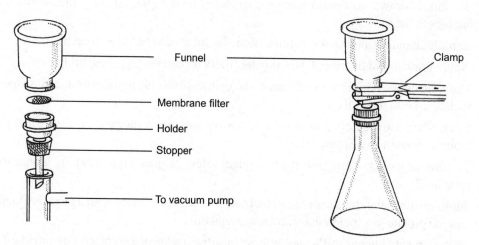

Fig. 2.19 Filtration assembly

The filtration assembly is made up of a buchner funnel, a glass holder, rubber stopper, rubber tubing, filtration flask and filter paper. The filtration flask has a side arm which is attached to the vaccum pump with the help of the rubber tubing.

Method:

- Turn on the vacuum.
- Check the vacuum by feeling for suction at the end of tubing.
- Vacuum should be strong enough to hold the tube on to your finger without falling off.

- Connect the tubing to the side arm of filter flask and check the suction at the top of the flask. Place the black rubber ring adapter in the top of the flask and then the Buchner funnel.

- Check once again for proper suction.

- Place gloved hand across the top of the funnel. If you do not feel strong suction, there is a poor connection and a leak somewhere in the system. Place a filter paper in the Buchner funnel and wet it with clean solvent.

- The paper should be sucked down against the holes in the funnel and the solvent should quickly pass through into the filter flask.

- Slowly pour the sample into the center of the filter paper.

- Use solvent to rinse the beaker, so that the entire solid is collected.

- Continue to draw air through the solid, to evaporate any remaining solvent in your sample. When finished, break the vacuum at the connection between the flask and the trap. Turn off the vacuum.

WATER BATH

This device is used to heat cultures or those solutions, which cannot withstand dry heat. It is made up of a metallic body containing a heating element. Water is filled in the container and heated. A thermostat maintains the water temperature. Water baths combined with a shaker are also available for use with materials/cultures, which require aeration (Fig. 2.20).

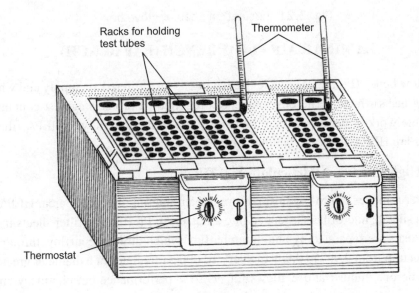

Fig. 2.20 Water bath

BIOLOGICAL SAFETY CABINETS

Biological safety cabinets, also known as tissue culture hoods, are one of the most important tools of microbiology. Their proper use prevents contamination and protects the operator from exposure. Fig.2.21 shows a wooden inoculation hood also called a glove box.

Fig. 2.21 Inoculation hood or glove box

LAMINAR AIR FLOW BENCH (LAF BENCH)

The laminar flow bench (Fig.2.22a) is a device having its own filtered air supply and which is used for performing tasks where sterile environment is required. It protects against contamination by ensuring that the work in process within the bench area is exposed only to HEPA-filtered air. Its main attributes are HEPA-filter and laminar air flow.

The HEPA (High Efficiency Particulate Air) Filter

The HEPA filter removes micro-organisms from the air entering the bench area. HEPA filters are made of boron silicate microfibers arranged into a flat sheet. Pleating of filter sheets increases the surface area. Pleats are separated by aluminum baffles, which direct the airflow through the filter. HEPA filter can trap from 9,997 to 9,999 of every 10,000 particulates of a diameter greater than and less than 0.3 microns. Industrial use allows for 99.97% performance but pharmacy, medical and other research laboratory applications, however, require a 99.99% filtration performance level. vapors or gases are not absorbed by HEPA filter.

Fig. 2.22a Laminar air flow bench

Laminar air flow

When the entire air within a confined area flows in unidirectional velocity along parallel flow lines it is termed as laminar airflow.

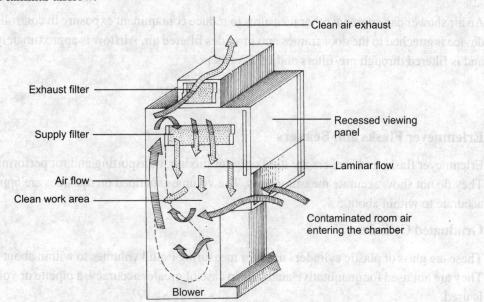

Fig. 2.22b Flow of air through a Laminar flow bench

Working of LAF bench

In an LAF bench air is drawn into the base of the hood by the blower/motor, through a washable, reusable pre-filter and then pushed up the rear plenum of the hood through the HEPA filter. Filtered air is directed horizontally across the work surface at a constant velocity of 100 FPM toward the user (Fig. 2.22b). Air quality in the laminar flow clean bench is maintained with accurate control of airflow volumes and velocities. Obstructions such as instruments placed in the bench, user etc. cause interruptions in airflow which allow particulates to enter the work area, sharply increasing particle counts inside the hood. It may also cause backwash i.e. entry of unfiltered room air into the work area. Backwash may be created when the user inserts a hand or materials into the work area. Some models of LAF bench are now equipped with high velocity return air slots, which are located along the front edge of the work surface and along the side walls of the work area. Air is drawn into these slots at 1,000 FPM ensuring that no unfiltered air enters the work area.

Air Curtain

Air curtain produces high-speed air jet across the whole width of the door. The high-speed air jet separates between a hot zone and a cold zone, prevents penetration of dust, smells, insects and other pollutants into the inner protected area. The effective operational range of the air curtain is an opening, up to 2.8 m high. In areas where the temperature is low, the air curtain can supply air that will be up to 10°C warmer than the ambient temperature. The motors in the air curtains have three speeds, which allows regulation of the air outlet speed.

Air Shower

An air shower can augment other measures to reduce contaminant exposure in controlled areas. The device is attached to the door frames and provides filtered air. Airflow is approximately 5000 ft/min and is filtered through pre-filters and HEPA filters.

GLASSWARE

Erlenmeyer Flasks and Beakers

Erlenmeyer flasks and beakers are used only for mixing, transporting and for performing reactions. They do not show accurate measurements. The volumes stamped on the sides are approximate and accurate to within about 5%.

Graduated Cylinders

These are glass or plastic cylinders used for measuring liquid volumes to within about 1% accuracy. They are not used for quantitative analysis. In case of greater accuracy, a pipette or volumetric flask is used.

Volumetric Flask

A volumetric flask is used to make up an accurate solution of fixed volume. It usually has an error of ± 0.2 ml. This is a relative uncertainty of 4×10^{-4} or 400 ppm.

In order to make up a solution, the solid material is first completely dissolved in a little amount of water. Then the flask is filled upto the required mark. Eye should be at the level of the mark on the neck of the flask and the circle around the neck should look like a line and not like an ellipse. Then distilled water is added drop by drop till the bottom of the meniscus lines up exactly with the mark on the neck of the flask. Liquid should not rise above the mark on the neck of the flask. Mix the solution thoroughly, by inverting the flask and shaking.

Burette and Titration

Burette is a glass instrument used to deliver precisely measured, variable volumes of one reactant during titration, into another until the precise end point of the reaction is reached. Titration is a method of analysis that allows to determine the precise endpoint of a reaction and therefore the precise quantity of reactant in the titration flask. A burette is used to deliver the second reactant to the flask and an indicator or a pH Meter is used to detect the endpoint of the reaction.

Method

- Condition the burette by washing it with the titrant solution and then fill it with titrant solution.
- Check for air bubbles and leaks.
- Take the initial volume reading and record it.
- Place the solution to be titrated, along with a few drops of indicator, in a clean flask or beaker and add titrant to within a couple of ml of expected endpoint.
- The indicator changes color when the titrant hits the solution in the flask, but the color change disappears upon stirring.
- This indicates that the end point is near.
- Titrate slowly till the end point is achieved. Record the volume of titrant used.
- Subtract the initial volume to determine the amount of titrant delivered.
- Use this, the concentration of the titrant, and the stoichiometry of the titration reaction to calculate the number of moles of reactant in your solution. For phenolphthalein, endpoint is the first permanent pale pink. The pale pink fades in about 10–20 minutes.

Titrating with a pH meter

Titration with a pH meter follows the same procedure as titration with an indicator, except that the endpoint is detected by a rapid change in pH, rather than the color change of an indicator. Arrange the sample, stirrer, burette, and pH meter electrode so that you can read the pH and operate the burette with ease. To detect the endpoint accurately, record pH v/s. volume of titrant added and plot the titration curve as you titrate.

Precautions while using the burette

Precautions to be followed while using a burette are:

- Never pour the reagent directly into the burette. Use a funnel to fill the burette.
- Check that the stopcock is closed before filling.
- The liquid in the burette should flow freely. Lift the funnel slightly to reduce the vacuum and to allow the solution to flow freely.
- Always condition the burette by filling and draining the titrant solution repeatedly so that all surfaces are coated with solution. This will insure that, the concentration of the titrant does not change by stray drops of water or any other previously used liquid.
- The tip should be clean and dry before you take an initial volume reading.
- Air bubble should not block the tip of the burette. If an air bubble is present during a titration, volume readings may be in error.
- To remove an air bubble, tap the side of the burette tip while the solution is flowing.
- Rinse the tip of the burette thoroughly with water from a wash bottle and dry it carefully.
- Check the tip for leakage.
- Once the burette is conditioned and filled, without air bubbles or leaks, take the initial volume reading.
- For accurate reading, use a burette reading card. The card has a black rectangle printed on it. Hold it behind the burette to see the meniscus of the liquid clearly.
- Keep the eye at the level of meniscus. Reading from an angle, or above or below the meniscus results in a parallax error. Read the lower level of the meniscus.
- Open the stopcock to deliver solution.
- The solution can be delivered quickly until a couple of ml from the endpoint, but the endpoint should be approached slowly, a drop at a time.
- Rinse, clean and dry the burette after use.

Quantitative Transfer

Quantitative transfer simply means that the complete volume of material should be transferred from one container to another. For e.g. every solid particle must be transferred from the weighing paper to the (clean) beaker.

Transferring a solid

Carefully tip the butter paper to pour the solid into the beaker. Tapping the paper with a spatula will knock particles into the beaker. Finally, the paper should be rinsed into the beaker, to remove all traces of the solid.

Transferring a liquid

When transferring a solution or heterogeneous mixture to another vessel, rinse the container with solvent to be sure that the transfer is quantitative. The rinsings should be transferred to the second vessel along with the rest of the mixture or solution.

INOCULATION NEEDLE AND LOOP

Inoculation needle is a metallic rod with a straight wire attached at one end with the help of a special holder. Usually a stainless steel or a platinum iridium wire is used. It is used for stab culture and for picking single colonies. In inoculation loop, the end of wire is closed to form a flat circular loop of 2–4 mm diameter. It is used for streaking as it can pick up large amount of solid culture or liquid (Fig. 2.23). Both should be sterilized by heating almost vertically in a flame. The tip should become red hot.

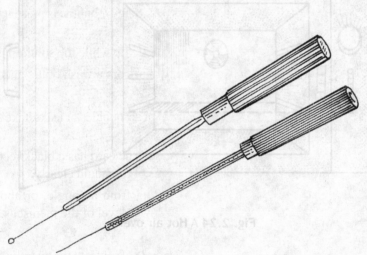

Fig. 2.23 Inoculation needle and loop

HOT AIR OVEN

Hot air oven is used to sterilize glassware, heat resistant material, oils, powders, waxes and other substances that cannot be sterilized by moist heat as they either get spoiled or are not sterilized effectively. Hot air oven sterilizes by dry heat, which removes moisture. It is not as effective as moist heat. For effective sterilization items have to be exposed to 150^0C –180^0C for 2–4 hours. It is made up of an insulated cabinet, in which heating is done by an electrical heating mechanism and a thermometer monitors temperature (Fig. 2.24). The shelves are perforated so that there is proper circulation of hot air.

The sterilization time and temperature relationship is as follows:

120⁰C	8 hours
140⁰C	3 hours
150⁰C	2 and 1/2 hours
160⁰C	2 hours
170⁰C	1 hour
180⁰C	20 minutes

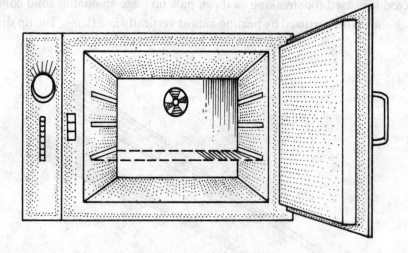

Fig. 2.24 A Hot air oven

INCUBATOR

Incubator is used to culture microbes at constant temperature conditions (Fig. 2.25). It is essentially similar to hot air oven except that temperature in this device is maintained at a constant level by an automatic device called thermostat. The thermostat cuts off the heating device when temperature reaches the desired level and once again switches it on when temperature drops below the desired level. Perforated shelves ensure proper ventilation. The incubator has double doors. The inner door is made up of glass to facilitate viewing of contents without opening the door and letting the outside air in. Thus in an incubator microbes can be cultured at optimum conditions required for growth. Since incubator maintains temperature by dry heat, a beaker filled with water should be kept to provide moisture and prevent dehydration of medium. In some incubators fluorescent light attachment is also provided for culturing microbes that require light for growth or sporulation.

Bacteriological Incubator

These structures are very sturdy double walled units with an outer chamber which is made up of a thick PCRC sheet duly degreased and treated for rust proofing and is painted with an attractive stove enamel, while the inner chamber is made up of a high-grade stainless steel sheet/anodized aluminium sheet. The gap between the walls is filled with a high-grade glass wool, for minimum thermal loss. Beaded elements of nichrome/kanthal A-1 wire are placed in the ribs at the sides and at the bottom of the chamber to provide uniform heating. The chamber is provided with four perforated shelves adjustable to convenient height. Temperature is controlled with an accuracy of ± 2° Celsius. An L-shaped prismatic thermometer is fitted to the unit for reading the temperature of the chamber. Air ventilators are provided at the sides.

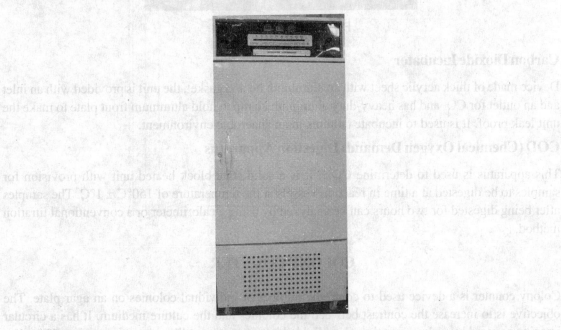

Fig. 2.25 Incubator

Temperature Range: 5°C above ambient to 80°C with three heat switches.

Anaerobic Culture Jar

It is a steel, glass or polycarbonate jar or with a rubber sealing ring and a cover held in place by a strong bridge clamp used to culture anaerobic microbes (Fig. 2.26). The cover is fitted with two needle valves for evacuation and a sachet containing a catalyst is placed inside it. The sachet is also known as gas pack and contains chemical which consume O_2 and generate anaerobic conditions. Once the inoculum is placed inside the sachet is opened and the lid is closed firmly.

Fig. 2.26 Anaerobic jar

Carbon Dioxide Incubator

Device made of thick acrylic sheet with an aluminum lid and gasket, the unit is provided with an inlet and an outlet for CO_2 and has heavy-duty aluminum clamp to hold aluminum front plate to make the unit leak proof. It is used to incubate cultures in an anaerobic environment.

COD (Chemical Oxygen Demand) Digestion Apparatus

This apparatus is used to determine COD. It is a solid state block heated unit with provision for samples to be digested at a time in reaction vessels at the temperature of $150°C \pm 1°C$. The samples after being digested for two hours can be analyzed by using a calorimeter or a conventional titration method.

COLONY COUNTER

Colony counter is a device used to count the number of individual colonies on an agar plate. The objective is to increase the contrast between the colonies and the culture medium. It has a circular opening covered by a glass counting plate divided into squares just like a hemocytometer. This plate can accommodate petri dishes of up to 100 mm external diameter (Fig. 2.27).

Light source is fitted behind the plate. A 4¼-inch thick lens having a standard 1.5X magnification is mounted on a sliding rod above the plate. The petriplate is placed on the counting plate. Light behind the glass plate illuminates the colonies so that they appear as brilliant spots or points on a subdued background. Magnifying glass improves the visibility so that very small colonies also can be seen clearly. In advanced versions readjustable electromagnetic counter provides instant, cumulative and accurate recording of colony counts. A switch in the electronic pen energizes the counter. Different ink colors can be used if different markings are desired. The marking pen can also be used in conjunction with the disposable membranes. Wet membranes are fixed at the back of the petri dish and after counting the sheets are indexed, dried and stored for reference (Fig. 2.27).

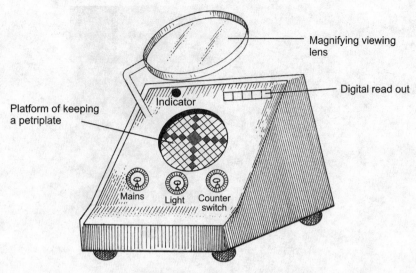

Fig. 2.27 Colony counter

LYOPHILIZER (FREEZE DRYER)

Freeze drying is the process of removing the moisture from a biological product while maintaining the integrity of biological and chemical products and their structure and activites.

Freeze drying technology is used in particular for the removal of moisture from the products usually of biological origin without causing any noticeable change in the original characteristics. Freeze drying is also the preferred process in the fields of pharmacy, chemistry and in conservation of food.

Bacterial growth and enzyme action do not take place in freeze-dried products. Proteins do not coagulate. Heat-sensitive biological materials may be freeze-dried without affecting their potency. Oxidation of many materials such as hemoglobin and vaccines is completely prevented when drying and sealing is done under high vacuum. The dried materials may be stored for years at room temperature without loss of their original characteristics. The freeze-dried material may be transported and shipped in small containers, and no provision is necessary for refrigeration en route.

MICRO-WELL PLATES

In order to assay several samples together or use small volumes of samples and reagents, special structures called micro-well plates are used. These plates are made up of plastic and contain 96 wells arranged in an 8 × 12 format with columns labeled 1–12 and rows labeled A–H. Each well has a definite name. Well shape and size differs according to the test. The plates cannot be autoclaved and are disinfected after use by soaking overnight in a 2% chlorine solution. They are then washed and rinsed three times in tap water, followed by rinsing three times in distilled water and then dried and reused.

3

TECHNIQUES OF MICROBIOLOGY

Study of microbes involves use of methods which enable us to observe them, culture them, measure their physiological properties and enumerate their biodiversity. Various methods and techniques used to study microbes are explained herein.

STERILISATION

Maintenance of sterile environment is absolutely essential in a microbiology laboratory. The working bench, laminar flow bench etc. should be wiped with spirit before and after use. Periodic fumigation of room, refrigerators, incubators etc. is essential. Wash hands with spirit before inoculation and wear sterile (autoclaved) gloves, masks and aprons while working. All the equipments, instruments etc. which come in direct contact with the microbes, need to be sterilised completely. The method of sterilisation chosen will depend on the nature of the material that needs to be sterilised. Various methods used for sterilisation are:

1. High Temperature

High temperature denatures proteins. It is used for sterilising equipments or substances that do not get damaged by heat. It is of following types:

- **Dry Heat Sterilisation:** Used to sterilise materials that can withstand high temperature as well as materials that get damaged by moisture such as:
    a. Inoculating loops, needles, tubes etc. which are sterilised by heating in flame before use and after use. (Do not touch the inoculum with hot needle, as it will scald the organism).

b. Glassware, instruments etc. are heated/dried in hot air ovens at 171°C : 1 h; 160°C : 2 h; 121°C : 16 h. for sterilisation.

- **Moist Heat Sterilisation:** Used to sterilise liquids and material which cannot withstand dry heat and get easily charred. This method is effective as it sterilises at a lower temperature as compared to dry heat, penetrates more quickly and does not dehydrate. Mostly used to sterilise gloves, masks, aprons, glassware, instruments like needles, forceps, loops, culture, media, water etc.

- **Boiling:** Medium/water is boiled at 100°C for a specific period. Only disadvantage is that boiling does not kill thermophiles and endospores.

- **Autoclaving:** Sterilisation of media, glassware, instruments, and water is done in an autoclave at 121°C, 15 psi, for 20 min. This method is more effective than boiling as it uses pressure to raise the temperature above that of boiling. It even kills the most resistant spores of *Bacillus stearothermophilus*.

- **Pasteurisation:** Material is heated at 63°C for 30 m. or 72°C for 15 seconds. It is named after Louis Pasteur who developed this method. It does not sterilise but limits growth and is generally used to sterilise food, milk, dairy products, wine, beer etc. It causes minimal damage to the product.

2. Low Temperature

This method is microbiostatic and is generally used to store media, solutions, reagents and cultures. It does not sterilise but slows down enzyme activity and metabolism.

- **Refrigeration** inhibits the growth of most pathogenic as well as disease-causing mesophiles but not psychrophiles: Exception: *Listeria* spp., which causes listeriosis (food poisoning).

- **Freezing** kills most bacteria, but the psychrophiles can remain alive for long periods in the frozen state. Compounds like DMSO, milk, or glycerol are added to protect proteins.

3. Radiation

Radiation denatures DNA. Shorter wavelengths have greater energy and thus are more lethal. Two types of radiation that kill bacteria directly are **UV (ultraviolet) Light** and **Ionising Radiation** i.e. **Xrays** and **Gamma rays**.

Sterilisation by UV: Tubes or flasks containing media, glassware, handling equipments etc. are exposed for 20 min to UV (germicidal tube) light.

Precautions: UV radiations are harmful. Never see the tube directly with naked eyes. Since UV radiations destroy DNA by causing **thymidine dimer formation**, never expose inoculum to UV light.

4. Filtration

This process sterilises by physical removal of cellular organisms except virus, as they are too small. It operates through exclusion rather than destruction of microorganisms. It is used to sterilise media, antibiotics and other such heat sensitive materials. It is gradually replacing pasteurisation in some cases, as it does not cause any damage e.g. cold filtered beers. It is safe for the user and is employed for sensitive liquids and gases as well. The filters are heat-sterilised before use.

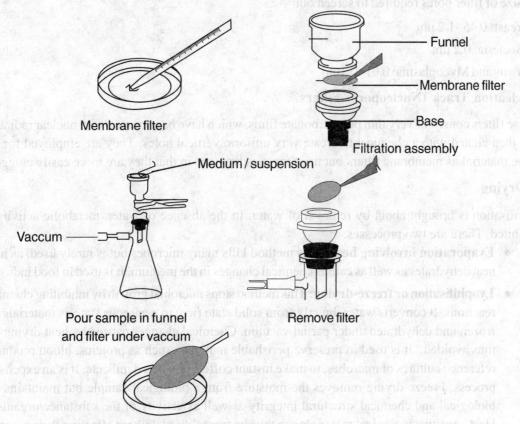

Membrane filter

Funnel

Membrane filter

Base

Filtration assembly

Medium / suspension

Vaccum

Pour sample in funnel
and filter under vaccum

Remove filter

Place filter in Nutrient medium and incubate

Fig. 3.1 Membrane filtration procedure

Three types of filters currently in use are:

Depth Filters

These are made of columns packed with fibrous materials such as glass wool or cotton wool. The twisting and turning fibers entrap particles and so act as filters; they show little resistance to flow and are used mainly for gases or as pre-filters for membrane filters, which are easily clogged.

Membrane Filters

These filters act by screening out particles. Their effectiveness depends on the size of the membrane pores and the electrostatic attractions present. The most commonly used filters in microbiology are usually made of cellulose acetate or cellulose nitrate. The sterile filters are placed in between the funnel and base of filtration assembly. The suspension / liquid which is to be filtered is poured in the funnel and filtered under vaccum (details are given in chapter 2). Filter is removed and placed on nutrient medium and incubated to check for bacterial growth and contamination (Fig. 3.1).

Size of filter pores required to screen out:

Yeast: 0.45–1.2 μm

Bacteria: 0.2 μm

Virus and Mycoplasma: 0.01–0.1μm

Nucleation Track (Nucleopore) Filters

These filters consist of very thin polycarbonate films, which have been treated with nuclear radiation and then etched with a chemical to create very uniform vertical holes. They are employed for the same material as membrane filters but have the disadvantage in that they are more easily clogged.

5. Drying

Sterilisation is brought about by removal of water. In the absence of water, metabolic activity is inhibited. There are two processes:

- **Evaporation involving heat:** This method kills many microbes but is rarely used, as high heat dehydrates as well as causes chemical changes in the medium. It is used in food industry.

- **Lyophilisation or freeze-drying:** This method stops microbial growth by inhibiting chemical reactions. It converts water directly from solid state (ice) to a gaseous state as materials are frozen and dehydrated under partial vacuum. Chemical changes caused by heat drying are thus avoided. It is used to preserve perishable materials such as proteins, blood products, reference cultures of microbes, to make instant coffee, powdered milk etc. It is an expensive process. Freeze-drying removes the moisture from a biological sample but maintains the biological and chemical structural integrity as well as activity of the substance/organism. Heat-sensitive biological materials can thus be freeze-dried without affecting their potency. The dried materials remain viable for long time and therefore can be stored for years at room temperature without loss of their original characteristics. The freeze-dried material can be transported and shipped in small containers without refrigeration. Bacterial growth and enzyme action do not take place in freeze-dried products. Proteins do not coagulate. Oxidation of many materials such as haemoglobin and vaccines is completely prevented when drying and sealing is done under high vacuum.

- **Osmotic Strength:** High concentration of salt or sugar causes dehydration of cells. Disadvantage is that once added, solutes (such as salt or sugar) cannot be easily removed. This method is used mainly for food preservation.

6. Chemical sterilisation

Chemicals are used for surface sterilisation of worktop, plant or animal tissues, etc. The commonly used chemicals are spirit, phenolic compounds, halogens, alcohols, etc. Chemicals bring about denaturation of proteins by hydrolysis, oxidation or by attachment of atoms or reactive groups, heavy metals, etc. They degrade the cell wall, affect nucleic acid structure and hinder metabolism.

Types of chemicals:

- **Phenols and phenolics:** These compounds denature proteins and damage cell membranes. They are microbicidal and are not affected by the presence of organic acids. E.g. Lizol, Phenyl, etc.

- **Alcohols:** These also affect proteins and cell membranes. Their effectiveness increases with increase in concentration. They are mainly used to surface sterilise skin, worktops, etc. (50–70% solution). They are unable to kill the endospores.

- **Halogens:** These inactivate enzymes. E.g. Iodine tincture, Betadine (iodophore i.e., iodine mixed with surfactant).

- **Surfactants:** These compounds break oily compounds by sticking to their surface. This results in formation of an emulsion which is then rinsed away. Microbial cells get washed away with the emulsion. Surfactants are usually combined with chemicals that help in penetration of fatty substances. Such surfactants are called wetting agents.

- **Quaternary ammonium compounds:** These are germicidal nitrogenous compounds in which four organic groups are attached to a nitrogen atom. They disrupt cell membranes. Soap hinders their activity.

- **Hydrogen peroxide:** This is an oxidizing agent, it denatures proteins. It is degraded by catalase and peroxidases into water and oxygen. Oxygen produced kills the anaerobes.

- **Heavy metals:** These denature proteins by reacting with the sulfhydryl groups. They are germicidal. E.g. Mercuric chloride, silver nitrate etc.

- **Alkylating agents:** They alkylate (attach short chains to) proteins and nucleic acids. They are carcinogenic.

- **Formalin:** It preserves, kills and inactivates microbes depending on the concentration used. It is effective against spores also.

- **Ethylene oxide:** Kills bacteria as well as spores. It is used to sterilise substances that cannot be heat sterilised.

- **Gluteraldehyde:** Is used if heat sterilisation is not possible.

- **Dyes:** Some dyes like crystal violet inhibit bacterial growth by affecting cell wall synthesis. It is specially effective in inhibiting Gram (+) cells and yeasts.

Comparing the effectiveness of germicides

Germicides effectivity can be tested by comparing their action with phenol (carbolic acid), the most commonly used germicide. Various methods used are:

1. Phenol coefficient: The ratio of the effective dilution of the chemical agent to the dilution of phenol that has the same effect, is the phenol coefficient.

2. Use-dilution test: The test microbe is added to different dilutions of the chemical agent. The highest dilution that remains clear after incubation indicates a germicide's effectiveness.

3. Disc method: Paper discs impregnated with chemicals are placed on agar plates pre-inoculated with test organism. Development of clear zone around the disc is an indication of inhibition of growth. Area of zone indicates the range of effectiveness.

Cleaning of slides

New slides: clean with a cotton swab dipped in spirit and heat over flame or immerse in absolute alcohol for 10–15 minutes. Used slides: Immerse in dichromic acid cleaning solution overnight. Wash and oven dry.

Dichromic acid solution

Potassium dichromate; 20g, distilled water; 100ml, Concentrated Sulphuric acid; 100 ml.

Dissolve the dichromate in water. Slowly and carefully add acid in small amounts. Cool the solution and store in glass bottles. Immerse glassware in it for at least five hours and wash thoroughly in running water. The solution can be used repeatedly till it turns dark.

OBSERVATION OF MICROORGANISMS

1. Observation of Live Microorganisms

Live microorganisms are observed to study their motility, interaction with one another, response to stimulus etc. For this, the mounting liquid should be of such a consistency that it allows the microbes to move freely. Observations can be continuous over a long period or short period. Short term or single observations can be easily made by mixing microbes in normal saline or distilled water and simply mounting the suspension on a clean glass slide, but for making a series of continuous observations, the medium used should be such that it provides enough nutrition to the cells allowing them to survive for the required time period. Usually nutrient medium is used for this purpose. The technique is known as micro-culture, where the microbes are grown on the slide and observed directly.

Various methods used for routine microbiological studies are:

- Wet Mounts
- Hanging drop preparations
- Agar block preparations

Wet Mounts

Drying, heat fixing and staining of bacteria kills them. In a wet mount, live and unstained bacteria are suspended in liquid, placed on a glass slide and covered with a small, thin piece of glass (or plastic) called cover slip. Most often bacteria are viewed in a wet mount to see if they are motile.

Method:

- Place a drop or one to two loopfuls of a liquid culture of microbial suspension in the centre of a clean, dry, grease proof slide.
- Cover it with a cover slip and observe.
- To prepare a wet mount from a culture on solid medium, place one or two loopfuls of water on the slide, remove a small amount of bacterial growth from the surface of the agar, and mix it well in the drop on the slide.
- Place a cover slip over it.

Hanging drop preparations

This technique is used to study motility in bacteria. A drop of liquid containing the microscopic cells/objects is suspended in a cavity. If bacteria shift and shake while floating, then movement is due to **Brownian motion** caused by the collision of water molecules with the bacterial cell. Currents in mounting medium may also sweep bacteria along in the same direction. This passive motion is referred to as **streaming.** Bacteria are being carried by the currents and are not actively swimming. Bacteria that move around rapidly and then quickly wriggle away are truly motile. True motility is dependent on the presence of one to many flagella on the cell surface.

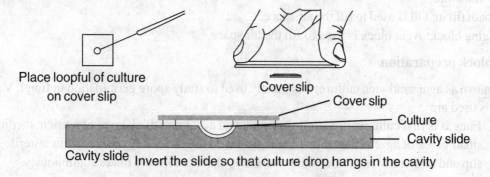

Place loopful of culture on cover slip

Cover slip

Cover slip

Culture

Cavity slide

Cavity slide   Invert the slide so that culture drop hangs in the cavity

Fig. 3.2 Hanging drop procedure

Method:

1. Take a clean, dry cover slip and place a loop full of suspension in the centre. Spread it a little with the help of a sterile needle or a loop.
2. Make a thin line of petroleum jelly around the edges of the cover slip.
3. Take a clean dry cavity slide (slide with a slight depression or a cavity in the centre) and place it over the cover slip in such a manner that the cavity is positioned directly on the suspension drop.
4. Next very gently invert the slide.
5. The suspension now hangs in the cavity. Observe under a light microscope (Fig. 3.2).

Advantages: There is enough space for the microbes to move freely. Bigger microbes do not get trapped between the slide and the cover slip. Nutrient and oxygen availability is limited but enough to support the microbes for little longer. Thus apart from motility, interaction between microbes especially conjugation, feeding behaviour, response to stimulus of protozoa, spore germination in fungi, generation time and lifecycle in bacteria etc. can be observed easily. Vaseline prevents quick drying of preparation. This method is relatively easier and quicker than agar slope method, it differentiates between true motility where cells keep changing place and Brownian movement where cells oscillate in one place.

Disadvantages: At times, movement ceases if the temperature of stage is lower than the body temperature of the microbe. To overcome this, a device called warm stage can be used. It heats the stage either electrically or by circulating warm water without disturbing the viewing area of the slide. Image clarity is compromised as interference with the pathway of light increases.

Precautions: Density of mounting medium should neither be too high nor too low. The cavity should be positioned directly on the drop. There should be no gaps in the vaseline lining the edges. Observations should be made at low intensity light and at the inner side of the drop edge. To slow down the speed of fast moving microbes such as protozoa, 2% carboxymethyl cellulose can be added to the medium.

Modifications: Several modifications have been suggested for this method. The commonly used modifications are:

Oil bath fill in: Oil is used to fill the air space.

Hanging block: Agar block is used to fill the air space.

Agar block preparation

Also known as agar sandwich culture; it is mostly used to study spore germination in fungi. Various methods used are:

a. Fungus is first cultured in a sterile petriplate on a 2–3 mm thick layer of nutrient medium. A small block of agar is then cut out and placed on the slide. It is covered with a sterile cover slip and allowed to incubate while a series of observations are made continuously.

b. Molten agar is poured over a sterile cover slip placed at an angle. When the agar dries to form a thin streak, a drop of inoculum is allowed to run down on it. Once the excess inoculum has drained off, a 3 × 3 mm size square is cut out of the agar streak and placed on a sterile slide, covered by a cover glass and sealed with paraffin. The preparation can be observed with oil immersion objective.

c. Moist chamber preparation: A 2–3 mm thick layer of medium is prepared in a sterile petridish. Out of this layer, a 5 mm block is cut and placed in the centre of a sterile glass slide. This block is then inoculated at the edge and covered by a cover slip. Complete assembly is transferred on a sterile U shaped glass rod placed horizontally in a petriplate containing 20% glycerol or water. The petriplate is covered by the lid and incubated at room temperature till adequate time. Complete assembly can be placed under the microscope for observation (Fig. 3.3).

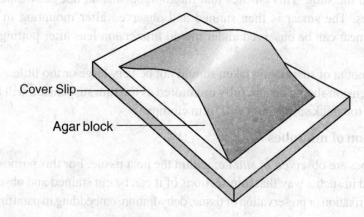

Cover Slip

Agar block

Fig. 3.3 Agar block

Disadvantages: The techniques are tedious and time consuming. Require skill and careful handling. Deposition of droplets of condensed water under the petriplate or cover glass reduces image clarity. Frequent contamination by other microorganisms also occurs.

Advantages: It allows the observation of fungi in continuous stage of growth.

Precautions: Maintenance of sterile conditions, incubation temperature, thickness of block etc. To overcome contamination, selective media can be used.

2. Observation of killed microorganisms

This is done to study the morphology, size, staining reactions etc. Microorganisms are observed by adhering or fixing them to the slide. Fixation kills them.

Specimen preparation for light microscopy

Microorganisms can be studied by light microscope in two forms:
- As a stained smear of the liquid containing them such as blood, water, soil solution, normal saline etc.
- *In situ* i.e. inside a stained section of the tissue containing them.

Preparation of a smear

Smear is just an even spreading of microbial suspension in the form of an uniform film on a clean, grease free glass slide or a cover slip. Smears are of two types: Thick films and Thin films.

Thick films are usually made by spreading the microbial solutions in the centre of a slide in a circle with an area of about 20 mm diameter. A drop of suspension is placed on the slide inside the marked circle and spread with the help of a sterile loop or needle.

Fixing and observing the smear

The smear is either air dried or gently heated on a spirit lamp or Bunsen burner for fixing the microorganisms on the slide. This ensures that microorganisms do not get washed off during the staining procedures. The smear is then stained and observed after mounting in canada balsam, alternatively the smear can be observed under the oil immersion lens after putting a drop of cedar wood oil on it..

Precautions: Amount of suspension taken should not be very large or too little.

Density of suspension should be carefully monitored as too thin suspension will tend to run away from the mark and too thick suspension will form clumps.

In situ observation of microbes

Sometimes microbes are observed *In situ* i.e. within the host tissue. For this purpose the host tissue has to be processed in such a way that thin sections of it can be cut stained and observed. The entire procedure involves fixation or preservation of tissue, dehydration, embedding in paraffin wax, sectioning, staining etc.

Preservation of plant material: In order to preserve the subcellular details fresh tissue is treated with Formalin Acetic Acid (FAA, Rectified spirit; 70 ml, Formalin; 5 ml, Acetic acid; 5 ml, DW; 20 ml) and then passed through 70% alcohol for 2 hours or immersed in 70% alcohol overnight. Dehydration of the tissue is then carried out by treating with Tertiary Butyl Alcohol (TBA) series in the following manner:

50% TBA series -------------	2 hours
70% TBA series -------------	overnight
85% TBA series -------------	2 hours
95% TBA series -------------	2 hours
100% TBA series ------------	2 hours
Pure TBA --------------------	2 hours
Pure TBA --------------------	2 hours
Pure TBA --------------------	2 hours
TBA + Paraffin oil (50-50) -	2 hours
Pure wax --------------------	2-3 changes till TBA smell disappears.

Preparation of TBA series

% of TBA	50%	70%	85%	95%	100%	Pure TBA	TBA+ paraffin
DW	50	30	15	–	–	–	–
95% Alcohol	40	59	50	45	–	–	–
TBA	10	20	35	55	75	100	50
Absolute Alcohol	–	–	–	–	25	–	–
Paraffin	–	–	–	–	–	–	50

Preservation of animal tissue: Properly washed tissue is cut into small pieces and then fixed in either formalin (4% formaldehyde) in normal saline for 48 h with one change or in warm Zenker formol fixative (Dissolve 5 g mercuric chloride, 2.5 g potassium bicarbonate, 1g sodium sulphate in 100 ml DW. Just before use add 5ml formalin/100ml solution) or Bouin`s fixative (Mix 75 ml saturated aqueous picric acid solution, 25 ml formalin and 5ml of glacial acetic acid) or Susas fixative (Dissolve 45 g mercuric chloride in 800 ml DW. Add 5g NaCl and 20g of Trichloroacetic acid and mix well. Finally add 40 ml of glacial acetic acid and 200 ml of formalin and mix well) for 24 h followed by washing in running water for next 24 h to remove traces of fixative. Formalin and Zenker formol fixative are general purpose fixatives. Bouin`s gives good results with virus inclusion bodies and Susa fixative is one of the best fixative for normal as well as pathological samples. Dehydration of the tissue is carried out as follows:

50% alcohol -----------------------------	2–5 h.
90% alcohol -----------------------------	2–5 h.
Absolute alcohol ------------------------	2 h (2 changes)
Chloroform - alcohol (1:1) -------------	overnight
Chloroform ------------------------------	6 h
Chloroform- paraffin wax (1:1) -------	1 h
Molten paraffin wax --------------------	2h (55 °C)

Embedding and sectioning of the tissue: Dehydrated tissue (plant/animal)is transferred to a container filled with molten wax. Paraffin is cooled immediately and appropriate sized blocks containing the tissue are cut out from this. Wax provides support to the tissue during sectioning and prevents damage by excessive drying. The ribbon containing the sections is then fixed on a slide to which egg albumin or haupt`s adhesive (1g pure gelatin is dissolved in 100 ml of DW at 30 °C. Few crystals of phenol and 15ml glycerol are added to it and mixed) has been applied. Slide is flooded with 4% formalin and kept on a slide warmer. Surface of the warmer should remain covered by water in order to prevent damage to cells by direct heat. Appropriate sized ribbon is placed on the warm slide, wax melts due to heat and the ribbon gets fixed on the slide.

Staining of sections: Before staining wax from the sections has to be removed otherwise it will not allow the aqueous stains to penetrate. Sections are treated with xylene to remove wax followed by down series of alcohol (Absolute, 95%, 70%, 50%. 30% alcohol) and finally water. The tissue is then stained with desired stains. After staining, the tissue is once again dehydrated by treating with up series (30%,50%, 70%, 95% and absolute alcohol), cleared in xylene and mounted in Canada balsam or **DPX** (5ml dibutyl phthalate, 35ml pure xylene and 10g Distrene 80).

STAINING OF MICROORGANISMS

Although microorganisms are chemically different from the medium in which they are growing, but their protoplast has the same refractive index as that of the medium. Thus, it is essential to stain the bacteria. They react with the dye or chemical by forming coloured complexes, which colour the cells or cell components. Staining of microbes therefore, provides a method for the study of morphology as well as general characteristics of a microbe; e.g. colour, shape, size, arrangement, structure, growth pattern etc. based on the reactions to the chemicals present in stains.

Chemical basis for structural staining

Stains get their colour due to dyes or chemicals called **Chromophores**, which in both positive and negative forms stain everything. The stain is made up of **chromogen** (an organic compound having benzene ring and a chromophore) and **auxochrome**, a compound that ionises the chromogen enabling it to form salts and bind to cellular structures. Staining takes place due to binding of cellular structures with the charge on **chromogen** surface. The most commonly used stains are salts e.g. Methylene blue chloride (methylene blue+Cl⁻]—disassociates in acid (HCl), yielding the staining component. If the microbe is incompatible with that specific dye it does not take up the stain.

Protein and other materials contain groups which bind with the staining chemical. The **surface** of **bacteria is acidic** due to the carboxyl groups of surface proteins. Active carboxyl (COO^-) or phosphoric (HPO_4^{2-}) groups impart a negative charge to the surface and are **able to bind covalently with** the **basic chromophoric ions**, of the dye used for example, methylene blue (tetramethylthionine chlorhydrate).

COOH (on cell surface) _____ ionisation to _____ $COO^- + H^+$

H^+ _____ replaced by _____ Na^+ / K^+ (in nature)

MBCl (methylene blue chloride) _____ ionisation to _____ $MB^+ + Cl^+$

$Na^+ / K^+ / H^+$ (on cell surface) _____ replaced by _____ MB^+

Formation of ionic bond between dye and cell surface results in a stained cell. At a lower pH, active amine (NH^{3+}) groups bind with acidic stains, e.g. potassium eosinate. This amphoteric character of proteins and pH thus determine the intensity and the type of stain, which will react with a particular protein. Raising the pH increases the staining by basic stains. Combined stains containing two or more ingredients to reveal more structures are used at or near pH 6 as some proteins are acidophilic; others and the nucleic acids are basophilic at this pH. Haemoglobin, granules of eosinophil, white blood cells continue to give an acidophilic reaction even at pH above 6. Oxyphil, oxyntic, eosinophilic, and acidophilic dyes are used synonymously for cells or components behaving in this way.

Use of mordants: Results in indirect union between tissue groups and radicals of the dye and are therefore used as an intermediary for some stains.

Types of Dye groups

a. **Acidic:** Stain bacteria having a negative charge on the cell surface. E.g. sodium, potassium, calcium, or ammonium salt of a coloured acid that yields a positive chromogen.

b. **Basic:** Stain bacteria having a positive charge on the cell surface. E.g. chloride or sulphate salt of coloured bases that yields a negative chromogen.

c. **Neutral:** Formed by a mixing of aqueous solutions of certain acidic and basic dyes.

Most bacteria tend to have a negative cell surface when the pH is either neutral (about 7) or slightly alkaline (about 8.3). This is the reason most bacteria are stained with basic dye.

Types of staining

- **Simple or Monochromatic staining:** Only one dye is used for staining.
- **Negative staining:** The cells are not stained but background is stained, therefore they can be seen easily against a dark background.
- **Orthochromatic staining:** The structure stains with the colour of the stain employed, e.g. collagen stains green with light green.
- **Metachromatic staining:** The dye, e.g. toluidine blue, combines with the sulphated proteoglycan in such a way that the dye molecules aggregate, causing a colour change from blue to reddish–purple.
- **Progressive staining:** Leaves the section in the dye until it is adequately coloured.
- **Regressive staining:** Overstains the tissue, then the excess stain is removed or differentiated out of the section by a solvent or oxidising agent.
- **Specific staining:** Stains just one structure or material.
- **Selective staining:** Preferentially stains one structure or material; others are stained, but less strongly.
- **Vital staining:** Involves the injection of materials into a living animal to reveal, say, macrophages.
- **Supra-vital staining:** Is applied to live cells held briefly in culture.
- **Differential staining:** Two colours are used separately or in combination to stain different cellular structures.
- **Cytological staining:** These are complex stains and stain a specific cell structure only such as capsule, endospore etc.
- **Microchemical staining:** This techniques involves use of complex stains which stain a specific cell component such as lipid, polyphosphates, glycogen etc.

Useful tips: Very thick layer of dye blocks the light and the dry smear develops cracks. A very thin film does not create a good contrast.

Important: All stains should be tested at appropriate intervals for their ability to distinguish positive and negative organisms and document the results. If no standard stains are available, known laboratory stains should be used as controls. The quality control procedure for stains needs to be performed on a weekly basis and also as and when a new lot of reagents for staining is procured/prepared.

Useful tip to blot dry: Place the slide in between the leaves of pad of blotting paper, close the pad on the slide and gently press down. The paper absorbs the water and leaves the slide almost dry.

SIMPLE STAINING

A single stain is used to colour microbes. Simple staining is mainly done to study the morphology and spatial arrangement of cells. Since the cell wall is negatively charged, basic dyes are generally used, as they are attracted towards the cell wall. E.g. Methylene blue, carbol fuchsin. crystal violet etc.

Procedure

- Heat fix the smear and flood with only one stain for e.g. Loeffler's methylene blue or 1% crystal violet or carbol fuchsin for 2–3 min.
- Wash the slide with water, blot dry and observe.
- Bacteria appear blue, violet or red, respectively, depending on the stain taken.

   Advantages: Easy, simple, quick, can stain almost all bacteria.

Composition of stains/reagents used:

   a. **Loeffler's methylene blue:** Dissolve 0.3 g methylene blue chloride in 30 ml of 95% ethanol. Add 100 ml of 0.1% KOH and filter. The stain can be stored at room temperature.

   b. **1% crystal violet:** 1 g Crystal violet is dissolved in 10 ml of 95% ethyl alcohol and final volume is made upto 100 ml with distilled water.

   c. **Carbol fuchsin:** Solution made for Ziehl–Neelsen staining is diluted and used.

NEGATIVE STAINING

Acidic stains are used to observe the cells. Since both the dye and cell surface have same charge, the dye is repelled away and the colourless cell is clearly visible against a coloured background.

Procedure

- Spread a loop full of bacterial suspension and a drop of 10% nigrosin or 2% congo red solution on a clean glass slide.
- Air-dry and observe under oil immersion.
- Since acidic dyes (eosin or nigrosin) have a negative charge, they cannot combine with the carboxyl groups on the cell surface. Thus, colourless cells are clearly visible against a dark background.

Advantages: Since cells are not heated, there is very little distortion of shape and size. Unstainable or difficult to stain bacteria can be easily observed by this technique.

Composition of stains/reagents used

   a. **10% nigrosin:** 10 g in 100 ml water.

   b. **2% congo red solution:** 2 g in 98 ml water.

DIFFERENTIAL STAINING

Gram Staining

This method was developed by Hans Christian Gram, a Danish bacteriologist as a means to differentiate pneumococci from *Klebsiella pneumoniae* in 1884. It is the most frequently used differential stain that divides bacteria into two major groups. It uses a primary stain, mordant, decolorising agent and a counter stain.Gram-positive cells (e.g. Cocci), appear purple or violet due to staining by **primary stain;** crystal violet and gram-negative cells (e.g. Coliforms and Pseudomonads) appear red due to staining by **counter stain;** safranin which gives a pink/red colour to bacteria.

    Principle: Gram staining is based on the differences in the cell walls of Gram-positive and Gram-negative organisms. Gram-positive organisms have a cell wall composed almost entirely of peptidoglycan where as Gram-negative organisms have a lipid rich outer membrane. The crystal violet complex is an insoluble complex that is difficult to remove especially after the use of a **mordant;** iodine, which results in the formation of a crystal violet–iodine complex. This complex intensifies the colour. **Decolorising agent;** alcohol dehydrates the Gram-positive cell wall and traps the complex inside the cell wall. However, it dissolves the lipid rich wall of Gram-negative organisms and therefore removes the stain from these cells. The colourless Gram-negative organisms are counter stained with fuchsin or safranin so that the Gram negative and Gram-positive organisms can be distinguished from one another.

Staining procedure

- Place a loopful of sample on a clean slide. Place another slide at an angle on it (Fig. 3.4a).
- Prepare a smear/thin film on a clean glass slide by dragging the slide over it (Fig. 3.4b).
- Air dry and heat fix the smear by passing through flame.
- Immerse the smear in crystal violet for one to one and a half minutes.
- Wash the slide with water (Fig. 3.4d) and then immerse in Gram's iodine for one to one and a half minutes. At this time all cells appear violet.
- Wash with water and decolorise by shaking the slide gently for 10–15 s in acetone/alcohol till the violet colour comes off the slide.
- Immediately wash with water and subsequently counter stain with safranin for 30 s.
- Once again wash the slide with water, blot dry and examine under the oil immersion lens of a microscope.
- The sequence of staining with alternate washing is as follows: Crystal violet / wash / Iodine/ wash / alcohol / wash / Safranin. (Fig. 3.4c).

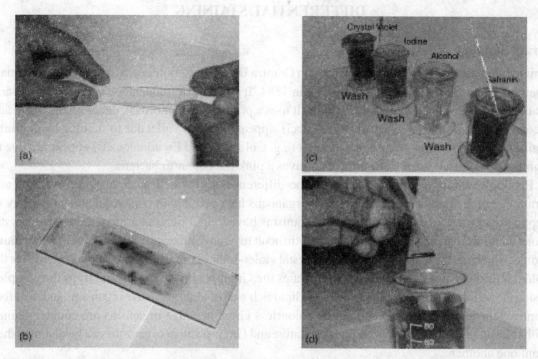

Fig. 3.4 Gram staining (a) Smear preparation; (b) Smear; (c) Proper arrangement of reagents; (d) Proper method of washing the smear.

Advantages: Differentiates cells on chemical basis without performing detailed procedures for chemical characterisation.

Precautions: Over-decolorisation can result in loss of stain from gram-positive cells also. Under-decolorisation will not remove colour from gram-negative cells. Thus, the results in both cases will be false. Always remove excess stain by thorough washing after every step by water. A dropper can be used for this purpose. This will prevent interference of the reagents with each other. Best results are obtained with 24 h-old cultures.

Composition of stains/reagents used

Ammonium oxalate–crystal violet solution

Solution A: 2.0 g crystal violet is dissolved in 20 ml of 95% ethanol.

Solution B: Dissolve 0.8 g ammonium oxalate in 80 ml distilled water. Mix solutions A and B. Store for 24 h before use.

Gram's iodine: 10 g iodine crystals and 20 g potassium iodide are dissolved in 1 litre distilled water. (Dry iodine and potassium iodide are ground in a mortar adding a few ml of water at a time until the iodine and iodide dissolve). Solution is filled in an amber glass bottle by rinsing the mortar pestle with the remainder of the distilled water.

Safranin stock solution: Dissolve 2.5 g Safranin O in 100 ml 95% ethanol. Filter before use.

Working solution: 10 ml of stock solution is mixed with 90 ml of distilled water.

Useful tips: It is best to use younger cells (24 h-old culture) because older Gram-positive bacteria are subject to break down of the cell wall by enzymes produced by the bacteria with age. Thus, gram-positive cells will also appear gram negative. Too much heating cooks the bacteria, also results in loss of gram-positive nature. Time taken for decolorisation also affects the result; hence, it should be standardized. Acetone is a better decolorising agent than alcohol.

Other counterstains which can be used instead of safranin are:

Dilute carbol fuchsin: dilute Ziehl-Neelsen's stain 10–20 times with water. Apply for 10–30 s.

Basic fuchsin: Dissolve 0.75g basic fuchsin in 25ml of 95% ethyl alcohol. Dilute 1: 9 before use. Apply for 10–30 s.

Neutral red: Dissolve 1 g neutral red in 2ml of 1% aqueous acetic acid and 1 litre of distilled water.

Modifications of Gram Staining

- Heat fixed smear is stained with Hucker's crystal violet solution for one minute. Crystal violet solution is washed off with Gram's iodine and then treated with fresh iodine for one minute followed by rinsing in running water and decolorisation by 1:1 acetone alcohol mixture for 10–15 s. Smear is once again rinsed with water and counterstained with safranin or ten times diluted carbol fuchsin for 30–60 s. Slide is blotted dry and observed under an oil immersion objective.

- Heat fixed smear is flooded with ammonium oxalate crystal violet solution for 30 s. Next it is washed with Gram's iodine followed by iodine–acetone mixture for 30 s. Smear is then washed with water and counterstained with 10 times dilute carbol fuchsin, washed, dried and observed under an oil immersion objective.

- Crystal violet solution is allowed to run over the heat fixed smear followed by a wash with undiluted Gram's iodine. The smear is immediately decolorised with acetone–alcohol mixture, quickly washed and counterstained with safranin for 5 s. It is observed after washing and drying under an oil immersion objective. The whole procedure takes about 15–30 s, hence, is also known as rapid method for Gram's stain.

Composition of the stains/reagents used

a. **Hucker's crystal violet:** 2 g of crystal violet chloride is dissolved in 20 ml of ethanol and mixed with 80 ml of 1% ammonium oxalate.

b. **Carbol fuchsin:** Mix 1 g basic fuchsin with 10 ml of 95% ethanol. Add 100 ml of 5% phenol. Let the solution stand for 7–14 days before use.

c. **Acetone–alcohol mixture:** Mix both in 1:1 ratio.

Decolorizers:

Acetone: fastest decolorizer as slide is dipped for only 2–3s.

Absolute alcohol: Needs a longer exposure i.e. one minute.

Acetone - alcohol: used in 1:1 ratio.

Acid-fast staining

The method was developed by **Paul Ehrlich**, a German physician in **1882** as a means of staining the tubercle bacillus, *Mycobacterium tuberculosis*. Ziehl and Neelsen modified the original method. The most commonly used Gram stain method was derived from the acid-fast stain.

Principle: Acid-fast staining is based on the high content of mycolic acids present in acid-fast organisms that are resistant to staining. A highly reactive dye, such as carbol fuchsin is used to stain the organisms. It penetrates the cell, combines with mycolic acids present in the cell, and binds tightly to them. Hence, cells containing high concentrations of mycolic acids are difficult to decolorise and are termed acid fast organisms, but non-acid-fast cells can be decolorised. These non-acid-fast cells can be counter stained by malachite green or methylene blue. 5% aqueous phenol is used as a chemical intensifier as it aids the penetration by primary colour. Heat acts as a physical intensifier.

Staining procedure

- Spread the suspension evenly on a clean glass slide and air dry for 15–30 min.
- Heat fix the smear by passing over the flame 3–5 times.
- Place the slide on the staining rack with the smeared side facing upwards and cover by 1% carbol fuchsin (**Primary stain**).
- Heat the slide until vapour starts rising. Continue the process for five minutes. Keep the slide flooded with the stain.
- Cool the slide for 5–7 min and then gently rinse with tap water to remove the excess carbol fuchsin stain.
- The smear appears red at this point.
- Decolorise by pouring acid alcohol or 25% sulphuric acid on the slide and leaving the acid for 2–4 min.
- Wash the excess stain. If the slide is still red, reapply sulphuric acid for 1–3 min and then rinse gently with tap water.
- Counter stain by pouring 1% malachite green solution or 0.1% Loeffler's methylene blue solution onto the slide and let it stand for one minute.
- Once again gently rinse the slide and air dry before examination under oil immersion.

Fig. 3.5 Steaming of smear for acid-fast staining

Advantages: Useful to identify cells which are not easily stained by Gram's stain especially *Mycobacterium* spp. due to the presence of very thick walls which are not easily penetrated by the stain. E.g. *M. tuberculosis*, *M. leprae*.

Useful tips: Unlike Gram-positive organisms, it is best to use older organisms for this staining technique because younger *Mycobacterium* spp. may not be as acid-fast as older ones since they have not accumulated as much acid. Do not let carbol fuchsin to boil or the slide to become absolutely dry.

Composition of the stains/reagents used

a. **1% Carbol fuchsin:** Mix 1 g basic fuchsin with 10 ml of 95% ethanol. Add 100 ml of 5% phenol. Let the solution stand for 7–14 days before use.

b. **25% Sulphuric acid:** 25 ml in 75 ml distilled water.

c. **0.1% Loeffler's Methylene blue:** Mix 0.3 g of methylene blue chloride and 30 ml of 95% ethanol. Add this to 100 ml of 0.1.1 KOH made in water.

d. **Acid alcohol:** 30 ml conc. HCl is added to 970 ml of 95% alcohol (kept in cold water).

Ziehl–Neelsen Staining

Ziehl and Neelsen developed this method in 1882–1883 independently. It also distinguishes acid-fast bacilli such as *Mycobacterium tuberculosis* and *M. leprae* from other non acid-fast bacilli. The procedure is essentially same as acid fast staining except instead of heat fixing a wetting agent

Turgitol is used in combination with carbol fuchsin.

Modifications of Ziehl–Neelsen Staining

Heat fixed smear is stained by basic fuchsin containing Tween 80 solution at room temperature for 5–10 minutes. Smear is rinsed with water and decolorised with 3% HCl in ethanol and rinsed once again. Counterstaining is done with Loeffler's methylene blue for three minutes. The slide is washed, dried and observed under oil immersion. Acid-fast organisms appear red and non-acid-fast organisms as well as the background appears blue.

Composition of the stains/reagents used

Basic fuchsin with tween 80 solution:

a. **Stock solution:** 4 g of basic fuchsin chloride is dissolved in 12 ml of phenol at 80°C. Twenty five ml of 95% ethyl alcohol is added to this with continuous stirring. The solution is made upto 300 ml with distilled water. The mixture is allowed to stand for 1–2 weeks before use.

b. **Working solution:** Filter and add 10 drops of Tween 80 to 100 ml stock solutions.

MICRO CHEMICAL STAINING

A. Staining of Nuclear Material

Since bacteria do not have a well-defined nucleus, the genetic material is stained using stains which will specifically stain only DNA or RNA. DNA of the nucleus can be seen by the Feulgen reaction in which mild acid hydrolysis unmasks aldehyde groups of the DNA, but leaves the RNA unchanged. Schiff's reagent then reveals these free aldehyde groups. A control is prepared by pretreatment with deoxyribonuclease enzyme, known to remove specifically DNA and then repeating the procedure.

1. Staining with Giemsa solution

- Fix the bacterial growth on agar block by osmic acid vapour for 2–3 min.
- Take a cover slip and press on the agar block to make an impression smear.
- Air-dry and keep the cover slip smear side down on warm Schaudin's fixative for five minutes.
- Rinse the cover slip with water (Cover slip can be stored till further use in 70% ethanol, before use wash with water.).
- Place the cover slip in 1 N HCl for 5–10 min at 60°C (This hydrolyses the nucleic acid).
- Wash the cover slip three times with tap water followed by distilled water.
- Stain for 30 min with Giemsa diluted 10 times with Sorensen's buffer.
- Rinse and mount in water.
- Immediately observe under oil immersion.
- DNA appears red against a colourless background.

Composition of the stains/reagents used

a. **Schaudin's fixative:** Thirty-three ml of absolute alcohol, 1 ml of glacial acetic acid, and 66 ml of saturated aqueous solution of mercuric chloride (6.9 g/100 ml at 20°C) are mixed together.

b. **70% ethanol:** 70 ml alcohol in 30 ml of distilled water.

c. **1 N HCl**

d. **Giemsa:** 1.6 g of Giemsa's stain R66 is dissolved in 177 ml of methanol and 42 ml of glycerol. Stirr constantly for few hours.

e. **Sorensen's buffer.**

Solution A: Dissolve 0.675 g of NaH_2PO_4 in one litre of distilled water.

Solution B: Dissolve 0.710 g of anhydrous Na_2HPO_4 in one litre of distilled water. To prepare buffer of pH 7, mix 52 ml of solution A and 48 ml of solution B.

2. Staining with crystal violet solution

- Prepare an impression smear without fixing with osmic acid.
- Air-dry the smear.
- Immerse it in boiling 0.2 N HCl for 5 s.
- Rinse with tap water several times.
- Stain with 0.1% crystal violet.
- Air-dry and immediately and observe under oil immersion.
- Nucleoid appears purple violet.

Composition of the stains/reagents used

a. **0.1% crystal violet:** Dilute stain prepared earlier with distilled water.

b. **0.2 N HCl:**

3. Staining with Feulgen solution

- Fix the bacterial growth on agar block by osmic acid vapour for 2–3 min.
- Take a cover slip and press on the agar block to make an impression smear.
- Air-dry and keep the cover slip smear side down on warm Schaudin's fixative for five minutes.
- Rinse the cover slip with water (cover slip can be stored till further use in 70% ethanol, wash with water before use).
- Place the cover slip in 1 N HCl for 5–10 min at 60°C (This hydrolyses the nucleic acid).
- Wash the cover slip three times with tap water followed by distilled water.
- Stain for one hour with Schiff's reagent at 15–20°C.

- Rinse and mount in water.
- Immediately observe under oil immersion.
- DNA appears red against a colourless background.

Composition of the stains/reagents used

Schiff's reagent: Dissolve 1 g of basic fuchsin in 400 ml of boiling distilled water. Cool, filter and add 1 ml of thionyl chloride. Leave the solution overnight. Shake the solution with activated charcoal to clear it, filter and store.

4. Staining with Acridine orange solution

- Fix the bacterial growth on agar block by osmic acid vapour for 2–3 min.
- Take a cover slip and press on the agar block to make an impression smear.
- Air-dry and keep the cover slip smear side down on warm Schaudin's fixative for five minutes.
- Rinse the cover slip with water (Cover slip can be stored till further use in 70% ethanol, wash with water before use).
- Place the cover slip in 3% HCl in alcohol for 5 minutes (This hydrolyses the nucleic acid).
- Rinse twice for 2 min each with citrate phosphate buffer (pH 3.8).
- Stain with 0.01% Acridine orange for 5–10 min (Acridine orange has a marked specificity for RNA and DNA).
- Rinse again twice for 2 min each with citrate phosphate buffer (pH 3.8).
- Mount in buffer, seal with nail paint and observe under fluorescence microscope.
- RNA component appears orange or red and DNA appears green.

Composition of the stains/reagents used

a. **3% HCl:** 3 ml HCl in 97 ml ethanol

b. **Citrate phosphate buffer (pH 3.8):** Mix 32.3 ml of 0.1 mol/lit (19.21g/l) citric acid solution with 17.7 ml of 0.2 mol/lit (28.39 g/l) Na_2HPO_4 solution and make up the final volume to 100 ml with distilled water.

c. **0.1% Acridine orange:** 1.0 mg in 100 ml distilled water, store in dark at 4 °C. Before use dilute with buffer to make 0.01% solution.

B. Staining of metachromatic granules

The bacterial cell stores phosphates, energy raw material for nucleic acid synthesis etc. as cytoplasmic inclusions known as volutin or metachromatic granules, also found in fungi and protozoa. These inclusions mainly contain polyphosphates, RNA and proteins.

Useful tip: These structures are more prominent in old cultures on the verge of starvation.

1. Staining with Albert solution

Metachromatic granules are acidic in nature. The acidified basic dye selectively stains them. The pH of the dye is adjusted (around 2.8) so that it is acidic to cytoplasm and basic to the granules. Toluidine blue-O stains the granules and methyl green stains the cytoplasm. Iodine acts as a mordant.

Staining procedure

- Cover the heat-fixed smear with Albert's stain I for 2–5 min.
- Wash with water.
- Cover with Albert stain II for two minutes, wash with water, blot dry and examine.
- Meta chromatic granules appear bluish black and the cell appears green or bluish green.

Composition of stains/reagents used

a. **Albert stain I:** Toluidine blue 0.15 gm, malachite green 0.20 gm, glacial acetic acid 1.0 ml alcohol (95%) 2.0 ml, distilled water 100 ml. The dyes are ground and first dissolved in alcohol. Water and acetic acid are then added to it. The mixture is allowed to stand for 24 h and then filtered.

b. **Albert stain II:** Iodine 2.0 gm, potassium iodide 3.0 gm, distilled water 300 ml. Iodine and potassium iodide are dissolved in water by grinding in a mortar with a pestle and then filtered through a filter paper. (Use lugol's iodine)

2. Staining with Methylene blue solution

- Cover the heat-fixed smear with Loeffler's alkaline methylene blue solution for 5 min.
- Wash with water. Air-dry and examine under oil immersion.
- Meta chromatic granules appear red and the cell appears pale blue.

Composition of the stains/reagents used

Loeffler's alkaline methylene blue: Same as Ziehl–Neelsen staining.

3. Staining with Methylene blue solution (modified)

- Cover the heat-fixed smear with Loeffler's alkaline methylene blue solution for 5 min.
- Wash with water. Decolorise with 0.1% H_2SO_4 for few seconds and wash again.
- Cover with Gram's iodine for one minute and wash with water.
- Counter stain with 1% Eosin Y for one minute, wash, blot dry and examine under oil immersion.
- Meta chromatic granules appear black and the cell appears pink.

Composition of the stains/reagents used

a. **Loeffler's alkaline methylene blue:** Same as Ziehl–Neelsen staining.

b. **0.1% H_2SO_4:** 0.1 ml H_2SO_4 in 99.9 ml of distilled water.

c. **1% Eosin Y:** 1 g in 100 ml.

C. Staining of Intracellular Lipids

Lipids have an affinity for Sudan dyes. They are present in outer membrane (cell wall), cell membrane and as carbon reserves in cytoplasm.

Staining with Sudan black B

- Cover the heat-fixed smear with Sudan black B solution for 10–15 min.
- Drain off excess stain, air dry the slide.
- Counter stain with 0.5% safranin or dilute carbol fuchsin for 5–10 s.
- Wash, blot dry and examine under oil immersion.
- Lipid granules appear dark blue and the cell appears light pink.

Composition of the stains/reagents used

a. **Sudan black B:** 0.3 g Sudan black B is dissolved in 100 ml of 70% ethyl alcohol. Leave overnight.

b. **0.5% safranin:** 5 mg in 99.5 ml.

D. Staining of Intracellular Polysaccharides

Polysaccharides are present in capsule, outer membrane (cell wall), and cell membrane and as carbon and energy reserves in cytoplasm.

1. Staining with Schiff's fuchsin sulphite reagent

- Cover the heat-fixed smear with periodate solution for 5 min.
- Rinse with 70% ethyl alcohol.
- Cover the smear with reducing rinse solution for 5 min.
- Rinse with 70% ethyl alcohol.
- Stain with fuchsin sulphite solution for 15–45 min.
- Wash repeatedly first with freshly prepared sulphite wash solution and then with water.
- Counter stain with 0. 002% malachite green for few seconds.
- Wash, blot dry and examine under oil immersion.
- Proteins and nucleic acid remain colourless, polysaccharide granules appear red and the cell appears green.

Composition of the stains/reagents used

a. **Periodate solution:** 20 ml of 4% aqueous periodic acid solution is mixed with 10 ml of 0.2 M aqueous sodium acetate and 70 ml of ethyl alcohol. Mix well and store in a brown bottle.

b. **Fuchsin sulphite solution:** Dissolve 2 g of basic fuchsin in 400 ml of boiling distilled water. Filter after cooling till 50°C. To this filtrate add 10 ml of 2 N HCl and 4 g of potassium metabisulphite. Incubate overnight in dark and cool place. Add 1 g of activated charcoal, mix thoroughly and filter. Apply the solution on glass and let it dry, note the colour. Keep adding 2 N HCl to this mixture till the solution on application on glass turns does not turn pink after drying.

c. **70% ethyl alcohol:** 70 ml alcohol in 30 ml of distilled water.

d. **Reducing rinse solution:** Dissolve 10 g of potassium iodide and 10 g of sodium thiosulphate pentahydrate in 200 ml of distilled water. Add 300 ml of ethyl alcohol and 5 ml of 2 N HCl with stirring. Allow the sulphur precipitate to settle.

e. **Sulphite wash solution:** Mix 0.4 g of potassium metabisulphite, 1 ml of concentrated hydrochloric acid and 100 ml of distilled water (Always use freshly prepared).

f. **0.002% malachite green:** 0.2 mg in 100 ml of distilled water.

2. Staining with Alician blue reagent

- Cover the heat-fixed smear with Alician blue solution for 1 min.
- Wash with water and air dry.
- Counter stain with 1:9 diluted carbol fuchsin.
- Wash immediately, air dry and examine under oil immersion.
- Polysaccharide granules appear blue and the cell appears red.

Composition of the stains/reagents used:

Alician blue reagent: Mix 1 g of alician blue with 100 ml of 95% ethanol. Before use, dilute ten times.

CYTOLOGICAL STAINING

A. Staining of endospore

Endospores are thick walled, resting spores produced by some bactria.

1. Staining with malachite green:

- Steam fix the smear by placing the slide, smear side up, on the edge of a beaker of boiling water.
- When the lower side of slide is covered with drops of condensed water, cover the smear with 5% aqueous malachite green.
- Keep the water boiling and warm the slide with the stain for one minute.
- Rinse with cold water.

- Stain with 0.5% safranin or 0.5% basic fuchsin for 30 sec.
- Wash, blot dry and examine under oil immersion.
- Spore will appear green and the cell appears red.

Composition of the stains/reagents used:

a. **5% aqueous malachite green:** 5 g in 100 ml of distilled water.

b. **0.5% safranin:** 5 mg in 100 ml of distilled water.

c. **0.5% basic fuchsin:** 5 mg in 100 ml of distilled water.

2. Staining with Carbol fuchsin:

- Air dry the smear and place a piece of blotting paper on it.
- Saturate the paper with carbol fuchsin.
- Heat the slide smear side up till steam rises. Keep the blotting paper moist with stain.
- Decolorise immediately with 95% ethyl alcohol to get a neat preparation, but if cells do not retain colour, then omit this step.
- Rinse with water.
- Spread a drop of aqueous Nigrosin over the smear with another slide.
- Dry the slide by gentle heating and examine under oil immersion.
- Spore will appear red, cell appears colourless and background dark.

Composition of the stains/reagents used:

a. **Carbol fuchsin:** Mix 10 ml of 95% ethanol and 1 g of basic fuchsin. Add 100 ml of 5% phenol. Let the solution stand for 7–14 days before use.

b. **Aqueous Nigrosin:** saturated solution of nigrosin in distilled water.

3. Staining by cold method:

- Heat the slide 20 times to fix the smear.
- Cool the slide and stain with saturated aqueous malachite green at room temperature for 10 min.
- Rinse with water and counter stain with 0.25% aqueous safranin for 2 min.
- Wash, blot dry and examine under oil immersion.
- Spore appears green and the cell appears red.

Composition of the stains/reagents used:

a. **Saturated aqueous malachite green:** saturated solution of malachite green in distilled water.

b. **0.25% aqueous safranin:** 0.25 mg in 100 ml of distilled water.

B. Staining of capsule

Capsule is made up of polysaccharides and is non-ionic, thus, it has a very low affinity for dyes. Therefore, capsule layer is best visualised by phase contrast microscope, where it appears as a halo around the cell.

Useful tips: For better results prepare the suspension in 6% glucose or 1% glucose in serum.

1. Staining by Copper sulphate

- Heat fix the smear.
- Stain with crystal violet (stains the cell).
- Flood the slide with 20% $CuSO_4$ (due to the osmotic difference created, stain diffuses towards the outer surface and little is retained in the capsule).
- Wash off the copper sulphate and dry the slide.
- Rinse with water, blot dry and examine under oil immersion.
- Capsule appears as a light purple layer around dark purple cell.

Composition of the stains/reagents used

20% $CuSO_4$: 20 g in 100 ml of distilled water.

2. Staining by Copper sulphate (modified)

- Do not heat fix the smear.
- Stain with 1% crystal violet for 2 min.
- Wash the slide with 20% aqueous $CuSO_4 \cdot 5H_2O$.
- Wash, blot dry and examine under oil immersion.
- Capsule appears as a light blue layer around dark purple cell.

Composition of the stains/reagents used

a. **1% crystal violet:** Dissolve 1 g of crystal violet chloride in 10 ml of 95% ethyl alcohol and make up final volume to 100 ml with distilled water.

b. **20% aqueous $CuSO_4 \cdot 5H_2O$:** 20 g in 100 ml of distilled water.

3. Staining by White's staining solution

- Mix a loopful of bacterial suspension and White's staining solution A and spread the mixture on the slide.
- Air dry and heat fix the smear.
- Stain with crystal violet.
- Flood the slide with 0.5% HCl, drain and dry by gentle heating.
- Stain with White's staining solution B for 15–30 sec.
- Drain, air dry and examine under oil immersion.

- Capsule appears colourless or light violet, cell appears blue and the background appears orange to golden.

Composition of the stains/reagents used:

a. **White's staining solution A:** Dissolve 5 g of Congo red in 100 ml of distilled water. Add 11 ml of serum and mix. Decant the supernatant and discard the residues.

b. **White's staining solution B:** Dissolve 1 g of methylene blue chloride in 100 ml of distilled water. Add 5 drops of glacial acetic acid.

c. **0.5% HCl:** 0.5 ml in 99.5 ml of distilled water.

4. Staining by Moller's fixative

- Air dry and heat fix the smear.
- Stain and flood the slide with 0.5% HCl, drain and dry by gentle heating.
- Stain with Moller's fixative for 15 sec.
- Cover with crystal violet for 1–3 min.
- Wash with saturated $CuSO_4 \cdot 5H_2O$ for 10 s.
- Blot dry and examine under oil immersion.
- Capsule appears light purple and the cell appears dark violet.

Composition of the stains/reagents used:

a. **0.5% HCl:** 0.5 ml in 99.5 ml of distilled water.

b. **Moller's fixative:** Dissolve 9 g of lead acetate in 280 ml of distilled water. Mix 20 ml of formalin (37–40% formaldehyde). Store in an airtight bottle.

c. **Crystal violet:** Dissolve 0.5 g of crystal violet chloride in 10 ml of 95% ethyl alcohol and make up final volume to 100 ml with distilled water.

d. **Saturated $CuSO_4 \cdot 5H_2O$**

5. Staining by Nigrosin

- Mix a loopful of bacterial suspension with 6% glucose and 8% aqueous nigrosin and spread the mixture on the slide.
- Air-dry and fix the smear by dipping in methanol for few seconds.
- Air-dry and stain with methyl violet.
- Wash, blot dry and examine under oil immersion.
- Capsule appears colourless, cell appears purple and background is grey coloured.

Composition of the stains/reagents used:

a. **6% glucose:** 6 g of glucose in 94 ml of distilled water.

b. **8% aqueous nigrosin:** 8 g water soluble nigrosin in 92 ml of distilled water.

c. **Methyl violet:** Dissolve 1 g methyl violet in 100 ml of distilled water.

6. Staining by Maneval's stain

- Mix a loop full of bacterial suspension and a drop of 1% congo red and spread the mixture on the slide.
- Air-dry and stain with Maneval's stain for 1 min.
- Drain, wash, air dry and examine under oil immersion.
- Capsule appears colourless and the cell appears red.

Composition of the stains/reagents used

a. **1% congo red:** 1 g in 100 ml of distilled water.

b. **Maneval's stain:** Mix 30 ml of 5% aqueous phenol solution, 10 ml of 20% acetic acid, 4 ml of 30% $FeCl_3 \cdot 6H_2O$ and 2 ml of 1% acid fuchsin.

7. Staining by India ink

- Mix a loop full of bacterial suspension and India ink and spread the mixture on the slide.
- Place a clean cover slip on it and press gently.
- Examine under oil immersion.
- Capsule appears as colourless halo surrounding a dark coloured cell.

C. Staining of flagella

Useful tips: First confirm motility by hanging drop method and then proceed.

1. Staining by Baileys/reagents solution

- Air-dry the smear and flood with Bailey's solution A for 3–4 min.
- Drain and immediately flood the slide with Bailey's solution B for 7 min.
- Wash with distilled water and without drying cover with Ziehl's fuchsin and heat, smear side up till steam rises for 1 min.
- Wash with distilled water, air dry and examine under oil immersion.
- Flagella and cell wall appear red.

Composition of the stains/reagents used

a. **Baileys Solution A:** Mix 6 ml of 6% aqueous $FeCl_3 \cdot 6H_2O$ with 18 ml of 10% aqueous solution of tannic acid.

b. **Baileys Solution B:** Add 0.5 ml of 0.5% basic fuchsin in ethyl alcohol, 0.5% ml of concentrated HCl and 2 ml of formalin to 3.5 ml of Bailey's solution A.

c. **Ziehl's fuchsin:** Mix 10 ml of 95% ethanol and 1 g basic fuchsin. Add 100 ml of 5% phenol. Let the solution stand for 7–14 days before use.

2. Staining by Leifson's staining/reagent

- Stain the smear with Leifson's staining reagent for 5–15 min till precipitate appears in staining solution.
- Wash and drain the water.
- Counter stain with 1–10 times diluted Loefller's methylene blue for 1 minute.
- Examine under oil immersion.
- Flagella appear red, cell appears blue.

Composition of the stains/reagents used

a. **Leifson's staining reagent**

Solution A: Prepare 3% aqueous tannic acid containing 0.2% phenol.

Solution B: Prepare 1.5% aqueous sodium chloride.

Solution C: Prepare 1.2% basic fuchsin in ethyl alcohol. Adjust pH to 5.

Mix solutions A, B and C in equal volumes.

3. Staining by Gray's staining solution

- Stain the smear with Gray's staining reagent for 10 min at 25 °C.
- Wash and drain the water.
- Counter stain with carbol fuchsin for 10 min.
- Wash, air dry and examine under oil immersion.
- Flagella appear light red, cell appears red.

Composition of the stains/reagents used

a. **Gray's staining reagent**

Solution A: Mix 5 ml of saturated solution of aluminium-potassium sulphate, 2 ml of 20% tannic acid and 2 ml of saturated solution of mercuric chloride.

Solution B: Prepare a saturated solution of basic fuchsin in ethanol (6 g/100 ml).

Mix 9 ml of solution A and 0.4 ml of solution B to prepare the staining reagent.

D. Staining of cell wall

1. Staining by tannic acid solution

- Air dry and heat fix the smear.
- Flood the slide with 5% aqueous solution of tannic acid for 30 min.
- Rinse with water and without drying stain with 0.5% aqueous solution of crystal violet chloride for 2–3 min.
- Wash and counter stain with 0.5% aqueous solution of congo red for 2–3 min.

- Blot dry and examine under oil immersion.
- Cell wall as well as new cross wall, septa etc. appear blue.

Composition of the stains/reagents used

a. **5% aqueous solution of tannic acid:** 5 g in 100 ml of distilled water.

b. **0.5% aqueous solution of crystal violet chloride:** 0.5 g in 100 ml of distilled water.

c. **0.5% aqueous solution of congo red:** 0.5 g in 100 ml of distilled water.

2. Staining by Bouin's fixative

- Place a cover slip in a petri dish.
- Place a 2 mm thick agar slab on the cover slip, bacterial growth side down.
- Immerse the slab in Bouin's fixative for 3 h.
- Remove fixative as well as the agar slab.
- Wash the cover slip by letting drops of water drip and flow out for 2 h.
- Cover with 7% aqueous tannic acid for 20–30 min.
- Mount in water and seal with nail enamel.
- Cytoplasm appears light violet, cell wall appears dark brownish black.

Composition of the stains/reagents used

a. **Bouin's fixative:** 75 ml of saturated aqueous picric acid solution is mixed with 25 ml of formalin and 5 ml of glacial acetic acid.

b. **7% aqueous tannic acid:** 7 g in 100 ml of distilled water.

3. Staining by Congo red

- Air dry and heat fix the smear.
- Cover with 3 drops of 0.34% aqueous solution of cetylpyridinium chloride.
- Add a drop of 5% congo red to this.
- Wash and cover with 0.5% methylene blue solution for 10 sec.
- Wash, dry and examine under oil immersion.
- Cytoplasm appears blue, cell wall appears red.

Composition of the stains/reagents used

a. **0.34% aqueous solution of cetylpyridinium chloride:** 34 mg in 100 ml of distilled water.

b. **5% congo red:** 5 g in 100 ml of distilled water.

c. **0.5% methylene blue:** 5 mg in 100 ml of distilled water.

ORGANISM SPECIFIC STAINING

1. Iodine staining for ova and cysts

- Mix sample with normal saline and then with iodine solution with the help of a wire loop or applicator and spread on the slide.
- Examine under 10X and 40X of the microscope for various ova and cysts.
- For staining *Azotobacter* cysts, suspend bacteria in staining solution.
- To observe old cysts, slightly warm the preparation.
- Vegetative cells appear light yellowish–green. In the early stages of cyst formation, receding cytoplasm appears dark green separated from the cell wall by brownish–red layer.

Composition of the stains/reagents used

Staining solution: Add 8.5 ml of glacial acetic acid, 3.25 g of anhydrous sodium sulphate, 200 mg of neutral red, 200 mg of light green and 50 ml of ethanol to 100 ml of water. Incubate for 15 min and filter through 0.5 μ membrane filter.

2. Staining of milk bacteria

- Spread 0.001 ml milk sample in a square area on a clean slide and heat fix.
- Remove cellular fat by treating the smear with xylene or chloroform.
- Fix the smear in 95% alcohol for 2–3 min and then stain with Breed's methylene blue for 2 min.
- Wash with 90% alcohol till smear becomes light blue, air dry and observe under oil immersion.
- Bacteria appear dark blue against a light blue background.

Composition of the stains/reagents used

Breed's methylene blue: 0.3 g of methylene blue chloride is added to 30 ml of ethanol. Add this solution to 100 ml of 2% phenol in distilled water.

3. Staining of Spirochaetes

a. By Fontana's fixative

- Prepare a smear and air-dry it.
- Apply Fontana's fixative three times successively for 30 sec each.
- Treat with absolute ethyl alcohol for 3 min.
- Drain off excess alcohol and dry the slide.
- Cover with Fontana's mordant and heat smear side up till steam rises for 30 sec.
- Cool, wash, dry and apply Fontana's stain and heat smear side up till steam rises and slide becomes brown in colour.
- Wash, dry and observe under oil immersion.
- Spirochaetes appear brown–black. Background appears yellow.

Composition of the stains/reagents used:

a. **Fontana's fixative:** (Ruge's solution) Mix 1 ml of glacial acetic acid and 2 ml of formalin in 100 ml of distilled water.

b. **Fontana's mordant:** Dissolve 1 ml of melted phenol crystals in 100 ml of 5% aqueous tannic acid.

c. **Fontana's stain:** Add slowly drop wise, 10% ammonia solution to 0.5% aqueous silver nitrate solution till precipitate formed dissolves. Slowly add more silver nitrate solution till precipitate reappears.

b. By Basic fuchsin

- Prepare a smear and air-dry it.
- Apply 1% formalin for 1 min and dry.
- Treat the smear with 5% $NaHCO_3$ followed by 10 drops of basic fuchsin for 3–5 min.
- Wash, dry and observe under oil immersion.
- Spirochaetes appear red.

Composition of the stains/reagents used

a. **1% formalin:** 1 ml in 99 ml of distilled water.

b. **5% $NaHCO_3$:** 5 ml in 100 ml of distilled water.

c. **Basic fuchsin:** Dissolve 0.75 g of basic fuchsin in 25 ml of 95% ethyl alcohol. Dilute 1:9 times before use.

c. Staining of Spirochaetes in tissue

- Fix the tissue in 10% formalin overnight.
- Wash in water for one hour.
- Dehydrate by placing it in 96–98% alcohol for 24 h.
- Treat with silver nitrate pyridine solution first for two hours at room temperature and then for 4–6 h at 50°C.
- Wash in 10% pyridine solution and quickly transfer to reducing fluid.
- Keep in reducing fluid for 48 h at room temperature in dark.
- Wash with distilled water, dehydrate in increasing series of alcohol.
- Embed the tissue in paraffin as described earlier in this chapter.
- Cut sections, remove paraffin with xylol and mount in Canada balsam.
- Spirochaetes appear red.

Composition of the stains/reagents used

a. **Silver nitrate pyridine:** Mix 10 ml of pure pyridine with 90 ml of 1% solution of silver nitrate.

b. **10% formalin:** 10 ml in 100 ml of distilled water.

c. **10% pyridine:** 10 g in 100 ml of distilled water.

d. **Reducing fluid:** Just before use, add 10 ml of pure acetone and 15 ml of pure (pyrimidine) to 100 ml of 4% formalin.

4. Staining of Mycoplasma

a. Staining of Mycoplasma colony

- Flood the petri plate with 1:9 diluted Diene's staining solution.
- Examine the plate under low power objective.
- Colony appears granular with aggregates of royal blue to greenish–blue coloured cells.

Composition of the stains/reagents used

Diene's staining solution: Add 0.25 g of Azure II, 0.5 g of methylene blue, 2 g of maltose, 0.05 g of Na_2CO_3, 0.04 g benzoic acid to 20 ml of distilled water.

b. Staining of Mycoplasma on cover slip

- Make a smear of undiluted Diene's staining solution.
- Place the agar block containg mycoplasma colony on the dried smear growth side down.
- Apply vaseline to a 15–20 cm diameter brass ring and fix it on to the slide by pressing.
- Invert the cover slip on it, block side down.
- Examine under low power objective.
- Colony appears dark royal blue and mycoplasma appear reddish or greenish blue.

c. Staining of Mycoplasma on slide

- Transfer agar block to the slide, growth side down.
- Remove the agar by treating with hot water.
- Stain the colonies with Dienes' staining solution or 1:1 Diene's staining solution and Giemsa stain.
- Examine under oil immersion.

Composition of the stains/reagents used

a. **Diene's staining solution:** Same as above.

b. **Giemsa stain:** 1.6 g of Giemsa's stain R66 is dissolved in 177 ml of methanol and 42 ml of glycerol. Stir constantly for few hours. Stock solutions for Giemsa are also available that are stable and produce consistent results. Giemsa stain must be prepared daily by mixing one volume of stock stain with 15 volumes of distilled water.

c. **Stock solutions for Giemsa:** Take 500 mg of finely powdered azure B eosinate, 100 mg of Azure A eosinate, 400 mg of methylene blue eosinate, 200 mg of methylene blue and mix well. Add these to 200 ml each of methanol and glycerol (solvent). Incubate at 50–60°C. Allow the mixture to stand for 2–3 days, shaking vigorously in between.

d. Staining with cresyl violet

- Stain agar block with cresyl violet before applying to cover slip.
- Transfer agar block to the cover slip, growth side down.
- Apply vaseline to a 15–20 cm diameter brass ring and fix it on to the slide by pressing.
- Invert the cover slip on it, block side down.
- Examine under low power objective.
- Colony and mycoplasmas appear reddish or purple.

5. Staining of Virus inclusion bodies

a. **Staining with Giemsa:** Same as for Fast Romanowsky-type Staining.

b. **Fast Romanowsky-type Stain:**

- Make a thick film and dry for one hour.
- Stain unfixed film with staining solution for 10 min.
- Rinse in distilled water, air-dry and observe.

Composition of the stains/reagents used

Staining solution: Mix 4 ml of Giemsa stock solution, 3 ml of acetone, 2 ml of buffer solution, and 3 litres of distilled water.

Buffer solution (pH 7): 5.447 g of anhydrous Na_2HPO_4 and 4.752 g of KH_2PO_4 are ground in a mortar pestle. One gram of this powder is added to 2 litres of distilled water.

c. Staining with methylene blue–eosin

- Fix tissue in Bouin's fluid or Zenker's fluid.
- Prepare paraffin sections in the usual manner.
- Rehydrate sections.
- Stain with Mann's methylene blue–eosin stain for 12 h at 37°C.
- Rinse with water and differentiate under microscope in 70% alcohol containing 0.5% orange G.
- Negri bodies appear red, while nuclei and central granules of negri bodies appear blue.

Composition of the stains/reagents used

a. **Bouin's fluid:** Mix 25 ml formalin, 5 ml glacial acetic acid and 75 ml of saturated aqueous solution of picric acid.

b. **Zenker's fluid or Zenker formol fixative:** Add 5 g of mercuric chloride, 2.5 g of potassium bicarbonate, 1 g of sodium sulphate to 100 ml of distilled water. Just before use, add 5 ml of formalin/100 ml of solution.

c. **Mann's methylene blue–eosin stain:** Mix 35 ml of 1% aqueous solution of methylene blue with 45 ml of 1% aqueous solution of eosin and make up to 180 ml with distilled water.

d. **70% alcohol containing 0.5% orange G:** 70 ml of alcohol, 0.5 g of orange G in 30 ml of distilled water.

d. Staining with Tetrazine

- Fix tissue in formol saline or formol corrosive solution.
- Prepare paraffin sections in the usual manner.
- Rehydrate sections.
- Stain the sections with 0.5% phloxine-B or 0.5% Rose Bengal in 0.5% $CaCl_2$ for 30 min.
- Rinse with water and differentiate by adding saturated solution of tetrazine in 2-ethoxy-ethanol (cellosolve) drop by drop.
- Rinse with 95% ethanol, dehydrate, clear, mount and observe.
- Acidophilic inclusion bodies are seen red against a yellow background.
- Nuclei appear blue and negri bodies do not take the stain.

Composition of the stains/reagents used

a. **Formol saline:** Mix 90 ml of 0.9% NaCl with 10 ml of 40% formaldehyde solution. Maintain pH 7 by adding $CaCO_3$ granules.

b. **Formol corrosive solution:** Mix 90 ml of saturated aqueous solution of mercuric chloride with 10 ml of 40% formaldehyde solution.

c. **0.5% phloxine-B or 0.5% Rose Bengal in 0.5% $CaCl_2$.**

d. **Saturated solution of tetrazine in 2-ethoxy-ethanol (cellosolve).**

6. Staining of Virus elementary bodies and Rickettsia

a. Staining with Gutstein's stain

- Prepare a thin film.
- Rinse with saline and distilled water.
- Fix with methyl alcohol for 30 min.

- Place two capillary tubes in a petri dish and place the slide on them smear side down.
- Add Gutstein's stain, cover with the lid and let the assembly stand for 20–30 min at 37°C.
- Remove the slide, rinse, dry and observe.
- This method is used to observe elementary bodies of variola-vaccinia group from scraping of skin lesions.

Composition of the stains/reagents used

a. **Gutstein's stain: Solution A:** 1 g of methyl violet is dissolved in 100 ml of distilled water.

b. **Solution B:** 2 g of sodium carbonate is dissolved in 100 ml of distilled water.

Just before use, mix equal volumes of A and B and filter.

b. Staining with Azure II

- Prepare a thin film.
- Air dry and fix in 1 N HCl for 2 min.
- Wash thoroughly and stain with Azure II solution for 20 min.
- Wash and counter stain with 0.25% safranin for 6–8 s.
- Wash in running water, blot dry and observe.
- This method is used to observe rickettsia and elementary bodies of psittacosis.
- They are stained blue but the cell and the nuclei appear red.

Composition of the stains/reagents used

a. **1 N HCl**

b. **Azure II solution:** Dissolve 1 g of Azure II in 100 ml of distilled water. Filter before use.

c. **0.25% safranin:** 0.25 g in 100 ml of distilled water.

c. Staining with Nicholau's staining solution

- Prepare a thin film.
- Fix by heating or by methanol.
- Cover with Nicholau's staining solution and heat from below till steam rises for 5 min.
- Do not boil the stain.
- Rinse, blot dry and observe.

Composition of the stains/reagents used

a. **Nicholau's staining solution:** 1 g of Isamine blue is dissolved in 10 ml of 95% ethanol. Add this to 100 ml of distilled water containing 3 g of phenol. Store in a brown bottle.

b. **Van Gieson's stain:** Mix 1–3 parts of saturated solution of acid fuchsin with 100 parts of saturated aqueous solution of picric acid.

d. Staining of Rickettsiae by Micchiavells method

- Prepare a thin film.
- Fix by air-drying.
- Warm the slide gently.
- Cover with 0.25% aqueous solution of basic fuchsin (pH 7.3) for 4 min.
- Rinse quickly with 0.15% citric acid followed by water.
- Counter stain with 1% aqueous methylene blue for 5 s.
- Blot dry and observe.

Composition of the stains/reagents used

a. **0.25% aqueous solution of basic fuchsin:** Dilute basic fuchsin prepared earlier for Spirochaete staining.

b. **0.15% citric acid:** 0.15 g in 100 ml.

c. **1% aqueous methylene blue:** Dissolve 1 g of methylene blue in 100 ml of distilled water.

7. Staining of Fungi

- Yeasts, yeast like fungi and larger yeast forms of dimorphic fungi can be stained by Gram staining, but the method is not very satisfactory.
- Giemsa can stain alcohol fixed preparation of very small yeasts.
- Fungal hyphae are also stained by cotton blue and mounted in lactophenol as wet mounts or as agar blocks.
- Actinomycetes (filamentous bacteria) are stained by gram stain or Ziehl–Neelsen acid-fast stain.
- Fungi can be stained inside the host tissue in the following manner: Rehydrate the paraffin sections by passing through down or descending alcohol series. Wash with water and immerse in periodic acid solution for 5 min. Once again wash thoroughly in tap water for 15 min followed by distilled water. Stain the sections with Schiff's reagent for 15 min and then wash 2–3 times with freshly prepared sulphite wash solution, taking 5 min for each wash. Follow this by washing in running water for 10 min and rinse with distilled water. Blot dry and counter stain with 0.002% malachite green or light green for 30–60 s. Rinse, dehydrate in absolute alcohol, clear in xylene and mount in Canada balsam or DPX. Fungal hyphae appear red and the cytoplasm as well as nuclei will appear green.

Composition of the stains/reagents used

a. **Gram stain:** see gram staining.

b. **Giemsa:** see staining of Mycoplasma spp.

c. **Lactophenol–cotton blue:** Mix 20 ml of lactic acid, 20 g of phenol, 40 ml of glycerol and

20 ml of distilled water. Heat gently on a water bath till the solution becomes homogenous, add 0.05 g aniline blue or 2ml of 1.1 cotton blue.

d. **Ziehl–Neelsen acid-fast stain:** See Ziehl–Neelsen acid fast staining.

e. **Descending alcohol series:**

f. **Periodic acid solution:** See staining of polysaccharides.

g. **Schiff's reagent:** see Feulgen staining.

h. **Sulphite wash solution:** See staining of polysaccharides.

i. **0.002% malachite green or light green:** 0.002 g in 100 ml of distilled water.

8. Staining of Protozoa

a. Staining with Wiegert's iron Hematoxylin

- Fix the smear with Schuadinn's fixative for about 5 min.
- Wash with 50% ethyl alcohol and stain with Gram's iodine for 2 min.
- Wash with ethyl alcohol once again till no more iodine comes out.
- Rinse with distilled water and stain with Wiegert's iron haematoxylin for 10–20 min.
- Wash with water. Counter stain with Van Gieson's stain for 15–30 s.
- Once again wash with water, air dry and observe.
- If staining protozoa in the host tissue, treat the section similarly, dehydrate with alcohol, clear in xylene and mount in Canada balsam or DPX.
- Protozoa are stained blue–black.
- Alternatively after fixing the film or tissue section with Schuadinn's fixative for about 5 min and washing, treat with 2% aqueous ammonium molybdate for 10 min. This acts as a mordant.
- Wash and stain with freshly prepared haematoxylin solution.
- Wash with distilled water and then keep in tap water till the film becomes blue.
- Mount in water and observe.
- Dehydrate sections with alcohol, clear in xylene and mount in Canada balsam or DPX.
- Protozoa appear blue.

b. Staining with Leishman's Stain

- Protozoa can be stained in blood film by treating unfixed film with Leishman's stain for 1–2 min.
- Flood the slide with double the volume of buffer solution or distilled water and mix the stain thoroughly with it. Keep the slide in the diluted stain for 10 min.
- Pour off the stain, wash thoroughly till the smear appears pink.
- Blot dry and observe.

- Sections are treated with xylene to remove paraffin, there after with alcohol followed by water.
- Stain with Leishman's stain diluted with buffer in 1:3 ratio for 5 min.
- Wash with buffer and place the sections in 1:1500 acetic acid till protoplasm of host cell appears pink and nuclei blue.
- Wash with buffer, blot the sections, dehydrate in absolute alcohol, clear in xylene and mount in DPX.

c. Staining of Malarial Parasites

- Giemsa staining is used to stain malarial parasites and trypanosomes.
- Romanosky stain can be used o stain malarial and other parasites.

d. Staining with Fields Stain

- Stain with Fields stain A for 1–2 seconds.
- Wash till colour ceases to flow and then stain with Fields stain B for 1–2 seconds.
- Rinse and drain dry.
- Oven dry the slide at 40°C.
- Gently warm the slide on flame.

Composition of the stains/reagents used

a. **Van Gieson's stain:** See staining of virus inclusion bodies.

b. **Wiegert's iron haematoxylin:**

   Solution A: 1 g haematoxylin is dissolved in 100 ml of absolute alcohol.

   Solution B: Mix 4 ml of 30% liquor- ferri perchlorate with 1 ml of concentrated HCl. Add this mixture to 100 ml of distilled water.

   Just before use mix equal parts of A and B.

c. **50% ethyl alcohol:** 50 ml of distilled water and 50 ml of absolute alcohol.

d. **Gram's iodine:** see Gram staining.

e. **2% aqueous ammonium molybdate:** 2 g Ammonium molybdate is mixed with 100 ml of distilled water.

f. **Leishman's stain:** Readymade stain is available. To prepare repeatedly grind 0.15 g of Leishman's powder in 100 ml of pure methanol (pH 6.5) till all the powder dissolves in methanol.

g. **Buffer solution:** See staining of virus inclusion bodies.

h. **Schuadinn's fixative:** See staining of virus inclusion bodies.

i. **Giemsa:** See staining of Mycoplasma spp.

j. **Fields stain A:** Dissolve 1.3 g of methylene blue and 5 g of anhydrous Na_2HPO_4 in 50 ml of distilled water. Boil the mixture to dryness in a water bath. Add 500 ml of freshly boiled water cooled to 45°C. Mix till all the powder dissolves. Incubate for 24 h, filter before use.

k. **Fields stain B:** Dissolve 5 g of anhydrous Na_2HPO_4 and 6.25 g of anhydrous KH_2PO_4 in 500 ml of freshly boiled warm water. Add 1.3 g of eosin and mix well to this solution, incubate for 24 h, and filter before use.

Some Important facts

- Sudan dyes dissolve in fat preserved by frozen sectioning and colour it. This does not involve ionic combination.

- Osmium tetroxide forms a black complex with unsaturated fat and also acts as a fixative.

- Glycogen, glycoproteins and proteoglycans are oxidised by Periodic Acid to -C (aldehyde) groups which restore colour to Schiff's reagent (hence PAS technique).

- Colourless Schiff's reagent/leucofuchsin can be prepared by bleaching basic fuchsin with sulphurous acid.

- To distinguish glycogen from glyco conjugates, pretreat with saliva. Enzyme amylase, present in saliva will destroy only glycogen.

- During EM study, ruthenium red makes some glyco conjugates electron-dense (not red) and visible.

- Porphyrins and vitamin A are naturally auto-fluorescent materials.

- Formaldehyde converts the catecholamines to fluorescent quinoline compounds. The UV microscopy of formaldehyde-fixed sections shows the distribution of norepinephrine, for example, in the sympathetic, post-ganglionic, nerve fibres and adrenal medulla.

- When certain substances are exposed to ultraviolet (UV) light, they fluoresce i.e. they emit light of longer and visible wavelengths. The fluorescence microscope illuminates the section with UV light, and its areas of fluorescence are viewed through eyepieces incorporating a UV filter to protect the eyes.

- Immunofluorescent visualisation involves preparation of a pure sample of the material (peptide or polysaccharide), whose distribution in the tissues is to be studied.

- Injecting peptide or polysaccharide into a rabbit, whose plasma cells will treat it as an antigen and produce antibodies against it, is used to produce antibodies aganst peptides or polysaccharide.

- Conjugation of the serum antibody with fluorescein isothiocyanate makes its position traceable when viewed in UV light due to fluorescence.

- Monoclonal antibody (MoAb) can also be produced by fusing antibody-forming and malignant mouse lymphocytes *in vitro* and cloning them.

- The strong bond between avidin and biotin is the basis for effective means of tagging reagents for immunohistology.

- Immunostaining, with its high specificity and sensitivity, is used in electron microscopy by conjugating the antibody not with fluorescein, but with ferritin; recognisable as granules; or peroxidase which on incubation with substrate, gives a visible reaction product; or gold particles, which do not react, are visible, and are of standardised sizes. This helps to test two materials at the same time, for any co-localisation, or separate distributions.

For all staining techniques, the following points need to be remembered

- Always use clean and grease free slides. Label the slides.
- Shake the broth culture before use so as to disperse the bacteria all over the medium.
- For observing cells from a liquid (or "broth") culture, place a loop full of culture on the slide and spread it around until it's about the size of a rupee coin.
- When observing cells from a slant or plate culture, first place a drop of water on the slide. Then, with a sterile loop, remove a small amount of microbial material and mix it well in the water and break up any clumps.
- If the drop looks cloudy, there are too many cells. Rinse the slide and start again.
- Spread the liquid once the cells are mixed in it.
- Allow the liquid to air dry before heating. If the slide is not air-dried, the cells will either not stick or get boiled in the liquid.
- Before making a smear, mark the smear side with a glass marker.
- For heat fixing the smear, slowly pass the slide through the flame a couple of times. This causes the cells to stick to the slide and they don't get washed off during staining and rinsing.
- While rinsing, hold the slide at an angle and pour or drip the solution onto the slide above the cells, so that the solution flows over them, rather than directly on them. This will prevent the cells from getting washed off the slide.

Specimens preparation for SEM viewing

Specimens for SEM viewing are prepared by first fixing them. Fixation is done to preserve the cellular structure in a form as close as possible to that of a living cell by various chemical processes. These chemicals are toxic and should be used with care. Before fixation the desired wanted part from the specimen is separated from the rest of the tissue. This can be done either by a manual dissection or by using chemicals, which eat away the soft body material leaving only the required hard bodied sections. First the specimens coming in from the field are cleaned either manually or by using a sonicator, which sends a high frequency vibration through the specimen shaking off the dirt. Various chemicals can be also be used to remove the waxy layer on the surface of the specimen or dissolve or loosen the unwanted surface grime.

Next as the SEM works under a vacuum the specimen must be dried otherwise the specimen will either collapse or blow up in the vacuumed chamber. Specimens can be dried by any of the following methods:

- **Air-drying:** hard bodied specimens such as insects are dried as soon as they are captured so that after cleaning they can be directly placed into the SEM.

- **Critical Point Drying:** Liquid in the cells is replaced with gas. This results in a completely dry specimen with minimal or no cellular distortion.

- **Chemical dehydration:** the wet specimen is dehydrated by passing through alcohol series which replaces the water with alcohol and then the alcohol is slowly evaporated off leaving a dried specimen.

Once the specimen has been fixed, cleaned and dried the next step is to mount the specimen on a small, flat, round stub of metal piece that has a stem. The specimen or bits of the specimen are glued to the stub which has been covered with double-sided sticky tape and a thin layer of foil by a special silver conductive glue. This stabilises the specimen in one place and ensures that the specimen (which is not conductive) will be grounded or earthed to the stub as well as also makes sure that electron charging of the specimen in the SEM chamber is reduced. It prevents dislodging of specimen while it is being gold coated for viewing and maneuvering in the SEM chamber.

Finally the specimen is coated with some electron-opaque substance like gold by a machine called gold sputter coater otherwise the electron beam would probably travel right through the specimen, there would be no image and most probably the specimen will be destroyed.

Environmental SEM can be used to examine non-conductive samples without coating them with a conductive material. Non-metallic samples such as bugs, plants, fingernails, and ceramics, should be coated so they are electrically conductive. Metallic samples can be placed directly into the SEM.

Specimen preparation for E.M.

The sample is first fixed in 2.5% glutaraldehyde in 0.1M sodium cacodylate buffer at 4°C, for a minimum of 4 hours. Larger samples must be cut into approximately 1mm cubes for fixing (fixed tissue can be stored in glass processing vials with plastic caps).

Fixed tissue is then washed in 0.1M buffer for 1 hour ($\times$ 2 or overnight at 4°C) followed by post fixing in osmium tetroxide in 0.2M buffer for 1 hour (Mix equal quantities of 2% aqueous OsO_4 and 0.4M buffer and use immediately.). Sample is then rinsed twice in 0.2M buffer for 5 minutes and then dehydrated in 70% ethanol for 20 minutes ($\times$ 2, sample at this stage can be stored overnight if necessary), 90% ethanol for 10 minutes ($\times$ 2) and 100% ethanol for 20 minutes ($\times$ 2). Tissue is then treated with propylene oxide (1.2 epoxy propane) for 10 minutes ($\times$ 2), propylene oxide/epoxy resin mixture (50/50) for 1 hour and then kept in epoxy resin overnight with caps removed from vials. This allows any remaining propylene oxide to evaporate. Finally tissue is embedded in labelled capsules with freshly prepared resin and polymerised at 60°C for 48 hours.

Embedding of cell samples in agar/resin

Cell samples suspended in fluid are centrifuged at 5,000 rpm for 5 minutes to form a pellet. The supernatant is removed and replaced with 2.5% glutaraldehyde in 0.1M sodium cacodylate buffer and left for a minimum of 4 hours at 4°C. After this the fixative is decanted and replaced with 0.1M sodium cacodylate buffer. Sample is re-suspended and left for 2 hours (or overnight) then re-spun. Simultaneously 1% solution of high strength agar in distilled water is made. The buffer is decanted from the sample tubes and each tube is filled with agar solution, which has been cooled to 60 °C. Samples are once again spun at full speed for 30 secs to 1 min. The tubes are cooled in a beaker of cold water to set the agar. Agar plug is removed with a mounted needle and the end containing the sample is cut off. 1mm cubes are cut and placed in 0.1M sodium cacodylate buffer. From here on the samples are treated as other specimens by starting with post fixing in osmium tetroxide in 0.2M buffer and following the entire procedure.

Staining the samples

To increase electron opacity samples are impregnated with heavy metals such as uranyl acetate, Reynold's lead citrate etc.

Specimens in epoxy resin are stained for 10 minutes with methanolic uranyl acetate or for 20 minutes with aqueous Uranyl acetate or for 5 minutes with lead citrate.

Specimens in acrylic resin are stained for 5 mins with aqueous uranyl acetate (only aqueous stain is used as alcohol softens the resin) or with lead citrate for 5 mins.

Composition of Stains used

Methanolic uranyl acetate: saturated uranyl acetate in 50% methanol.

  Aqueous uranyl acetate: saturated uranyl acetate in distilled water.

  Make and store in a brown glass bottle. Can be used for approximately 3 months.

Reynold's lead citrate

Mix 1.33g lead nitrate, 1.76g sodium citrate, 30ml distilled water and shake for 1 minute. Allow the mixture to stand for 30 mts. shaking it occasionally. After 30 mts add 8ml 1M NaOH (Analar) and mix throughly. Dilute the solution to 50ml with distilled water. Final pH should be 12. Can be used for approximately 6 months.

Important

Before use, all staining solutions should either be filtered through millipore filters or centrifuged. Washing between stains should be done with filtered distilled water or 50% methanol then distilled water if using methanolic uranyl acetate. Breathing on lead citrate should be avoided during staining as a precipitate of lead carbonate may form and contaminate the sections.

Sectioning for EM viewing

Semi-thin sectioning allows selection of the appropriate tissue area before proceeding to E.M. viewing. 1m thick sections are cut using glass knives and an ultramicrotome. These sections are placed on a glass slide and dried on a hotplate at 80°C and then heated over a flame for a few seconds to ensure adhesion. Next the sections are stained with 1% toluidine blue in 1% borax solution for 1 minute at 80°C. The stain is rinsed off with distilled water and the sections are dried and covered with a glass coverslip using a synthetic mounting medium such as D.P.X.

Thin sections are cut using a diamond knife, with the ultramicrotome set to cut at around 100nm using heat advance. The sections are picked up onto 300 mesh (300 squares), thin-bar, copper grids unless they are for immunocytochemistry, in which case gold or nickel grids are used.

For viewing bacteria straight from the culture plate

One drop of 1% ammonium acetate is put onto a clean slide. bacteria from the plate are picked up using a sterile glass "hockey stick" and added to the ammonium acetate on the slide and mixed. One drop of -ve stain (preferably 1% ammonium molybdate) is then added to the slide and mixed. A small amount of this mixture is put onto a formvar grid and left for one minute. The edge of the grid is blotted to remove excess mixture, dried at room temperature and viewed under E.M.

MEASUREMENT OF MICROBIAL SIZE

Size of microscopic objects and organisme is measured by using a micrometer. **(See details of micrometer in chapter 2).**

- Insert and calibrate the oculometer (see chapter 2 for details).
- Take a clean glass slide.
- Spread the suspension on it.
- Air dry the slide and stain bacteria with carbol fuchsin and fungi with cotton blue for better visualization.
- Replace the stage micrometer with this slide.
- Focus the cells and superimpose the oculometer scale on them.
- Remove the stage micrometer and place the slide carrying the object or organism.
- Count the number of ocular meter divisions traversing the length / breadth / diameter of the organism. Take more than one readings from different areas.
- Calculate the average number of divisions and multiply with calibration factor (use the same calibration factor as the magnification used for observation). This gives the size of organism.
- Tabulate the data.

Axial Linear Measurements

Specimen depth can be measured by measuring the axial linear distance along the optical (or **z**) axis of the microscope. Microscopes having graduated fine focus knobs can measure specimen depth. Size of each division on the graduated focus knob is usually provided with the model.

This is the calibration factor. In case it is not marked or given it can be calculated as follows:

Method:

- Determine the size of each division on the graduated focus knob (if not provided) using a glass cover slip.

- Thickness of the cover slip is measured with a machinist's micrometer or dial caliper.

- Make a mark on each side of the cover slip with glass marker and place it on a glass slide.

- Focus the mark placed on the upper surface of the cover slip and note the position of the fine focus knob graduations with respect to a reference point.

- Next focus the lower surface mark and once again note the fine focus knob graduations with respect to a reference point.

- The axial distance corresponding to each division of the fine focus knob is equal to the thickness of the cover glass (in micrometers) divided by the total number of knob graduations traversed from the top to the bottom of the cover slip. **(Calibration factor)**

- For measuring the depth of the specimen, locate the feature of interest in the view field and focus the bottom surface of the specimen.

- Note the exact position of the focus knob.

- Now repeat the procedure with upper surface.

- Subtract this value from the one obtained from imaging the bottom surface.

- Specimen dimensions can be calculated by multiplying the number of focus knob increments by the calibration factor discussed above or supplied by the manufacturer.

Measurements may not be accurate because

- Readings fluctuate due to fluctuations of specimen depth.

- Refraction artifacts and spherical aberration also affect the results.

MEASURING OF MICROSCOPIC NUMBERS

Microbial population or concentration in a suspension can be measured directly or indirectly:

1. **Direct Measurements:** There are several methods which help in direct counting of microbes. These methods give more accurate measurements of numbers of microbes.

   - **Direct Counting:** can be done by using Coulter Counter, an electronic device. Counts can be made rapidly and accurately only if bacterial cells are the only particles present in the solution. Only disadvantage is that it gives a total count of live and dead cells.

   - **Standard Plate Count or Quantitative plating:** Bacterial solution of different dilutions is first made by serial dilution technique (see chapter 4 for details). One ml inoculum from different dilutions is then streaked on well labelled respective petriplates containing nutrient medium and incubated. Well-separated colonies are viewed through magnifying glass against a colony-counting grid called a Quebec colony counter. The petri plate is inverted on the counter. Each colony is marked to prevent recounting. Number of cells is calculated by multiplying the number of colonies by the dilution factor. Dilution factor is expressed mathematically as the reciprocal of dilution e.g. dilution 10^{-6} of has dilution factor 10^6. It gives a viable count. e.g.

50 colonies × dilution factor of 10^{-3} = 50 × 10^3 = 500.00 cells/ml.

- **Filtration:** A known volume of liquid or air is drawn through a membrane filter by vacuum. The pores in the filter are too small for microbial cells to pass through. Therefore, cells are retained on the filter, which is then pressed on the surface of petri dish containing appropriate solid medium. The medium is incubated and after 48 h and number of colonies is counted. This is the number of viable microbial cells in that volume of liquid that was filtered.

- **Counting chambers** are used for counting particles and cells in a specific volume of liquid. Counting chambers consist of a thick glass slide having a central polished and ruled platform. The platform is placed beneath twin polished cover slip supports to create a chamber, which is filled with a known amount of liquid. A clean glass cover slip is placed over this chamber and positioned centrally on the polished supports. The gap between the ruled counting platform and the cover slip equals to 100 micrometers, and the ruled (engraved) face is divided into squares of exact dimension. As a result, the volume of the liquid placed in the chamber can be easily calculated to yield an accurate analysis of the number of particles (cells) per unit volume in a suspension. The most common type of counting chamber is **hemocytometer,** which is used for counting blood cells. For details of counting method see chapter 2.

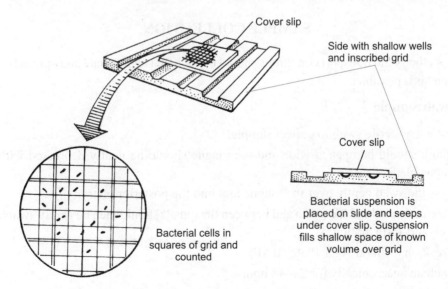

Fig. 3.13 Counting chamber

2. Indirect Measurements: measuring a property of cells and then estimating the cell number.

- **Turbidity:** A spectrophotometer can measure how much light a solution of microbial cell transmits; the greater the mass of cells in the culture, the greater its turbidity (cloudiness) and the less light that will be transmitted.

Method:

- Make serial dilutions of sample.
- Label culture tubes according to dilutions and add 15 ml broth in each tube.
- Inoculate culture tubes containing broth with respective dilutions.
- A tube containing only nutrient broth is used as blank.
- Set spectrophotometer to 100% transmission with blank.
- Now determine the OD of undiluted as well as diluted samples.
- Determine the number of cells by comparing with the reference or standard curve.
- Alternatively, streak the undiluted as well as diluted samples on respective labelled agar plates.
- Incubate for 48 hrs.
- Determine the plate count and plot densities of various dilutions against corresponding counts.

Disadvantages: Not sensitive in terms of numbers of bacterial cells and not useful for detecting minor contamination.

Metabolic Activity: By measuring the rate of formation of metabolic products, rate of utilization of a substrate and rate of reduction of certain dyes.

SAMPLE COLLECTION

Collection methods and transport conditions should be such that appropriate and optimal recovery of microorganisms is possible.

Throat Swab Sample

1. Use a dry sterile swab to collect sample.
2. Mouth should be opened wide and the tongue should be gently depressed with a sterile tongue blade.
3. Press the swab gently over the tongue and into the posterior pharynx.
4. The mucosa behind the uvula and between the tonsils should then be gently swabbed with a back-and-forth motion.
5. Streak on a Blood Agar Plate (BAP)
6. Incubate anaerobically for 24–48 hours.
7. Observe plates for evidence of hemolysis.

Tooth Swab Sample

1. Use a dry swab to collect a sample.
2. Swab the tooth surface, gum line, folds surrounding the molars, at the back of the mouth in a brushing the teeth motion.
3. Streak on a Blood Agar Plate (BAP).
4. Incubate for 24–48 hours in the presence of CO_2.
5. Observe plates for evidence of hemolysis.

Sputum Sample

- Early morning samples first are recommended.
- Collect the expressed sputum in a sterile container or dry sterile cotton swab.

Disadvantages:

- Contamination with **normal oral flora**.
- Effective only for upper respiratory infections.
- Recovery of lower respiratory pathogens not possible.

Water Sample:

Proper collection of water is very important. Samples such as drinking water, water from various sources such as lakes, swimming pools, wells and rivers must be collected in sterile, leak proof plastic or glass bottles containing sodium thiosulfate. Sludge samples can be collected in sterile plastic bags or wide-mouth sterile jars. Never collect samples from pipes or faucets that leak.

Collecting samples from tube wells, hand pumps, taps etc.: Let the water run for five minutes. Remove the cap carefully and ensure that it is not touched or contaminated. Hold the bottle from the base and fill it up to the brim. Avoid splashing.

Collecting samples from lakes, ponds, pools, etc.: Carefully remove the cap of the bottle and grasp it from the base. Immerse the bottle mouth down in the water to avoid surface scum. Place the mouth of the bottle into the current away from the hand. Fill the bottle up to the brim and replace the cap.

TITRATION

Titration is a method of analysis that allows determination of the precise endpoint of a reaction and therefore the precise quantity of reactant in the titration flask. A burette is used to deliver the second reactant to the flask and an indicator or pH Meter is used to detect the endpoint of the reaction.

- Condition and fill the burette with titrant solution.
- Check for air bubbles and leaks, before proceeding.
- Take an initial volume reading and record it in your notebook.

- Always calculate the expected endpoint volume before beginning a titration.
- If sample is a solid, make sure it is completely dissolved.
- Put a magnetic stirrer in the flask and add the indicator.
- Use the burette to deliver a stream of titrant to within a couple of ml of expected endpoint.
- When indicator changes color when the titrant hits the solution in the flask, but the color change disappears upon stirring.
- Approach the endpoint slowly and watch the color of liquid in flask carefully.
- For phenolphthalein, the end point is the first permanent pale pink. The pale pink fades in 10 to 20 minutes. This signals that end point is near.
- When you reach the endpoint, record the volume reading and add another drop.
- Subtract the initial volume to determine the amount of titrant delivered.
- Use this, the concentration of the titrant, and the stoichiometry of the titration reaction to calculate the number of moles of reactant in the solution.

Titrating with a pH meter

Titration with a pH meter follows the same procedure as a titration with an indicator, except that the endpoint is detected by a rapid change in pH, rather than the color change of an indicator. Arrange the sample, stirrer, burette, and pH meter electrode so that you can read the pH and operate the burette with ease. To detect the endpoint accurately, record pH vs. volume of titrant added and plot the titration curve as you titrate.

4

CULTURE OF MICROORGANISMS

Microorganisms are very small in size and therefore it is not possible to study their physical and physiological characteristics individually. Microbial morphology, biochemistry, physiology and reproduction etc. are therefore studied in a collective form by artificially culturing them. Pure cultures of these organisms are developed for this purpose by provided them with all the necessary optimum conditions required for growth such as nutrients, pH, temperature etc.

The earliest culture media were liquid, which made isolation and development of pure culture of bacteria very difficult. The procedure followed was introduction of bacteria in medium and then subsequent dilution of medium possibly till only one organism remained in the medium. This process was not only tedious but was also plagued with a major problem of contamination.

Brefeld introduced pure culture. He developed fungal cultures and used gelatin to solidify the medium. **Lister 1878** developed first pure culture from Pasteur's lactic acid ferment by using **Dilution method**. Bacterial suspension was diluted with sterile medium. **J. Schroeter** suggested growth on potato starch, starch paste, bread, egg albumin etc. But the credit for culturing bacteria on solid medium and efficiently developing pure culture goes to **Robert Koch** 1881 who established the pure culture methods, which are still being used.

Initially Koch tried to use sterile cut surfaces of potato. The potato was sliced with a knife sterilized in flame and inoculated with bacteria with a needle and then streaked out over the surface. The slices were incubated under bell jars in order to prevent contamination. Although, the isolated cells did grow into pure colonies, but unfortunately this method was not successful as potato is not a very good nutrient medium and many bacteria do not grow well on potato slices. Secondly it was difficult to distinguish between individual colonies of each bacterium as bacteria would soon spread all over the moist surface.

During this period, Frederick Loeffler who was Koch's associate developed a medium containing meat extract and peptone. Koch tried to solidify this medium by adding gelatin to it. He developed the **streak plate method** by spreading the meat extract, peptone medium containing gelatin on a glass slide and inoculated bacteria in the same way as he did on potato slice that is by streaking the inoculum over it. He purified the culture by repeated streaking and finally transferred the culture to tubes containing solid medium. He plugged the tubes with cotton and placed them in a slanting position. This led to development of **slant culture**. Later he found that if instead of streaking, bacteria were mixed with liquid gelatin than on solidification, cells get immobilized within and develop into individual colonies. This gave rise to **pour plate method**.

He found that microbes degraded gelatin as it was protienaceous, secondly gelatin liquefied at 28°C so it could not be incubated at 37°C, the required temperature for culturing pathogenic microbes. **Fannie Eilshemius Hesse** wife of Koch's associate **Walter Hesse** suggested the use of Agar a solidifying agent used to prepare jams and jellies in those days. In 1882 Koch replaced gelatin by agar, a complex polysaccharide isolated from red algae *Gelidium*. This medium became an instant success and has not been replaced since then.

Agar is a good solidifying agent because:

- It remains liquid till 44°C, hence it is easier to use.
- Requires heating to 100°C for melting.
- Solidifies at room temperature therefore there is no need to refrigerate it.
- Cannot be degraded by microbes.

In order to develop pure cultures, microorganisms are separated from mixed population. Pure culture of this organism is then developed to get similar cells. Once the organism has been isolated, it is cultured. The complete process is known as **Isolation, purification and cultivation.** Satisfactory growth can be achieved if right nutrient medium, optimal environmental conditions such as temperature, pH, oxygen tension and sterile conditions are maintained.

CULTURE MEDIA

In order to grow bacteria in laboratory we need to provide all those conditions required for growth by bacteria in mature. The most important thing is to provide all the nutrients. Culture medium is an aqueous solution to which all the necessary nutrients have been added. **Nutrient medium** designed by Koch was based on the fact that microbes grow within the host, so meat infusions and extracts were chosen as basic nutrients. Basic medium contained 0.5% peptone (enzyme digest of meat), 0.3% meat extract (concentrated water soluble contents of meat) and 0.8% NaCl (same salt concentration as meat).

Depending on the type and combination of nutrients, different categories of media can be made. The mixture of necessary nutrients can be used as a liquid medium, or a solidifying agent can be added. "Agar Agar" is a natural polysaccharide produced by marine algae and is the most commonly used solidifying agent added to media (end concentration usually 1.5 % w/v). If hydrolysis of the agar is suspected, silica gel can be used as a replacement solidifying agent. Solid media are useful for identifying bacteria by colony characteristics. Isolation of pure culture can be done by using solid media whereas liquid media yield mixture of all types of bacteria present in the sample.

COMPONENTS OF MEDIA

1. **Nutrients:** Carbohydrates, serum, whole blood, and ascitic fluid, yeast extract, peptone, beef extract.
2. **Solidifying agent:** Agar
3. **Water:** Tap water with low mineral content, gas distilled water or demineralised water.
4. **Additives or supplements:** Dyes, vitamins, amino acids, growth factors and antibiotics.

ROLE OF SUBSTANCES ADDED TO THE MEDIA

Carbohydrates: Provide energy, act as a carbon source and also indicate fermentation reactions.

Serum, whole blood, and ascitic fluid: Promote growth of less hardy organisms.

Dyes: Selective growth inhibitors for certain bacteria; indicate pH change due to acid formation. E.g. Phenol red is red in alkaline medium and yellow in acidic. Gentian violet inhibits gram-positive bacteria.

Agar: Solidifies medium. It is made from red algae *Gelidium* (Rhodophyta). Chemically it is mostly galactose and very few microbes can degrade it. It remains liquid at 100°C (easy to pour) and solidifies at 40°C (incubation temperatures).

Yeast extract: Aqueous extract of yeast cells, available as powder. It is a rich source of B vitamins and organic nitrogen as well as carbon compounds.

Peptone: It is a product resulting from digestion of protein rich materials like meat, casein, gelatin, etc. It is a source of nitrogen as well as vitamins and growth factors.

Beef or meat extract: It is an aqueous extract of lean beef tissue concentrated to a paste. It contains water-soluble substances of animal tissue which include carbohydrates, organic nitrogenous compounds, water-soluble vitamins and salts.

Casein hydrolysate: Contains aminoacids obtained by hydrolysis of milk protein casein. It can be used instead of peptone. Its constitution is more clearly defined than other peptones.

Malt extract: Consists of soluble substances extracted from barley sprouts at 55°C. The obtained liquid is concentrated by heating at 55°C to form a thick brown viscous material containing maltose, starch, dextrin, glucose, 5% proteins and protein breakdown products, mineral salts, growth factors and inositol.

Media Supplements: Substances such as vitamins, amino acids, growth factors and antibiotics are added to media for a definite purpose. E.g. actidione (cyclohexamide): antibiotic added after filter sterilisation, L (+)-arabinose, arginine* absorbs CO_2, asparagine*, biotin , cysteine*, dextrin, Ehrlichs reagent, fuchsin, galactose, glucose, glycerol, glycogen, lactose, maltose, mannitol, ornithine*, phenylalanine*, resazurine (indicator), ribose, citric acid, sorbitol, starch, sucrose, EDTA (disodium salt used to complex iron in media)

TYPES OF MEDIA

Media are classified in various ways. On the basis of molecular oxygen and reducing substance in the media, they may be classified as aerobic and anaerobic culture media.

Based on physical state, they may be solid or liquid.

Various types of media are:

1. **Broths**: liquid media.

2. **Semi-solid:** containing <1% agar.

3. **Solid:** containing > 1% agar. Solid medium may be of following types:

   - **Slants:** Test tubes filled with liquid agar and allowed to solidify at an angle. Slants are used to maintain pure cultures.

   - **Deeps:** Test tubes filled with liquid agar and allowed to solidify while level, often used deeps to identify gas production.

   - **Plates:** petri dishes filled with agar. Provides a large surface area for microbial growth. Plates are used to isolate pure cultures. Plates should always be incubated in an inverted position to prevent condensation from dripping on the medium surface (water droplets on the surface can spread bacteria and ruin the streak).

4. **Complex:** Support growth of most heterotrophic organisms.

5. **Defined:** Support growth of specific heterotrophs and are often mandatory for chemoautotrophs, photoautotrophs and for microbiological assays.

6. **Reducing:** Support growth of obligate anaerobes.

7. **Simple or Basal media:** Nutrient broth is an example of basal media which contains Peptone—1%, Meat extracts—1%, NaCl—0.5%, and distilled water. It has a pH of 7.4–7.5. When 2–3% agar is added to nutrient broth it is called nutrient agar.

8. **Natural or empirical medium:** Media in which the concentration of various nutrients is not exactly known. e.g. milk, urine, diluted blood, vegetable and fruit juices, meat extracts and infusions, etc.

---

* pH of the medium is also important. It will rise if medium is hot.
* All stock solutions of amino acids can be autoclaved at 120°C for 20 minutes.

9. **Semi synthetic media:** Media, which contain both natural and chemical nutrients. e.g. potato dextrose agar.

10. **Synthetic media:** Media made up of entirely synthetic substances. e.g. sabourad agar.

11. **Dehydrated media:** These are commercially available, powdered, ready to use media. It only requires the addition of specified amount of distilled water in specified amount of powder.

12. **Living culture media:** Obligate parasites are cultured in living cells, callus, organs or animals or plants. e.g. chick embryo is used to culture viruses.

13. **Minimal media:** A defined medium, which has just enough ingredients to support growth. Types of ingredients vary with species.

14. **Special media:** When certain ingredients are added to a basal medium to study special characteristics or to provide special nutrients required for the growth of the organism, it is called complex medium. Virtually all special media are complex media, these are further divided as:

- **Enrichment media:** Similar to selective media but designed to increase the numbers of desired microorganisms to a detectable level without stimulating the rest of the bacterial population. When some special nutrients such as blood or serum, egg or meat pieces are added to basal media, the latter are converted into enriched media. It favours the multiplication of a particular species of bacteria by incorporating special substances, which selectively favour its growth or inhibit the growth of competitors. e.g. Selenite broth, Alkaline peptone water, Blood agar, Loeffler's serum media.

- **Selective media:** Suppresses growth of unwanted microbes, or encourages growth of desired microbes. In addition to basal media they contain certain substances such as bile salt or deoxycholate citrate, which inhibits all bacteria except those of a particular type or group. Isolation of a particular species of bacteria from a mixed inoculum is possible by the help of selective medium. e.g. MacConkey agar, Deoxycholate citrate agar (DCA).

- **Indicator media:** When certain indicator (neutral red or bromothymol blue) or reducing agent (potassium tellurite) is incorporated in the culture medium it is called **indicator medium**. The colour of the medium changes with change in pH due to bacterial growth.

- **Differential media:** These distinguish colonies of specific microbes from others. When a culture medium containing certain substance helps to distinguish differing properties of different bacteria, it is called **differential media**. e.g. MacConkey agar: It is also an indicator medium. It contains peptone, agar, lactose, sodium taurocholate and neutral red. The lactose fermenters form pink colonies while non-lactose fermenters produce colourless or pale colonies. Blood agar serves both as an enriched as well as indicator medium. It shows different types of haemolysis.

- **Transport media:** When specimens e.g. faeces, throat swab, etc. are to be sent to laboratory from distant places, the bacteria may not survive the time taken for transit or may be over grown by nonpathogenic bacteria. For transporting specimens, special media have been devised which are called **transport media**. These media preserve the viability of the organism. The medium is non-nutrient and therefore other commensal bacteria are not able to grow in it. e.g. Stuart's transport medium for urethral discharge (Gonococci), deep semisolid thioglycollate medium for anaerobes, etc.

- **Sugar media:** The standard media used for biochemical tests contain 1% sugar in peptone water along with indicator (Andrade's indicator). A small tube (Durham's fermentation tube) is kept inverted in the large tube containing sugar media. With production of acid by bacteria the colorless medium turns pink and gas production is indicated by accumulation of gas bubbles on the top of the inverted tube. Glucose, sucrose, lactose, and mannitol are routinely employed for sugar fermentation tests.

- **Assay media:** Media of prescribed composition used for assay of vitamins, amino acids and antibiotics.

- **Maintenance or storage media:** It is a medium which maintains the viability and physiological characteristics over a period of time without accelerating growth. It contains just basic nutrients. Substances which enhance growth or substances which when degraded result in the production of harmful substances such as acids, etc. are omitted.

- **Selective-differential media:** Some media have both properties. e.g. MacConkey agar. It is selective as it contains crystal violet and bile salts which inhibit the growth of bacteria other than coliforms. It is differential as it contains neutral red and lactose. Lactose is degraded by coliforms to acid which is detected due to change in pH. At acidic pH colourless neutral red becomes red and the colonies develop red colour. *Shigella* and *Salmonella* colonies remain colourless therefore can be easily distinguished.

Chemically defined media

These are made up of completely synthetic and inorganic substances added in known quantities.
Examples:

1. **Inorganic synthetic broth:**

Sodium chloride	5.0 g
Magnesium sulfate	0.2 g
Ammonium dihydrogen phosphate	1.0 g
Dipotassium hydrogen phosphate	1.0 g
Distilled water	1000 ml

2. **Glucose broth:**

Inorganic synthetic broth	1000 ml
Glucose	5.0 g

Complex media

Such media are made up of complex animal or plant materials. The exact chemical composition of the constituents is not known. Examples:

1. **Nutrient broth** (basic medium, peptone provides nitrogen and beef extract is a source of organic nitrogen, carbon, inorganic salts as well as vitamins).

Peptone	5.0 g
Beef extract	3.0 g
Distilled water	1000 ml

2. **Yeast extract broth** (enriched medium, yeast provides additional organic nitrogen and carbon as well as vitamin B)

Peptone	5.0 g
Beef extract	3.0 g
Yeast extract	5.0 g
Distilled water	1000 ml

Composition of several other media is given in annexure 5.

INTERACTION OF MEDIA COMPONENTS

Medium composition plays an important role in culture of microbes. Interactions between media components can affect growth. e.g. divalent Hg, Pb, Ag and Cu ions are bound by yeast, peptone and amino acids. Non-selective metal binding by glucose can cause metal deficiency of essential ions or decrease the toxicity of toxic ions. Deficiency of essential trace metals is caused by EDTA, which forms chelates. This effect can be reversed by the addition of other metals e.g. Cu and Fe. NaCI increases Zn toxicity by forming a Zn-chloral complex. Sn toxicity increases by formation of a soluble toxic tin agar complex. Silica gel decreases toxicity of Sn, Cd, Pb, Ni and Zn divalent ions. Precipitation* of insoluble phosphates decreases toxicity of toxic ions like Cd and Pb or causes deficiency of essential ions like Fe. Precipitation of insoluble carbonates decreases Pb toxicity. Citrate causes non-selective metal binding and Trisnon causes selective metal binding.

Useful tips: use selective and/or differential media to isolate specific microbes.

PREPARATION OF NUTRIENT MEDIUM

- To prepare one litre of nutrient broth, warm 500 ml water in a round bottom flask
- Dissolve beef extract, peptone and other ingredients in it by continuous shaking.
- Add remaining 500 ml water and pour into tubes or conical flasks.

* Can be avoided by addition of a HEPES buffer to reduce precipitation by phosphate buffer.

- Check the pH of the medium. Adjust the pH with NaOH or acid according to the requirement.
- Plug with cotton plug and autoclave at 121°C and 15psi for 20 minutes.
- To make solid medium add desired quantity of agar to the broth and boil till all the agar dissolves.
- In case you are using commercially available dehydrated medium, weigh suggested amount of the dry powder and add to warm water. Add agar if required in the same manner as described above.

Useful tips:

- Never add all the agar powder at once. Sprinkle agar powder slowly on the surface of hot water and stir continuously till all the powder gets dispersed. Adding too much together will result in formation of lumps.
- Never add too much medium in the culture tube. Always leave two-third of culture tube empty for effective aeration. Approximately 15–20 ml medium or broth is sufficient for this.
- Never use the medium immediately after autoclaving. Incubate the medium after autoclaving at 30°C for at least three days before using. (This rules out contamination due to improper sterilisation).
- The medium should never touch the cotton plug. To ensure this, place the culture tubes and flasks in an upright position in the autoclave. Tie the tubes in groups.
- Make sure that the cotton plugs are tight because after autoclaving the plugs shrink a little and become loose. Such plugs come out very easily and may lead to accidents while taking out from the autoclave. Loose plugs also lead to contamination.
- It is better to autoclave the medium in smaller flasks. This way, you can use the medium as and when required without exposing the complete medium to atmosphere.
- Certain ingredients like sugars, vitamins, hormones, etc. should be added to the medium after autoclaving as heating destroys them. Such substances can be sterilised separately (The methods are discussed in Chapter 3).
- Before autoclaving always check the pH of the medium. At times the medium does not solidify due to very high or very low pH.
- Brown colour of medium after autoclaving is due to caramelising of the sugars due to over heating.
- Any change in colour, consistency or growth in medium after incubation means some kind of contamination.
- Always autoclave used or contaminated media containing any kind of growth before discarding it.

CULTURE TECHNIQUES

All microbes except viruses and phytoplasmas can be cultured artificially on nutrient medium. Obligate parasites cultured either in the host or dual culture of host and the parasite is done together. For example bacterial cultures or cell monolayers are used for culture of viruses. Microbial culture is done under absolute sterile conditions preferably in a laminar flow bench in order to minimize the chances of contamination.

The basic procedure involves formation and sterilization of nutrient medium. Nutrient medium is sterilized by any of the methods described in chapter 3. The working area of LAF bench is disinfected with a surface disinfectant. Before inoculation, date of inoculation, temperature of incubation, duration of incubation, name of organism, name of person etc are written on the bottom of the plates or on tubes and then the microorganism is inoculated by any of the standard techniques.

Before inoculation usually the bacterial suspenion or sample is diluted to reduce the number of organisms present in approximately 1 ml of medium. The procedure is called **serial dilution**.

Serial dilution

Since millions of bacteria can be found in a single drop of water, bacterial suspension is diluted serially (Fig. 4.1) so as to reduce the bacterial population by few cells/ml.

- Arrange test tubes in the test tube stand. Label them in a serial manner. Fill all but the first tube with 9 ml sterile distilled water. First tube is to be filled with the sample solution.
- Prepare sample by adding known amount of substance containing bacteria (soil, water, tissue etc.) to known volume of sterile distilled water and homogenize or mix thoroughly. Pour 10 ml of this suspension in the first tube. This tube thus contains concentrated bacterial suspension. It has maximum number of cells/ml.
- Take 1 ml of the sample suspension from the first tube and add it into the second tube containing 9 ml of sterile distilled water. This gives 1:10 or 10^{-1} dilution.
- Now take 1 ml of this dilution and add to next tube already containing 9 ml of sterile distilled water. This gives 1:100 dilutions or 10^{-2} dilution.
- Once again take 1 ml of this dilution and add to 9 ml of sterile distilled water present in the next tube. This gives 1:1000 dilutions or 10^{-3} dilution.
- In this manner go on preparing dilutions upto 10^{-10}.
- Isolate bacteria by standard methods.

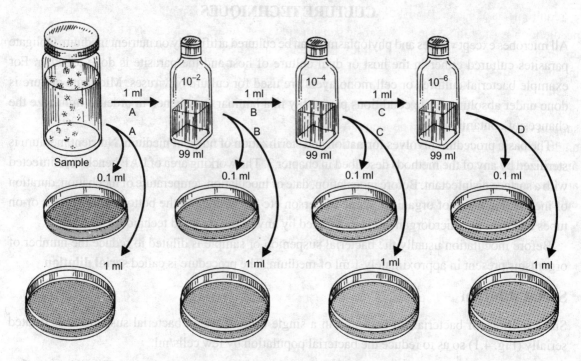

Fig. 4.1 Serial dilution Technique

Inoculation

The **process of introducing microbes (inoculum) in the nutrient medium** is called **inoculation.**
Autoclaved media, glassware, inoculation instruments, spirit lamp etc. are placed in the LAF bench
and exposed to UV light for 20 minutes followed by inoculation. Inoculation is done with the help of
an inoculating needle or loop. The inoculating loop or needle is sterilized by dipping it in absolute
alcohol and heating in flame till it glows red. The hot tip is first touched to the surface of the medium
to cool it and then dipped in the broth or scraped over the colony to pick up inoculum and results in
charging of needle. This **charged needle** (needle carrying inoculum) is then used to transfer the
inoculum from the suspension to tubes or petriplates containing nutrient broth or agar.

Culture of bacteria in culture tubes

Bacteria can be cultured in liquid as well as solid medium in culture tubes. The various methods used
for this purpose are:

   Broth culture: Culture in liquid medium is known as **broth culture** and is usually done in culture
tubes. Since bacteria absorb nutrients in liquid form and from general body surface, nutrient absorption
is better in broth culture as all the nutrients are present in soluble form. It is mostly used to increase
the initial concentration of bacteria in the sample. It is one of the easiest methods to find out oxygen

requirements of bacteria and is also used in growth experiments, biochemical assays etc. Inoculating needle/loop is first dipped in bacterial suspension and then immersed in the broth. Inoculated tubes are then incubated at desired temperature. Before inoculation nutrient broth is transparent and after inoculation and growth of bacteria it becomes turbid (Fig. 4.2).

Slant culture: Bacteria can be cultured in a culture tube on solid medium also. Molten medium is poured in culture tubes and the tubes are arranged in slanted position at an angle of 45° against a support or on special stands. The medium solidifies to form **agar slant** (Fig. 4.2) in tubes kept at an angle. This method of preparing culture tubes increases the surface area of the medium. For preparing **agar deep tubes,** after filling the tubes are kept erect and the medium solidifies as it is. These tubes are usually prepared to study oxygen requirements and for storage of bacteria. Inoculation is done by either of the following methods:

a. **Stroke culture:** It is employed for providing a pure growth of the bacterium for slide agglutination and other diagnostic tests. Stroke culture is made in tubes containing **agar slope (slant)** by the help of inoculating needle charged with culture. The charged needle is dragged in a zig zag manner across the medium (Fig. 4.2).

b. **Stab culture:** It is employed for maintenance of stock culture and study of oxygen requirements as well. An inoculating needle charged with culture is pushed deep inside the **agar deep tube or agar slant** in a straight line and rapidly withdrawn without breaking the medium. Anaerobic organisms grow inside the medium and aerobes grow on the surface.

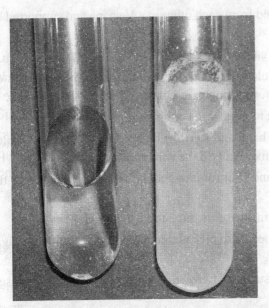 

Fig. 4.2 Uninoculated nutrient broth and inoculated nutrient broth, inoculated agar slant, uninoculated agar slant

Preparation of culture tubes

Culture tubes are autoclaved along with the medium. 15 to 20 ml molten media is poured in culture tubes and plugged with cotton plugs/screw caps. 5–7 tubes are tied together and covered with aluminium foil. The tubes are placed in the autoclave with other glassware and autoclaved at 121°C, 15 psi for 20 mts. Autoclaved tubes containing solid medium are placed in a slanted position to form agar slants. Autoclaved tubes are stored in refrigerator till further use.

Inoculation Procedure

- Clean the working top of the laminar flow bench or the inoculation tube with spirit.
- Arrange the culture tubes containing the medium, beakers containing the medium, autoclaved petri plates, beaker containing inoculating instruments dipped in spirit on the working top.
- Close the door of the laminar flow bench or the hood and expose the tubes and the instruments to 20 minutes of UV light.
- Wear sterile apron, gloves and mask, wash your hands with spirit, and wait till spirit evaporates before lighting the burner.

Inoculation in a culture tube

- Hold the stock culture tube (inoculum) and the tube to be inoculated in the palm of your left hand near the flame (Fig.4.3 a).
- The tubes are held from the base, separated from each other in a V shape and supported by the thumb (Fig.4.3 a).
- Remove the cotton plugs or unscrew the caps with the right hand.
- Do not place them on the working top. Hold the caps in the palm of the right hand. Cotton plugs are inserted between the little finger, ring finger and the middle finger (Fig.4.3 d).
- Pass the necks of the tubes through flame to minimize contamination.
- Hold the loop or the needle in your right hand and sterilize by holding it in the inner blue portion, which is the hottest part of the flame. Wait till the wire becomes red hot (Fig.4.3 b).
- Never put the loop on the working table and wait for it to cool down before picking up the inoculum.
- Touch the tip of the loop or needle to the inside of the tube to further cool it.
- Depending on the medium, use a loop or needle to pick up inoculum and inoculate the agar slants/deeps/broth (Fig.4.3 c).
- Pick up inoculum by lightly scraping the surface of growth. Do not stab the agar   or insert the needle in the medium.
- Insert the loop containing the inoculum into the culture tube.
- Shake the needle in the broth to dispense the cells. Drag the needle in a zigzag manner over solid surface in agar slant.

- To inoculate agar deep tube, insert the needle into the medium in a straight line and rapidly withdraw it in the same line without breaking the medium.
- Place the needle or loop in a container filled with spirit.
- Reflame the necks, close the cap or insert the cotton plug tightly.
- Label the tubes and incubate.
- Incubate at 24–25°C in an incubator for 48 hours.

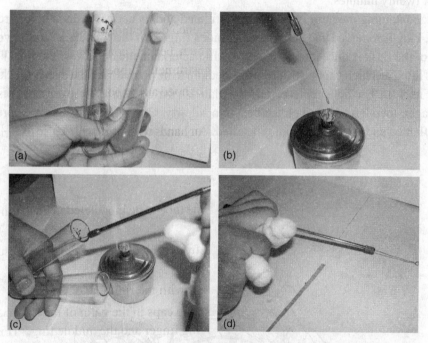

(a) Hold the tubes between fingers and thumb; (b) Flame the needle;
(c) Transfer inoculum from first tube to second;
(d) Proper method of holding cotton plugs and needle during inoculation.

Fig. 4.3 Inoculation technique

Culture of bacteria in agar plates

A sterile petri dish containing agar plus nutrients and used to culture bacteria or fungi is called **Agar plate.** Culture in agar plate is usually done on solid medium. At times semisolid medium containing 2% agar is also used to study bacterial motility.

Preparation of agar plates

Medium is poured in agar plates just before inoculation unlike tubes which are filled with medium and then autoclaved. Clean empty agar plates are autoclaved in batches and stored till further use.

Autoclaved medium is poured in agar plates in complete sterile condotions in LAF bench.

Agar plates are prepared in absolute sterile conditions in a LAF bench. Before pouring the medium one should cover the hair with a sterile cotton cap and wear a sterile mask. Hands and the LAF bench should be sterilized. Hands are first washed with antimicrobial soap and hot water and then rinsed with ethanol. The LAF bench is also wiped with ethanol or some other disinfectant. Autoclaved plates as well as the flasks or tubes containing medium are placed on the bench, with their lids still on along with burner, inoculating instruments and container filled with spirit. All these are exposed to UV light for twenty minutes.

The flask containing medium is held in the left hand and its neck is flamed before removing the cotton plug. Plug is removed and held little and index fingers of the right hand. It is never put down. The mouth of the flask is once again flamed to kill bacteria on the outside of the rim. The lid of the petri plate is lifted just high enough to allow the medium to be poured, and the dish is quickly half filled with agar (Fig. 4.4). The lid is replaced and the plate is swirled gently to ensure even distribution of the molten agar, then left to stand on the bench to solidify. Once all the plates are poured, the flask mouth is reflamed and the cotton wool reinserted. Any unused agar is still sterile.

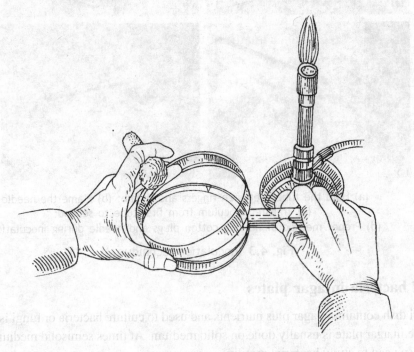

Fig. 4.4. Preparation of agar plate

Inoculation in agar plate

There are various methods by which agarplates can be inoculated. These are:

a. Streak plate method:

- Hold the prepared agar plate in the left hand between the fingers and the thumb near the flame.
- Hold the loop in the right hand, flame it and cool it.
- Dip it in the bacterial suspension after touching it to the sides of the culture tube.
- Open the lid of the petriplate and inoculate the medium (Fig. 4.5 a).
- There are various methods of streaking.

  a. **One-way streaking:** Place a loopful of the inoculum on agar surface. Without lifting the loop, drag it across the medium in a zigzag manner all over the medium in one motion (Fig.4.5 b).

  b. **Two way streaking:** Petriplate is divided into two equal halves. The culture is first streaked in first half. Petri dish is rotated by 180° and culture from the end of first streak is dragged across the second half (Fig.4.5 b).

  c. **Three way streaking:** Petriplate is divided into three areas. Culture is first streaked in first area, then second and finally in the third by rotating the Petri dish (Fig.4.5 b, c).

  d. **Four way streaking or quadrant manner:** Petriplate is divided into four quadrants or areas. The culture is first streaked in first quadrant. The loop is reflamed, petridish is rotated at 90^0 and streaking is done in second quadrant starting from the end point of first streak. Turn the dish, reflame and streak the third quadrant. Once again turn the dish and this time streak without lifting the loop and reflaming it. In this manner each time the concentration of organism in each quadrant gets reduced as compared to the earlier quadrants, organisms get separated and isolated colonies are developed.

  e. **Spiral streaking:** Culture is placed in the center of the plate and drawn into spiral manner across the medium.

  f. **Christmas tree streaking:** This pattern of streaking is used to culture urine samples. A single streak is made down the middle of the agar plate with a loop dipped in the sample. Loop is then moved in and out going through the streak multiple times at right-angles to the first streak to form a christmas tree like.

During streaking bacterial cells get isolated and when the plate is incubated, the resulting colonies develop from just one bacterium. Streaking ensures that with each streak the concentration of organism reduces as compared to the earlier quadrants. In this manner the organisms get separated and isolated colonies are developed. Incubate cultures at 24–26°C for 48 hours.

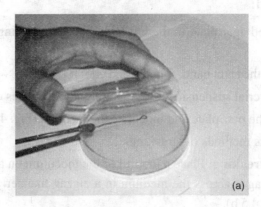

Fig. 4.5 (a) Streaking of an agar plate

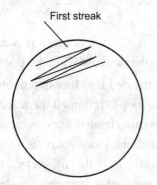
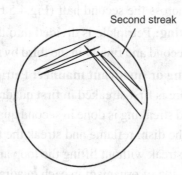

First streak

Second streak

Third streak

Fig. 4.5 (b) Streaking Technique

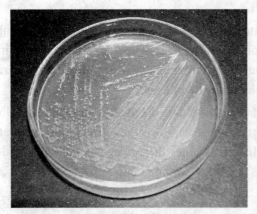
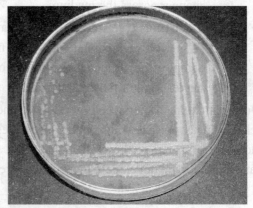

Fig.4.5 (c) Three way streaking of Bacillus on Nutrient Agar

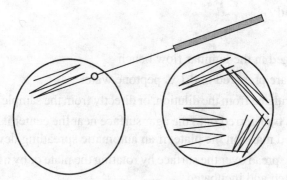

Fig. 4.5 (d) Four way streaking

b. Pour plate method

The Pour Plate technique can be used for culture and enumeration of bacteria. Conditions vary depending upon the type(s) of bacteria being enumerated.

1. Agar is melted and kept at 44–46°C.
2. Serial dilutions are prepared using 0.1 % peptone water. Following incubation, one of the dilutions will yield growth of 30 - 300 colonies the agar plate.
3. 1.0 ml of the sample or dilution is transferred to a sterile, empty petri dish containing approximately 15 ml of molten agar medium (Fig. 4.6, 4.7).
4. The sample and agar are mixed thoroughly by rotating the plate several times, clockwise, then counterclockwise.
5. When the media has solidified, the plates are inverted and incubated.
6. After incubation, colonies are counted using a Quebec counter and hand tally.
7. Otherwise in a simpler method, take one ml of bacterial suspension and add it into culture tube containing molten agar. Mix well and pour it into a petriplate. Allow it to solidify. Incubate at 24-26° C for 48 hours.

Fig. 4.6 Pour Plate Inoculation Technique

c. Spread plate method

- Sterile petri
- plates are arranged in the laminar flow bench
- Serial dilutions are prepared in 0.1 % peptone water.
- The plate is inoculated from the dilution, or directly from the sample using a 0.1 ml inoculum.
- The inoculum is transferred onto the agar surface near the center if the plate is hand spread, or at a designated mark on the plate, if an automatic spreading device is spreading it.
- The inoculum is spread over the surface by rotating the plate or by a bent glass tube (Fig.4.7). Plates are inverted and incubated.
- Following the appropriate length of incubation, colonies are counted using a Quebec counter and hand tally.

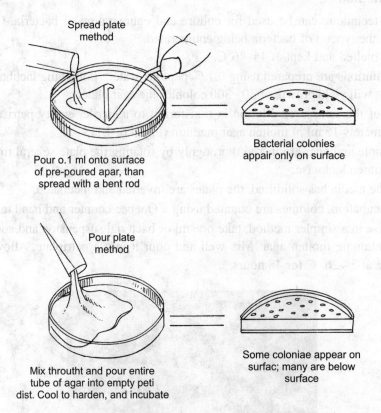

Spread plate method

Pour o.1 ml onto surface of pre-poured apar, than spread with a bent rod

Bacterial colonies appair only on surface

Pour plate method

Mix throutht and pour entire tube of agar into empty peti dist. Cool to harden, and incubate

Some coloniae appear on surfac; many are below surface

Fig 4.7. Spread and Pour Plate Inoculation Technique

d. Single cell cultures

single cells are microscopically captured by a fine capillary pipette and deposited on the medium. Colony developed from these cells is made up of identical cells, which are progeny of a single cell.

Single cells are picked up with the help of a micromanipulator. Micromanipulator helps to control the movement of the micropipette or a microprobe.

e. Lawn Culture or Carpet Culture

An even and complete spread of growth all over the agar plate is called a lawn. It is often used for testing sensitivity to antibiotics, or for work with bacteriophages. To prepare a lawn a 0.5 **McFarland suspension** of bacteria in saline (saline is slightly turbid) is used. Specimen of liquid culture is poured on the surface of a culture plate and kept for a minute and then excess material is thrown out. Alternatively the surface of the plate inoculated by a sterile swab soaked in liquid culture (Fig. 4.8). The charged swab is moved from side to side down the whole agar plate so all the area is covered. The plate is rotated 90 ° and the swab is once again moved side to side perpendicular to the first direction. This is done once more with the swab rotated at 45°.

Fig.4.8 Lawn culture; Uniform lawn bacteria growing on surface of agar plate

Incubation of agar plates

All plates are incubated upside down to prevent drops of condensation from collecting on the inoculated surface. **Plates containing human pathogens** are usually incubated at 37° C in a 5% CO_2 (the temperature and conditions that most of the body's bacteria will grow) atmosphere in incubators which maintain these conditions.Fungi, and some bacteria (e.g. *Yersinia* sp.) are incubated slightly cooler usually at 30°C aerobically. Bacteria present in grow at much higher temperatures (~45°C). Certain bacteria need special agar plates and microaerophilic environment e.g. *Campylobacter*.

Disposal of agar plates

Used plates must be discarded after sterilizing them in an autoclave. They are placed in an autoclave bag and then sterilised by autoclaving at 121°C, 103 kPa (15 psi) for 15 minutes. Bag is used as the disposable plastic plates will melt. The bag can safely be thrown away after cooling. In absence of autoclave ordinary domestic pressure cooker or a hospital or professional lab incinerator should be used. Other equipment should be decontaminated by being placed in a suitable disinfectant for 24 hours.

Pure Culture of Bacteria

Pure cultures contain only one kind of cells, which are assumed to be the progeny of a single cell. Such cultures are free from contamination by any other kind of living form. Thus, they are clones of each other. **Mixed cultures** contain more than one type of organisms. Mixed culture containing only two kinds of organisms is called **dual culture.**

In order to develop pure cultures, microorganisms are separated from mixed population. Pure culture of this organism is then developed to get similar cells. Once the organism has been isolated, it is cultured. The complete process is known as **Isolation, purification and cultivation.** It can be done in several ways:

1. Development of pure culture by serial dilution method

- Prepare bacterial suspension by mixing the sample in distilled water or saline.
- Prepare serial dilutions of the suspension.
- Take 1 ml of first dilution and inoculate a sterile petriplate containing nutrient medium with it.
- Repeat this with all dilutions individually and separately on petri plates containing solidified nutrient medium.
- Incubate, and then observe the type of colonies. Discard plates containing overlapping or diffused colonies.
- Select plates showing different types of distinct, individual well separated colonies.
- Pick up inoculum from any one colony and mix in 10 ml of sterile distilled water. This gives a second sample.
- Repeat the procedure from the beginning till all colonies are of same color, type etc. This gives pure culture of one organism.
- Repeat the procedure with all different types of colonies to separate all bacteria present in the original sample.

Enrichment culture of bacteria

Enrichment culture provides nutrients and a suitable environment for only a specific microbe but not for others. It uses a selective medium that is supposed to increase the number of microbes in a sample. **Sergei Winogradsky** was one of the early scientists to use the concept of an enrichment culture to isolate and grow a particular microbe, ***Beggiatoa*** and said that it uses inorganic H_2S as an energy source and CO_2 as a carbon source. In 1890 he used an enrichment culture to isolate nitrifying bacteria in soil. In 1888, **Martinus Beijerinck** used enrichment culture without nitrogenous compounds to obtain a pure culture of the root nodule bacterium ***Rhizobium,*** showing that enrichment culture creates the conditions optimal for growing desired bacterium. In 1931, C.B. van Niel used enrichment cultures to show that photosynthetic bacteria use reduced compounds as electron donors without producing oxygen.

Difference between enrichment culture and enriched media is that enriched media has extra nutrients that allows microbes to comfortably grow; it does not select for a particular microbe. It simply provides an environment for all microbes to grow.

Often the sample from source material is directly inoculated into a broth medium, which will encourage the proliferation of the desired organism. This is called **enrichment. Selective enrichment** can also be done to suppress the growth of undesired organism and increase the probability of development of colonies of the desired organism upon subsequent streaking. For example for isolation of endospore former *Bacillus*, the soil suspension is boiled. This kills the vegetative cells as well as reproductive spores leaving only the endospores. Hence the colonies, which will develop subsequently, will be due to germination of endospores.

Following enrichment the organisms are plated on **isolation medium.** Usually a selective medium is used for this purpose. Selection can be achieved by either adding a selective agent which will kill undesired organisms while allowing proliferation of desired organism or by making the medium restrictive by using only one type of nutrient which can be used by only the desired organism or by leaving out the nutrient required by undesired organism. For example carbon source can be left out of medium for selecting autotrophs, which can obtain carbon from atmosphere or nitrogen source can be omitted from the medium for selecting nitrogen fixers. Non nitrogen fixers will not be able to grow in such a medium as they require external source of nitrogen where as nitrogen fixers can absorb it from atmosphere.

Method

- Take sample from natural source such as soil, water, etc and add directly to nutrient broth.
- Incubate overnight and take 1 ml from this overnight culture and add to isolation medium.
- Isolation medium can be chosen according to the organism requirement.
- Incubate and observe for development of colonies.
- Identify by standard methods.

Standard plate count

Standard plate count procedure is also known as the **heterotrophic plate count (HPC).** It is mostly used for estimating the number of live heterotrophic bacteria in any sample and to monitor water quality for presence of coliforms in water. Colonies arising from pairs, chains, clusters or single cells are included in the term **"colony-forming units" (CFU)** and all HPC are reported as colony forming units cfu/ per ml or per g. This method is mostly used to estimate number of coliforms in water sample.

Tips: High concentrations of the general bacterial population may hinder the recovery of coliforms.

Method

- Pour molten agar in plates and allow them to dry before inoculating.
- Prepare serial dilutions using 0.1% peptone water.
- Inoculate the plate from the one of the dilutions or directly from the sample using a 0.1 ml inoculum,
- Plates are inverted and incubated 35°C ± 0.5°C for 48 – 72 ± 2 hours.
- Following incubation count the number colonies

Replica plating

Replica plating is mostly used to **screen and select** mutants by comparing the master plate and secondary plates for a selectable phenotype. This technique was first described with refence to the **Luria-Delbruck experiment** in 1943. This technique has a number of uses, such as selection of auxotrophic and antibiotic resistant mutants etc. it is also used in genetic engineering to identify transformed cells.

By this technique microorganism colonies are rapidly transferred from the master plate, in an exact spatial pattern, to a number of secondary plates. In this manner exact copies of the master plate are formed. The secondary plates contains a different solid (agar-based) selective growth media which may either be lacking in some specific nutrient or may contain some chemical growth inhibitors such as antibiotics etc. When these plates are inoculated with the same colonies of microorganisms from a primary plate the original spatial pattern of colonies is reproduced. A colony which was present on the master plate but fails to appear at the same location on the secondary plate shows that the colony was sensitive or resistant to the specific marker substance missing/added in the secondary plate. For example if the secondary plate contains an antibiotic ampicillin and if all but few colonies on secondary plate die then the remaining colonies are obviously resistant mutants and pure cultures from these can be developed by picking up these colonis.

Similarly negative selection can also be done by this technique. As both, the primary and secondary plates have the same spatial patterns the colonies which die are sensitive to ampicillin and these sensitive colonies therefore can be selected from the primary plate.

Method

The primary plate is first developed by streaking mixed inoculum on agar surface and incubating it till colonies develop. A pad of sterile cloth, velvet, gauze, or even paper towel is tied to a wooden block the same size as the plate. This block is pressed on the surface of primary plate in such a manner that some cells from each colony growing on the primary plate stick to the cloth. The cloth is then pressed onto the surface of secondary plate. Few cells from the cloth will get deposited on the agar surface and colonies will grow in exactly the same positions on the new plate as on primary plate. Incubate the plates and observe for appearance or disappearance of colonies. Select the mutants accordingly (Fig. 4.9).

Advantages

A large number of individual isolated colonies can be rapidly screened for as many phenotypes as there are secondary plates by using different selective growth media in each plate. This technique is also used in genetic engineering to identify transformed cells. It saves the bother of streaking out individual colonies also. Usually large numbers of colonies (roughly 30-300) are replica plated together.

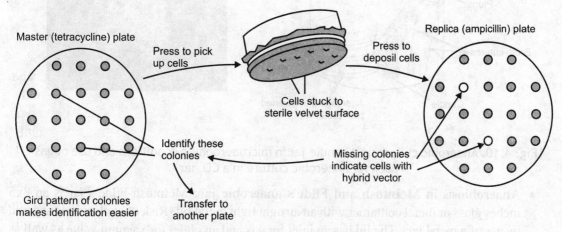

Fig. 4.9 Replica plating method

Anaerobic culture of bacteria

Some obligate anaerobes require oxygen free conditions. Anaerobic culture of such organisms is done by following methods:

1. By displacement of oxygen: oxygen is displaced from a sealed jar loaded with inoculated media by:

- **Boiling the medium:** before inoculation the medium is boiled in a water bath at 800°C for half an hour to drive out oxygen. For strict anaerobiosis the surface of the medium may be covered with a 1cm layer of sterile liquid paraffin

- **Passing inert gases:** like hydrogen or nitrogen through the container. Disadvantage: requires repeated evacuation and refilling. Complete evacuation of oxygen not possible.

- **Use of candle:** A lighted candle is kept in the container loaded with inoculated culture media plates. Candle supposedly uses all the available oxygen inside before it gets extinguished. But some amount of oxygen is always left behind (Fig. 4.10).

- **Absorption of oxygen by chemicals:** Chemicals generate hydrogen and carbon dioxide inside the jar with water. A disposable packet of aluminium foil containing pellets of sodium borohydride, cobalt chloride, citric acid and sodium bicarbonate called "Gas-Pak" with water is kept inside and the lid is tightly closed (Fig.4.10).

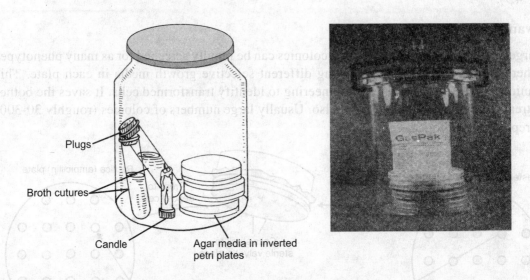

Plugs

Broth cutures

Candle

Agar media in inverted petri plates

Fig. 4.10. Anaerobic culture:in a candle jar in microaerophilic conditions and use of gas pack for anaerobic culture in a CO₂ jar

- **Anaerobiosis in McIntosh and Filde's anaerobic jar:** McIntosh-Filde's jar is an 8x5 inches glass or metal container with an airtight tight metal lid (Risk of explosion is negated by use of a metal jar). The lid has an inlet for gas and an outlet for vacuum valve as well as two electric terminals. Electric current is used to heat the catalyst i.e. Alumina pellets coated with palladium, which is suspended by wires connected to the electrical terminals. Air inside the jar is removed by closing the inlet tube and connecting out let tube to the vacuum pump. Pressure within is reduced to 100 mm Hg (i.e.660mm below atmosphere). Next, the out let tap is tightly closed and the inlet tap is connected to hydrogen supply. The reduced pressure is brought up to 760mmHg. (Atmospheric) which is monitored on the vacuum gauze as zero. After filling the jar with hydrogen, electric terminals are switched on for heating the catalyst. Spongy palladium or platinum brings about slow combination of hydrogen and oxygen to form water. Reduced methylene blue is used as indicator for verifying the anaerobic condition in the jar. It is a mixture of NaOH, methylene blue and glucose. In anaerobic conditions it is colorless and in presence of oxygen regains blue color.

2. By incorporating reducing agents in media: Oxygen from culture media can be removed by adding substances like 1% glucose, 0.1% thioglycollate, 0.05% cysteine, 0.1% ascorbic acid and cooked meat pieces. Unsaturated fatty acids present in meat utilize oxygen for auto oxidation. Certain reducing substances such as sulphydryl compounds, glutathione and cysteine also utilize oxygen.

- **Thioglycollate broth** - nutrient broth and 1% thioglycollate.
- **Robertson's cooked meat medium** – nutrient broth and pieces of fat free minced cooked meat of ox heart. It permits the growth of strict anaerobes and preserves delicate organisms.

3. By dilution shake culture: Take a tube of molten and cooled agar. Inoculate it and mix well. Transfer approximately 1/10th of the contents into 2nd tube containing molten medium and mix well. Repeat the procedure till 6-10 successive dilutions have been made. Cool the tubes rapidly and seal by pouring a layer of sterile petroleum jelly and paraffin on the surface. Prevent access of air to the medium. Incubate. Colonies will develop anaerobically in the agar column.To study the organism, remove the jelly and extract the agar column in a petridish by gently blowing a stream of gas through the capillary pipette inserted between the medium and culture tube wall.

4. By roll tube method: R E Hungate developed this method for use in field. Tubes containing pre reduced liquid media are tilted till agar almost touches the cotton plug and rolled in this position against a block of ice. This coats the entire inner surface of the tube with medium and provides a large surface area. The tube is closed by butyl using a rubber stopper, as ordinary rubber stoppers allow micro quantity of air, which also can be lethal. Whenever the tube is opened, passing a stream of oxygen free gases such as carbon di-oxide or nitrogen into the tube prevents entry of air.

5. Creation of anaerobic conditions by using pyrogallol and NaOH: Inoculate the agar slant and plug the tube immediately. Cut the upper portion of cotton plug in such a manner that the cut end is at level with the mouth of the tube. Push the plug into the test tube till it almost touches the medium. Fill the space between the plug and the mouth of culture tube with pyragallol and NaOH. Close the tube with a rubber stopper or screw cap. Pyragallol and NaOH act as oxygen scavengers and create anaerobic conditions. They also do not allow entry of fresh air in the tube. In this manner anaerobic environment is created in the culture tube (Fig. 4.12).

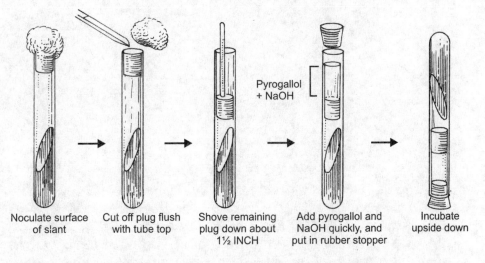

Pyrogallol + NaOH

Noculate surface of slant   Cut off plug flush with tube top   Shove remaining plug down about 1½ INCH   Add pyrogallol and NaOH quickly, and put in rubber stopper   Incubate upside down

Fig. 4.12. Creation of anaerobic conditions by using pyrogallol and NaOH

5

ISOLATION, PURIFICATION AND IDENTIFICATION OF BACTERIA

ISOLATION AND PURIFICATION OF BACTERIA

Several microbes are found in soil, water and air, etc. In order to study a specific microbe, it is necessary to first isolate (separate) it from the other bacteria growing in the same medium and develop a pure culture containing only the desired bacterial cells. The isolated bacterium is then identified according to Bergey's Manual specifications by examining its macroscopic (colonial) characters, microscopic morphology, antibiotic susceptibility and biochemical characters. Finally genome analysis of the isolate is carried out and the observations are compared with available data in specified data banks and culture collection centres to find out the genus, species and strain status. Isolation is done in a sequential manner as follows.

- **Broth culture:** The first step for isolation and purification of bacteria involves development of 24 hour culture containing viable cells. The sample is added to 15ml of sterile nutrient broth and incubated at 37°C for 24 hours. The advantage is that all cells become metabolically active including the spores, which germinate to produce viable vegetative cells. It also increases the number of inherent microbes.

- **Serial dilution:** Since the earlier step results in increase of number of cells, next the bacterial suspension is diluted by serial dilution technique. One ml of this 24 hour culture is added to

9.0 ml of sterile DW to get 10^{-1} dilution. From this tube, one ml suspension is once again added to 9.0 ml of DW. In this manner, by stepwise repeated dilution bacterial suspensions ranging from 10^{-1} to 10^{-10} are obtained. (See chapter 3 for details).

- **Culture of microbes:** The bacteria are inoculated on nutrient medium by **Streak plate method**. 20 ml of liquid medium is poured into petri dishes and cooled till it solidifies. One ml of bacterial suspension from successive dilutions is streaked onto the surface of solidified agar with the help of a transfer loop and incubated at 37 °C for 48 hrs. The plates are observed for colony formation. Cells from a single colony are picked and mixed in sterile broth/normal saline/sterile distilled water and the whole procedure is repeated till pure cultures develop i.e. all colonies in the petri plate are identical (See chapter 3 for details). Identification of these in there done by standard methods.

IDENTIFICATION OF BACTERIA

Identification of the pure isolates is done by methods involving direct microscopic examination, study of cultural characteristics as well as physiological and biochemical properties of the isolate. Identification is done according to Bergey's Manual of Systematic Bacteriology.

The following characteristics are studied for identification

1. Morphology
2. Staining reactions
3. Growth on differential media
4. Colony characteristics
5. Biochemical properties
6. Antibiotic susceptibility
7. Serology
8. Pathogenicity
9. Flow cytometry
10. Phage typing
11. Protein and lipid analysis
12. Genome analysis
13. Comparison of nucleotide sequences

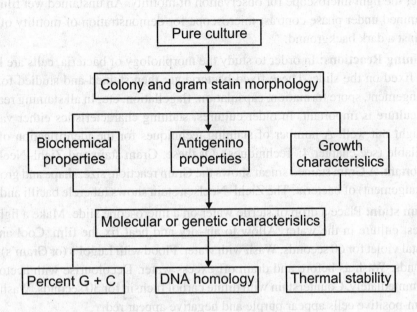

Fig. 5.1 Scheme for Bacterial identification

These characteristics can be divided into the following categories:

1. Direct microscopic examination
2. Study of cultural characteristics
3. Study of biochemical and physiological properties
4. Antibiotic susceptibility testing
5. Pathogenicity tests
6. Serological properties
7. Genome analysis

Few of these are described here. (See chapter 3 for details of staining and reagent composition as well as annexure V for media composition)

1. Direct microscopic examination

Bacteria are visualised under the compound light (bright field) microscope, for observing size, shape, and presence or absence of motility, spore formation and capsulation.

    A. Motility: Movement is studied by hanging drop method. Vaseline is applied around the depression of the cavity slide. One drop of unstained, live bacterial isolate is aseptically transferred to a clean cover slip and the cavity slide is inverted over it. The slide is then turned right side up so as to suspend the liquid containing bacterial isolate in the well for microscopic observation. An unstained wet film or hanging drop preparation is examined under the light microscope for observation of motility. An unstained wet film may also be examined under phase contrast microscope for demonstration of motility of spirochaetes against a dark background.

B. Staining Reactions: In order to study the morphology of bacteria, cells are killed by heat and fixed on the slide. These fixed bacteria are then stained and studied for size, shape, arrangement, spore formation, capsulation, flagellation, etc. In all staining reactions age of the culture is important. In older cultures, staining characteristics either vary or are not brought out well. A number of staining techniques for the identification of bacteria are available (see chapter 4: Techniques). Of these, Gram stain and Ziehl–Neelsen are most important. A Gram stained smear shows the Gram reaction, size, shape and grouping pattern (arrangement) of bacteria. The Ziehl–Neelsen stain shows tubercle bacilli and lepra bacilli.

- **Gram stain:** Place a drop of sterile water on a microscope slide. Make a light suspension of test culture in the water. Allow to air–dry and heat fix the film. Cool and flood with crystal violet for 60 seconds. Wash with water. Flood with Lugol's (or Gram's) iodine for 60 seconds. Wash as before and drain off excess water. Decolourise with acetone and wash off immediately. Counterstain with dilute carbol fuchsin for 30 seconds. Wash and blot dry. Gram-positive cells appear purple and negative appear red.

- **Ziehl–Neelsen Stain:** Flood a fixed slide with strong carbol fuchsin solution. Heat the slide until it steams and keep steaming for 5 minutes but do not boil. Do not allow the slide to dry out. Wash the slide with sterile water. Treat with 3% acid alcohol for 10 minutes or until only a suggestion of pink remains on the film. Wash the film with water. Counterstain for 15–30 seconds with methylene blue. Wash and blot dry. Acid-fast bacilli appear bright red while tissue cells and other bacteria stain blue.

- **Endospore staining:** Several days old culture of bacteria is stained and examined for endospore formation. Endospore staining is done by preparing a smear on a clean slide and heat fixing it. The smear is flooded and kept saturated with 5% malachite green while heating continuously for 5 minutes. Subsequently, smear is rinsed with water and counter stained with safranin for 30 seconds. The slide is once again washed, dried and investigated for endospore formation. Cells with endospore appear green while vegetative cells appear red.

- **Spore Staining by modified Ziehl–Neelsen method:** Stain the fixed film with strong carbol fuchsin for 3–5 minutes. Heat until steam rises. Wash with water. Treat with 0.25% sulphuric acid for 15–60 seconds. Wash with water. Counterstain with 1% aqueous methylene blue for 5 minutes. Wash and blot dry. Bacterial spores are seen as red structures, vegetative cells stain blue.

- **Capsule staining:** Bacterial cultures are stained with India ink. Place the cover slip over the test culture and remove excess ink. Slide is then observed for capsule stain. Capsule forms a clear halo around the cell against a dark background (Fig 5.2).

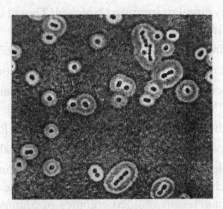

Fig. 5.2 Negative staining of capsule which appears as a clear halo around the coloured cell

- **Capsule staining (Relief staining with eosin):** Place a drop of broth culture on one end of a microscope slide. Add one drop of eosin solution and leave for one minute. Take a second slide and draw its edge back to contact the stained suspension. Holding the second slide at 45 degrees spread a thin layer of fluid along the first slide by a continuous forward movement. Allow the film to air–dry then examine under oil-immersion. Background material and cells stain red. Capsular material appears as an unstained halo around the cells.

C. Measurement of cell size: is done by using a micrometer. (See details in chapter 3).

2. Cultural characteristics

Colony characters of three-days old pure cultures are studied on solid medium.

1. **Colony size:** Pinpoint/ Extremely small/ Fraction of an mm/ Pinpoint to small/ 5–10mm (Fig. 5.3).

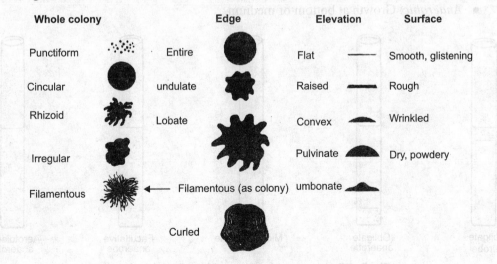

Fig.5.3 Colony Characteristics

2. **Colony shape:** Evenly circular/ Irregular/ Notched/ Thread like or Filamentous/ Rhizoidal/ Wavy/ Punctiform/ Circular (Fig. 5.1A)

3. **Colony margin:** Entire/ Undulate/ Lobate/ Curled/ Serrate/ Filamentous/

4. **Colony elevation:** Flat/ Raised/ Convex/ Pulvinate/ Umbonate (Fig. 5.1A)

5. **Colony surface:** Smooth; Shiny glistening/ Rough; dull, granular or matte/ Wrinkled; folded/ Dry: powdery, brittle/ Mucoid; slimy or gummy (Fig. 5.1A)

6. **Colony consistency:** Butyrous or buttery/ Viscous/ Stringy/ Rubbery/ Dry brittle or powdery

7. **Optical features:** Opaque/ Transparent/ Translucent/ Opalescent

8. **Chromogenesis or pigmentation:** Production and retention of water-soluble intracellular pigments due to which colony is colored. e.g pink colonies of *Flectobacillus major*, red colonies of *Serratia marsecens*, golden colonies of *Staphylococcus aureus*.

9. **Production and diffusion of water-soluble pigments:** which colour the medium. e.g. bluish green colour of medium due to secretion of pyocyanin by *Pseudomonas aeruginosa*.

10. **Production and diffusion of sparingly water-soluble pigments:** which accumulate as coloured crystals near the colony. e.g. green crystals of chlororaphin secreted by *Pseudomonas chlororaphis*.

11. **Production and diffusion of water-soluble and fluorescent pigments:** due to pigment medium glows white or blue green when exposed to UV rays. e.g. secretion of fluorescent pyoverdin by *Pseudomonas aeruginosa*.

12. **Amount of growth in broth:** Scanty/Moderate/Abundant.

13. **Oxygen requirement:** Bacteria are inoculated into nutrient broth to study the oxygen requirement of pure isolates (Fig. 5.4). Depending on growth they are grouped as:

    *   *Aerobic*: Growth on surface of medium.
    *   *Microaerobic*: Dispersed growth.
    *   *Anaerobic*: Growth at bottom of medium.

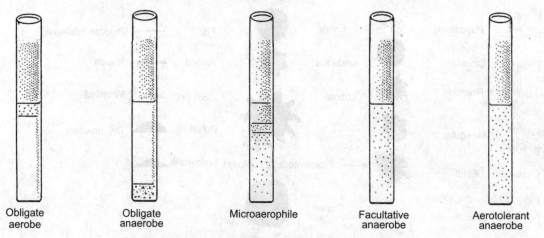

Obligate aerobe	Obligate anaerobe	Microaerophile	Facultative anaerobe	Aerotolerant anaerobe

Fig.5.4 Oxygen requirements of bacterial isolates

14. **Hemolytic behaviour:** Hemolysis of RBC by bacteria is studied by culturing them in blood agar. Based on the hemolytic behaviour they are classified as:
- *Alpha hemolytic*: complete hemolysis of RBC
- *Beta hemolytic*: partial hemolysis of RBC
- *Gamma hemolytic*: non hemolytic

3. Biochemical and physiological properties

Bacteria and other unicellular microbes lack extensive digestive system. Their nutrition is dependent on substances absorbed by the cell membrane. Since, macromolecules cannot pass through the membrane they have to be broken down into simpler substances by enzymatic activity outside the cell. The microbes therefore secrete extracellular enzymes to degrade polysaccharides, proteins, lipids etc. The breakdown products are smaller in size, have low molecular weight and can be easily absorbed by the general body surface and later used for metabolic activity within the cell. All organisms have a preference for a particular substrate and secrete enzymes accordingly, therefore detection of exo-enzyme secretion is a good method to identify and differentiate microbes. Presence or absence of enzyme activity is studied by providing the substrate in the medium. Digestion of the substrate is an indication of positive reaction. In some cases the reaction is clearly visible as change in either colour or property of medium, but in cases where the reaction is invisible indicator dyes are introduced in the medium. The dyes change colour according to pH of medium. Any change in biochemical composition of the medium is then reflected by change in colour of the dye.

Some common indicator dyes

Andrade's indicator: becomes reddish pink at about 5.5 pH.

Bromocresol purple: is yellow at 5.2 pH and becomes violet purple at 6.8.

Phenol red: yellow at 6.8 pH and becomes purple pink 8.4.

Bromothymol blue: yellow at 6.0 pH and becomes blue at 7.6.

Preparation of inoculum for test media

The desired colony is first plated out on general purpose medium and an isolated colony from the secondary medium is used as inoculum. This ensures that there is no contamination of the desired colony. Medium is also checked for sterility by keeping it at room temperature for at least 72 hours before inoculation.

Control tests

One tube plate in each experiment is inoculated with stock culture known to give positive results and one with stock culture known to give negative results. These are incubated along with the experimental tubes/plates. These controls are called **positive** and **negative controls** respectively.

Important points to remember:

- Carbohydrates undergo alteration due to heat or in presence of other compounds therefore such compounds must be sterilised separately and added to the medium after autoclaving.

- Carbohydrates are especially susceptible to heat in alkaline medium, addition of 1–2 drops of phosphoric acid ensures that the medium remains acidic.

- Such test compounds should be either sterilised by membrane filtration or by steam sterilisation at 100°C.

- Medium for fermentation tests is inoculated by stabbing the medium with a straight needle carrying the inoculum or by liquefying the medium and then inoculating it.

- Starch solution should be freshly prepared as it easily gets hydrolysed to glucose which can be fermented by almost any bacteria.

- Most test compounds can be stored in capped bottles in the refrigerator as 10% aqueous solution after sterilisation.

- Rarer compounds can be stored in bottles having a rubber washer and perforated screw caps. Solution can be withdrawn with the help of a sterile syringe as and when required without opening the bottle and exposing all the contents.

Biochemical tests performed to identify the pure isolates are grouped into four categories:

- Test for utilisation of specific substrate
- Test for enzyme activity
- Test for specific breakdown products
- Test for proteolytic activity

The various biochemical tests performed are:

1. IMViC Tests

Indole, MR, VP and Citrate tests are done in routine for the classification of gram-negative enteric bacteria. They are commonly referred to as IMViC tests.

a. Indole production

If bacteria possess enzyme tryptophanase, they degrade amino acid tryptophan to indole, pyruvic acid and ammonia.

$$\text{Tryptophan} \xrightarrow{\text{Tryptophanase}} \text{Indole} + \text{Pyruvic acid} + \text{Ammonia}$$

- Inoculate the test organism into peptone water broth.
- Incubate it at 37 °C for 48–96 h.
- Add 0.5 ml of Kovac's reagent and shake gently.
- **Pink/Red colour** in the alcohol layer indicates a **positive** reaction.

- Soak a filter paper in saturated solution of oxalic acid and dry it.
- Hang it inside the culture tube with the help of cotton plug for 24 hrs.
- Positive test is indicated if the paper turns Pink/Red.

Reagents used:

Kovac's reagent:

Paradimethylaminobenzaldehyde	10 g
Amyl or Isoamyl alcohol	150 ml
Conc. HCl	50 ml

Peptone water broth: See Annexure V.

b. Methyl red test

Bacteria produce acid by fermentation of glucose, which changes pH of the medium. It falls and is maintained below 4.5. This test detects the production of acid.

1. Inoculate the test organism in glucose phosphate broth.
2. Incubate at 37°C for 2–5 days.
3. Add five drops of 0.04% solution of alcoholic methyl red solution.
4. Mix well and read the result immediately.
5. **Positive** tests give **bright red** colour.
6. **Yellow** colour indicates **negative** test.
7. If the test is negative after 2 days repeat it after 5 days.

Reagents used:

Methyl red solution: 0.2 g of methyl red, 500 ml of ethyl alcohol, 500 ml of distilled water. Dissolve methyl red in alcohol, add water and filter.

Glucose phosphate broth: See Annexure V

c. Voges–Proskauer test

Fermentation of carbohydrate by several bacteria results in the production of acetyl methyl carbinol (acetoin). In the presence of alkali and atmospheric oxygen, acetoin is oxidised to diacetyl, which reacts with peptone of the broth to give a red colour.

- Inoculate test organism in 5 ml glucose phosphate broth.
- Incubate at 37°C for 48 h.
- Add 1 ml of 40% potassium hydroxide (KOH) containing 0.3% creatine and three ml of 5% solution of α-naphthol in absolute alcohol.
- A **positive** reaction is indicated by the development of **pink** colour in 2–5 min.

Reagents used:

40% potassium hydroxide (KOH): 40 ml of KOH and 60 ml of distilled water.

0.3% creatine: 3 mg in 100 ml of distilled water.

5% solution of α-naphthol in absolute alcohol: 5 g in 100 ml of distilled water.

Glucose phosphate broth : See Annexure V.

d. Citrate utilisation

This test is used to study the ability of an organism to utilise citrate present in Simmon's media, as a sole source of carbon for growth.

- Inoculate a colony directly on Simmon's citrate agar.
- The media also contains bromothymol blue as indicator.
- **Positive** test is indicated by the appearance of growth with **blue** colour.
- **Negative** test shows no growth with original **green** colour.

Simmon's citrate agar: see Annexure V.

2. Nitrate reduction test

The enzyme nitrate reductase reduces nitrate to nitrite, ammonia, nitrous oxide, nitrogen, etc. This test is used to detect the production of nitrate reductase. Nitrate reducing bacteria are found in soil.

$$\text{Nitrate} \xrightarrow{\quad\text{Nitrate reductase}\quad} \text{Nitrite}$$

- Inoculate the test organism in a five ml nitrate broth, containing potassium nitrate, peptone and distilled water.
- Incubate it at 37 °C for 96 h.
- Add one ml of α-naphthylamine reagent and one ml of sulphanilamide reagent.
- Development of **red colour** within a few minutes indicates the presence of nitrate and hence the ability of test organism to reduce nitrate to nitrite.

Reagents used:

Nitrate broth: See Annexure V.

0.5% α-naphthylamine reagent: Dissolve five grams of reagent in less than 1000 ml of 5 N acetic acid (30%) by heating gently. Filter through washed adsorbent cotton. Store in a glass stoppered brown bottle.

Sulphanilamide reagent: Dissolve eight grams of 0.8% sulphanilic acid in 1000 ml of 5 N acetic acid. Store in a glass stoppered brown bottle.

3. Urease test

This test detects the ability of an organism to produce urease enzyme.

- Inoculate test organism on the agar slant of urea agar.
- Incubate at 37°C.
- Observe after four hours and after overnight incubation.
- Development of **purple–pink colour** indicates production of urease.
- Urease in the presence of water converts urea into ammonia and carbon dioxide.
- Ammonia makes the medium alkaline and phenol red indicator changes to purple–red in colour.

$$\text{Urea} \xrightarrow{\text{Urease}} \text{Ammonia} + CO_2$$

Urea agar: See Annexure V.

4. Carbohydrate digestion (Sugar fermentation) test

The ability of an organism to ferment various sugars/digest carbohydrates is indicated by the production of acid and or gas as follows:

- Inoculate the test organism in peptone water broth containing 1% solution of desired sugar. (Total amount of sugar should be 10% of medium volume)
- Add phenol red as indicator of acid production.
- Insert an inverted Durham tube in the culture tubes. (**Inverted tube method**)
- Incubate at 37°C.
- Production of **acid** is indicated by the change of the colour of the medium to **red or pink**.
- If gas is produced it collects in Durham's tube, which rises up in the culture tube (Fig.5.5).
- Alternatively impregnate filter paper discs with sugar solution and place them on inoculated agar plates (**Disc method**).
- Observe after 48 h. Presence of acid will be indicated by appearance of yellow halo around the disc.
- Inoculate oxidation-fermentation (OF) slants by stab culture with 0.1% by volume sugar.
- Incubate one tube anaerobically and another aerobically.
- Observe for colour change.
- Under anaerobic conditions, sugar will be fermented to acid and medium will turn yellow.
- Under aerobic conditions, there will be no acid production.
- Homogenous colour indicates motility.

Reagents used:

Sugar solution: one percent solution of sugar. Autoclave at low pressure and add to peptone water broth to make a final concentration of 10% by volume.

Phenol red indicator: 0.1 g of phenol sulphonaphthalein, 14.1 ml of 0.02 N NaOH, 250 ml of distilled water. Mix all and autoclave and add to basal medium in order to make a final concentration of 0.4% by volume.

Bromothymol blue: 0.1 g of dibromo-cresol sulphanaphthalein, 8 ml of 0.02 N NaOH. Make up volume to 250 ml. It is yellow at pH 6 and blue at pH 7.6. Add to medium in order to make a final concentration of 0.4%.

Peptone water broth, oxidation–fermentation (OF) medium; See Annexure V.

Fig. 5.5 Formation of gas during fermentation reaction. a; No accumulation of gas in Durham tube shows negative reaction. b; Accumulation of gas in Durham tube shows positive reaction

5. Hydrogen sulphide production test

Some organisms degrade sulphur containing amino acids to produce H_2S.

- Inoculate the test organism in TSI medium or nutrient agar containing lead acetate or ferrous acetate.
- Production of H_2S is indicated by the change of colour of the medium to **black** or **brown**.

Regents used:

TSI Agar: See Annexure V.

6. Catalase test

Catalase degrades hydrogen peroxide and releases oxygen which is detected as effervesence. Place a drop of three percent H_2O_2 on colonies of nutrient agar. Prompt **effervescence** indicates catalase production. Culture media containing blood are unsuitable for the test as blood contains catalase.

7. Oxidase test

This test indicates the presence of cytochrome oxidase which catalyses oxidation of reduced cytochrome by oxygen. It indicates the ability of microbes to oxidise amines. Negative test is an indication of presence of enteric bacteria.

- Put a drop of freshly prepared one percent solution of oxidase reagent on a piece of filter paper.
- With a sterile loop pick a test colony and rub it on the paper in the area impregnated with oxidase reagent.
- If organism is oxidase **positive**, paper will become **deep purple blue** in colour in a few seconds.
- Alternatively, pour oxidase reagent over the colonies of the test organism on the culture plate.
- The colonies of oxidase positive organisms rapidly develop a deep purple blue colour.

Oxidase reagent: One percent solution of tetra methyl-*p*-phenylene diamine hydrochloride.

8. Litmus milk reaction

- Inoculate litmus milk and observe.
- There may be no change in the medium or acid or alkali may be produced; clotting of milk, peptonisation or saponification may occur. The clot may be disrupted by the gas produced (stormy fermentation).
- Milk will turn red if acid is produced and blue if alkali is produced.

Litmus milk: 150 g of skimmed milk powder, 1000 ml of distilled water, a few drops of aqueous solution of azolitmin.

9. Amylase test

This test detects the ability of an organism to produce amylase enzyme for starch hydrolysis.

- Inoculate test organism on the starch agar plates (agar containing starch as substrate).
- Incubate at 37°C for 24 h.

- Flood the plates with Gram's/lugol's iodine.
- Development of purple/blue–black colour indicates the absence of starch degradation.
- Presence of clear zone around the colonies indicates secretion of amylase and hydrolysis of starch (Fig. 5.6).
- Maintain a control of uninoculated starch agar plate.

$$\text{Starch (blue-black colour with iodine)} \xrightarrow{\text{amylase}} \text{Sugar (colourless)}$$

Starch agar: See Annexure V.

Gram's/lugol's iodine: See Chapter 3, Gram staining procedure.

Fig.5.6 Amylase test. Area around bacterial colony shows positive reaction rest of the agar becomes blue black on addition of iodine

10. Gelatinase test

This test detects the ability of an organism to produce proteolytic enzymes using gelatin as substrate.

- Prepare gelatin agar.
- Pour medium into culture tubes and allow it to solidify.
- Inoculate test organism in gelatin agar.
- Maintain a control with uninoculated gelatin agar tube.
- Incubate at 37°C for 2–7 days.

- Test for gelatin hydrolysis by chilling the tubes with ice or putting in the freezer.
- Solidifying of gelatin is a negative test.
- Failure of solidification indicates secretion of gelatinase and hydrolysis of gelatin.

Gelatin agar: See Annexure V.

11. Bile solubility test

- Emulsify a few colonies of the test culture in one ml of saline to form a smooth suspension.
- Add one drop of 10% sodium deoxycholate solution.
- Incubate at 37°C.
- Examine for clearing at 15 min, 30 min and 1 h.
- Clearing should occur within 30 min.
- In a mixed culture, place one drop of 10% sodium deoxycholate solution onto the test colony.
- This should lyse within 30 min.
- This method is not entirely reliable, and it is better to purify any suspected colony.

10% Sodium dioxycholate solution: 10 g of in sodium deoxycholate 100 ml of distilled water.

12. Coagulase Test

The test can be done by two methods:

a. Slide method

- Place three separate drops of saline on a clean slide.
- Suspend a loopful of test colony in two of these, and a loopful of control organism (*Staphylococcus aureus*) in the third.
- With a sterile loop, add a drop of citrated rabbit plasma to one test and the control suspension.
- Occurrence of clumping within 10 s indicates a positive result.
- The saline control should remain evenly suspended.
- This test detects the presence of "clumping factor" and is not a true coagulase test.

b. Tube method

- Prepare 1/10 dilution of plasma in 0.85% saline.
- Emulsify a few colonies of control (*S. aureus*) and the test isolate into appropriately labelled tubes containing a 1/10 dilution of plasma in 0.85% saline.
- Incubate at 37°C.
- Examine for coagulation at 1, 3 and 6 h.
- Conversion of the plasma into a soft or stiff gel, seen on tilting the tube to a horizontal position indicates a positive result.

13. DNAse Test

DNAse enzyme produced by the isolate hydrolyses the DNA present in the medium.

- Inoculate sections of tryptose agar medium containing DNA (DNAse agar) with material from test colonies.
- Inoculate *S. aureus* and *S. epidermidis* as positive and negative controls.
- Incubate the plate at 37°C for 18–24 h.
- Flood the plate with 1 N HCl that precipitates DNA and turns the medium cloudy.
- Presence of a zone of clearing round the area of growth indicates positive test.

Tryptose agar medium containing DNA (DNAse agar): See Annexure V.

14. Lancefield Grouping (Streptex Method):

- Emulsify a loopful of the test culture in 0.4 ml of extraction enzyme.
- Incubate at 37°C for 1 h.
- Add one drop of latex reagent to the appropriate circle of a black tile.
- Next, add one drop of extract to each circle and mix, using a wooden stick.
- Mix gently for one minute. Clumping indicates a positive reaction.

15. Nagler Test

This test indicates the production of *a*-toxin (lecithinase or phospholipase C) an exotoxin by *Clostridium perfringens*.

- Divide an egg yolk agar plate into two equal sections.
- Spread a loopful of *C. perfringens* antitoxin over half the plate and allow to dry.
- With a single streak, inoculate the plate with a loopful of the test culture, beginning on the untreated side of the plate.
- Incubate at 37°C under anaerobic conditions.
- Lecithinase activity results in the production of an opaque zone of precipitation around the area of growth.
- This precipitation should not be present on that side of the plate previously inoculated with specific α-antitoxin.

Egg yolk agar: See Annexure V.

16. Optochin Test

- Divide blood agar plate into three equal sections.
- Inoculate one with a known *Streptococcus pneumoniae*, another with a viridans *Streptococcus viridans* and the third with the test isolate.
- Keep the cultures separately.

- Place a 5 μg Optochin disc (ethylene hydrocupreine hydrochloride) in the centre of the plate.
- Incubate at 37°C overnight and observe the zone of inhibition.

Blood agar: See Annexure V.

17. Caesinase test

This test detects the ability of an organism to produce proteolytic enzymes using casein as substrate.

- Inoculate test organism on the milk agar plates (agar containing skimmed milk as substrate).
- Incubate at 37°C for 24 h.
- Presence of clear zone around the colonies indicates hydrolysis of casein.
- Maintain a control of uninoculated milk agar plate.

Milk agar: See Annexure V.

18. Lipase test

This test detects the ability of an organism to produce lipolytic enzymes using lipid as substrate.

- Inoculate test organism on the tributytrin agar plates (agar containing skimmed milk as substrate).
- Incubate at 37°C for 24 h.
- Presence of clear zone around the colonies indicates hydrolysis of lipid.
- Maintain a control of uninoculated tributytrin agar plate.

19. Egg yolk agar method for lipase detection

- Inoculate egg agar plates.
- Incubate at 37°C for 24 h.
- Appearance of opalescent precipitae is a positive test.

Tributytrin agar, Egg agar: See Annexure V.

20. Gluconate test

This test differentiates Enterobacteria.

- Inoculate gluconate broth.
- Incubate at 48°C for 24 h.
- Add equal volumes of Benedict's reagent and heat in boiling water bath
- Observe for precipitation.
- Brown precipitate indicates gluconate oxidation.
- Development of a deep blue colour is a negative test.

Gluconate broth: See Annexure V.

Benedict's reagent: 18 g of copper sulphate, 100 g of sodium carbonate, 200 g of sodium citrate, 125 g of KCNS, 5.0 ml of 5% potassium ferrocyanide solution and 100 ml of DW.

21. Deamination of amino acids

This test detects the ability of an organism to remove amino groups from amino acids.

- Inoculate test organism in 4% peptone broth.
- Incubate at 37 °C for 24 h.
- Test the tubes for presence of ammonia by adding Nessler's reagent to a drop of broth.
- Development of a deep yellow colour reflects deamination.
- Maintain a control of uninoculated broth.

$$\text{Amino acid} \longrightarrow \text{keto acid + ammonia}$$

Peptone broth: See Annexure V.

Nessler's reagent: Take 50 g of potassium iodide and dissolve in sufficient quantity of distilled water. To this add a saturated solution of mercuric chloride till a slight precipitate is formed. Add 400 ml of 50% clarified solution of NaOH. Dilute to 1000 ml by distilled water. Let it stand for one week. Decant the clear solution and keep in air tight bottles in dark.

22. Decarboxylation of amino acids

This test detects the ability of an organism to remove carboxyl groups from amino acids.

- Inoculate test organism in tyrosine broth.
- Incubate at 37°C for 2 days.
- Observe test tubes for gas formation.
- Test for acid production with pH paper.
- Maintain a control by inoculating tyrosine broth.

$$\text{Amino acid} \longrightarrow \text{amine + carbon dioxide}$$

Tyrosine broth: See Annexure V.

23. Hydrolysis of fats

This test detects the ability of an organism to decompose fats. This causes rancidity in foods due to the formation of fatty acids.

- Prepare nutrient agar plates containing 1% fat (oil/butter/ghee) and neutral red indicator.
- Streak the test organism in these plates.
- Incubate at 37°C for 2 days.

- Observe plates for the presence of red globules under the colonies.
- Acid production is indicated by change in the colour of indicator from yellow to red.
- Maintain a control with uninoculated.

24. Dehydrogenase activity

This test detects the ability of an organism to produce dehydrogenases. This causes removal of hydrogen from the substrate. The indicator dye reacts with this hydrogen and changes colour.

- Prepare six tubes containing 1.0 ml of inoculum, 1.0 ml of M/10 phosphate buffer of pH 7 and 0.5 ml of resazurin solution.
- Add 1.0 ml of water to tube number 1.
- Add 1.0 ml of M/10 glucose to tube number 2.
- Add 1.0 ml of M/10 lactose to tube number 3.
- Add 1.0 ml of M/10 sodium lactate to tube number 4.
- Add 1.0 ml of M/10 sodium succinate to tube number 5.
- Add 1.0 ml of M/10 sodium fumarate to tube number 6.
- Mix the contents thoroughly.
- Incubate at room temperature and observe for colour change from blue to pink every 15 min for one hour. Record the time of colour change.

4. Animal Pathogenicity

Various experimental models used in diagnostic microbiology laboratory are guinea pig, rabbit, armadillo and monkey. Various routes of inoculation are: intradermal, subcutaneous, intramuscular, intraperitoneal, intracerebral and intravenous. Oral and nasal routes can also be used. The identification of the organism is carried out on the basis of clinical and postmortem findings and cultural characteristics.

5. Antibiotic Sensitivity Testing

The sensitivity of the organism isolated from the patient to antimicrobial agents is tested by placing filter paper discs impregnated with the antibiotic on culture plates preceded with the organism to be tested and by judging the degree of sensitivity by the size of inhibition zones resulting after overnight incubation, during which adequate interaction between the bacterium and the drug takes place.

Two commonly used methods are: Kirby-Bauer method and Stoke's method.

1. Kirby–Bauer Method

This test can be performed in two ways:

1. By using antibiotic disc
2. By cup well method

- Inoculate bacteria into peptone water.
- Incubate at 37°C for 3–4 h.
- Take a sterile nutrient agar plate and dry it in the incubator till surface moisture become invisible.
- Flood the plate with growth in peptone water or with the help of a sterile swab rub the surface of agar. This will lead to development of a lawn of the bacterial growth on the plate.
- Place the antibiotic impregnated discs on the surface of culture medium at adequate distance of 2 cm or more.
- Incubate the plates overnight at 37°C in an incubator.
- Measure the zone of inhibition around the antibiotic discs.
- Alternately prepare a bacterial lawn as described above.
- Make a well in this plate with the help of a cork borer.
- Fill the well with antibiotic solution of desired concentration and amount.
- Incubate the plates at 37°C .
- Observe for formation of Zone of inhibition around the well.
- Measure the size of the zone.

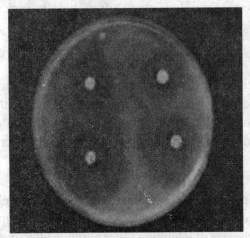

Antibiotic disc method   Cup well method

Fig. 5.7 Antibiotic sensitivity test

2. Stokes Method

In this method the test and control organisms are compared against the same antibiotic discs, on the same medium and under the same physical conditions.

- Divide the plate into three equal areas.

- Seed the middle one-third part of the culture plate with the test strain grown in peptone water.
- Inoculate the standard strain on either side, leaving a little gap between the inoculated areas.
- The discs are placed in the gap and incubated overnight at 37°C.
- A zone radius that is the same size or is larger than the control or is not smaller than 3 mm is reported to be sensitive to that particular antibiotic.

Standard Strains: *Staphylococcus aureus* (for testing of Gram-positive cocci), *Escherichia coli* (for testing of Gram-negative bacilli other than Pseudomonas), and *Pseudomonas aeruginosa* (for testing of Pseudomonas).

6. Serological Methods

Antibodies are highly selective in terms of the surface proteins to which they bind. Thus by studying antibody–antigen binding it is possible to distinguish one bacterial species from the other, or even one strain from the other on the basis of their surface proteins. Serological methods detect antibodies by the following methods–Agglutination tests, ELISAs, and Western blots.

7. Flow Cytometry

This technique uses methods, which analyze cells suspended in a liquid medium by light, electrical conductivity or fluorescence as they pass through a small orifice.

8. Phage Typing

Bacteriophages are species-specific. Infection by phages is used to distinguish bacterial species and strains.

9. Protein Analysis (gel electrophoresis, SDS-PAGE, establishment of clonality)

Identification is done on the basis of the size and other differences between proteins among different organisms by separating proteins using gel electrophoresis.

10. SDS-PAGE

Sodium dodecyl sulphate (also known as sodium lauryl sulphate)–Polyacrylamide gel electrophoresis can detect small differences between isolates and also establish clonality.

11. Comparison of Nucleotide Sequences

Methods like Southern blot, nucleic acid hybridization, RFLP, DNA fingerprinting are used to identify bacteria by determining the exact nucleotide sequence in the genome of different species, strains etc.

RAPID METHODS FOR IDENTIFYING MICROBES

Conventional methods to detect bacteria are time-consuming and depend on growth in culture media, followed by isolation, biochemical identification and sometimes serology. Advances in technology make detection and identification faster, more convenient, more sensitive, and more specific than conventional methods. Rapid methods are generally used as screening techniques, with negative results accepted as is, but positive results requiring confirmation by the appropriate method is necessary.

Rapid methods include miniaturised biochemical kits, antibody- and DNA-based tests, and assays that are modifications of conventional tests to speed up analysis. Most miniaturised kits are designed for enteric bacteria, but kits for the identification of non-Enterobacteriaceae are also available, including for *Campylobacter*, *Listeria*, anaerobes, non-fermenting Gram-negative bacteria and for Gram-positive bacteria.

Automatic analysers incubate the reactions and monitor biochemical changes to generate a phenotypic profile, which is then compared with the provided database stored in the computer to provide identification. Some systems identify bacteria based on compositional or metabolic properties, such as fatty acid profiles, carbon oxidation profiles and other traits.

DNA and antibody-based assays for numerous microbes or their toxins are now available commercially. DNA-based assay formats, such as DNA probes, PCR and bacteriophage have been developed for detecting food-borne pathogens. The highly specific interaction of phage with its bacterial host has also been used to develop assays in which a specific bacteriophage is engineered to carry a detectable marker (e.g. ice nucleation gene in *Salmonella*). The phage confers the marker to the host, which then expresses the phenotype to allow detection. Detection of antigen–antibody binding comprises the largest group of rapid methods being used.

Probabilistic Identification of Bacteria for Windows (PIBWin)

The PIBWin programme provides probabilistic identification of unknown bacterial isolates against identification matrices of known strains.

The programme has three major functions:

- The identification of an unknown isolate
- The selection of additional tests to distinguish between possible strains if identification is not achieved
- The storage and retrieval of results

IMPORTANT FEATURES OF SOME COMMON PATHOGENS

Neisseria: Aerobic coccobacilli, Non-spore-forming, gram-negative, non-motile, oxidase+, growth on chocolate agar containing antibiotics that inhibit growth of gram-negative bacteria, gram-positive bacteria, and molds, glucose fermentative, two species: *N. gonorrhoeae* and *N. meningitidis*.

Escherichia coli: Growth on EMB agar+, Lysine+, Citrate–, Indole+, Acetate+, Lactose+.

Shigella: and Serotype A:*S. dysenteriae*, Serotype B:*S. flexneri*, Serotype C:*S. boydii*
Serotype D—*S. sonnei*. Lysine–, non-motile, –/+TSI reaction (no gas), Acetate–, Lactose–

Edwardsiella tarda: Lysine+, Hydrogen sulfide+, –/+TSI reaction (with gas), Indole+, Citrate–

Salmonella: Lysine+, Hydrogen sulfide+, –/+TSI reaction (with gas), Indole+, Citrate+,
ONPGI–, Malonate–

Citrobacter: Lysine–, Hydrogen sulfide+ (*C. freundii*), +/+TSI reaction (with gas), Citrate+, Slow
urease.

Klebsiella: Citrate+, Indol–, +/+TSI (with gas), Non-motile, Ornithine–

Enterobacter: Lysine+ (except *E. cloacae*), Citrate+, Indol–, +/+TSI (with gas), Motile, Ornithine+

Serratia: Lysine+, Citrate+, Indol–, +/+TSI (No gas), DNase+

Proteus: Lysine–, Hydrogen sulfide+, Motile, Urease+

Morganella: Indole+, Ornithine+, Citrate+

Yersinia entercolitica: Urease+, Ornithine+, +/+TSI reaction (No gas), Motile at room temperature.
Y. pestis: Urease–, Ornithine–, Non-motile at room temperature.

Pleomorphic Gram-negative rods

Haemophilus: Growth on blood agar+, Coccobacilli, classified by their capsular properties into six
different serological groups, (a–f). Type b capsule are virulent.

Bordetella pertussis:. Growth on Blood Free Peptone–, Urease–, Nitrate Reduction–, Motility–,
Citrate–.

B. parapertussis: Growth on Blood Free Peptone+, Urease+, Nitrate Reduction –, Motility–, Citrate+.

B. bronchiseptica: Growth on Blood Free Peptone+, Urease+, Nitrate Reduction+, Motility+, Citrate+.

Brucella: Aerobic, gram-negative, Coccobacillus.

Pasteurella: gram-negative, non-motile, facultative anaerobic, Coccobacilli, can be cultured on
chocolate agar and produces foul odour. Oxidase+, non-motile, Catalase+, Non-haemolytic (some
are beta-haemolytic), Indole+, Ornithine decarboxylase+

Legionella: gram-negative, growth on buffered charcoal yeast extract agar off-white in colour and
circular in shape. Motile, Urease–, Catalase+, Nitrate–, Gelatinase+.

L. pneumophila: Beta-lactamase+, Hippurate hydrolysis+.

L. micdadei: Beta-lactamase–, Hippurate hydrolysis–, Acid fast.

Miscellaneous gram-negative rods

Campylobacter: gram-negative, microaerophiles, motile, microaerophiles, which means that they can survive in a low oxygen environment.

Vibrio: Motile, gram-negative, obligate aerobes, curved shape, single polar flagella.

V. cholerae: Oxidase+, Catalase+, Indole+, Lysine decarboxylase+, Ornithine deaminase+.

Helicobacter: gram-negative, micraerophilic, motile, catalase-positive, curved cell bodies. Causes peptic ulcers and chronic gastritis, growth on sheep blood agar, Urease+, Catalase+.

Non-fermenters

Pseudomonads: Motile, gram-negative rods, utilise glucose oxidatively, resistant to most antibiotics and they are capable of surviving in conditions that few other organisms can tolerate, produce a slime layer resistant to phagocytosis, cause of hospital acquired (nosocomal) infections.

Pseudomonas aeruginosa: Can grow at 42°C, produces a bluish pigment (pyocyanin) and a greenish pigment, characteristic fruity odour, Oxidase+, Beta-haemolytic, Motile.

Burkholderia (*Pseudomonas*) *cepacia*: Lysine positive.

Stenotrophomonas maltophila (formerly known as *Xanthomonas maltophila*): Motile, susceptible to trimethoprim sulfamethoxazole, lysine+, DNAse+.

Acinetobacter: Oxidase-negative, non-motile, gram-negative coccobacilli in pairs, good growth on MacConkey agar, Penicillin resistance (most strains).

Flavobacterium: Ubiquitous, produce indole when grown in tryptophan broth, metabolise glucose oxidatively, some really slow fermenters, motile and oxidase positive.

Alcaligenes faecalis: Oxidase positive, motile, can grow on MacConkey agar. Characteristic fruity odour, which distinguishes it from other *Alcaligenes* species.

Gram-positive anaerobes

Clostridium: gram-positive, spore-forming rods, anaerobic, motile, ubiquitous, appear as long drumsticks with a bulge located at their terminal ends, optimimum growth on blood agar at human body temperatures, produce **spores,** exotoxins are responsible for tetanus, botulism, and gas gangrene.

C. tetani: Motile, terminal spores, non-aerotolerant.

C. botulinum: Motile, no terminal spores, non-aerotolerant, lipase+ .

C. perfringens: Non-motile, no terminal spores, non-aerotolerant, double zone haemolysis.

C. difficile: Motile, no terminal spores, non-aerotolerant, gram-positive non-spore-forming bacilli.

Actinomyces: Filamentous, gram-positive, obligate anaerobes, Indole–, Catalase–, Lipase–, DNase–.

***Bifidobacterium*:** Anaerobic, gram-positive bacilli, appear to be bone shaped, obligate anaerobes, non-motile, Catalase–, forms branching filaments.

***Eubacterium*:** Biochemical testing can distinguish *Eubacterium* from the other gram-positive, anaerobic rods. These bacteria tend to form clumps under microscopic observation. Indole–, Catalase–, Hydrogen sulphide–.

***Propionibacterium*:** gram-positive, anaerobes, clump up and may show a slight tendency to branch, show uneven staining patterns following a Gram-stain procedure, anaerobic or microaerophilic, grow on blood agar, Indole+, Catalase+, glucose fermentation.

Gram negative anaerobes

***Bacteroides fragilis*:** Anaerobic, inhabit the digestive tract, anaerobic counterpart of *E. coli* except they are somewhat smaller, grow well on blood agar, non-spore-forming, produce a very large capsule, Indole–, Catalase+, Esculin hydrolysis+, Glucose fermentation, Lactose+.

***Fusobacterium*:** Anaerobic, Gram-negative bacilli, strong resemblance to certain *Bacteroides* species. Normally spindle-shaped cells with sharp ends, Indole+, Hydrogen sulphide–, Catalase–.

***Veillonella*:** gram-negative, cocci, anaerobic counterpart of *Neisseria,* non-motile diplococci, part of the normal flora of the mouth, negative for almost every biochemical test with the exception of an occasional strain being positive for catalase.

6

MAINTENANCE AND STORAGE
OF BACTERIAL STRAINS

Bacterial strains may be stored indefinitely at low temperatures such as −20°C and −80°C in 15–40% glycerol. Once the bacterial have been identified a stock culture is fist developed which can be preserved by various methods.

- **Preparation of stock culture**

    1. Inoculate a culture tube containing 5 ml of LBM (Luria Bertani Medium) or LBM+ antibiotic selective medium with fresh inoculum.

    2. Incubate at 37°C for 5 hours to overnight. Culture is now in late log or stationary phase.

Various methods of preservation are as follows:

1. Storage at low temperature

a. Storage at −80°C

    1. Take a sterile cryovial and add 225 μ of sterile 80% glycerol into it.

    2. Add 1.0 ml of the bacterial culture. The frozen stock will be 15% glycerol.

    3. Mix well and place tube at −80°C.

b. Storage at −20°C

    1. Mix equal volumes of 80% glycerol and bacterial culture in a polypropylene tube.

    2. Place the tube in a −20°C freezer.

Recovery of cells

- From the −80°C glycerol stock: With a sterile toothpick scrape off the ice and streak the cells on the appropriate medium. e.g. LBM + ampicillin.
- From the −20°C working stocks: pipette out 50 to 100 µl and use as inoculum.

Precautions

- If mixing is not proper, ice crystals will form decreasing the viability of the cells.
- Check the viability of the cells after 1 week.
- Periodical checking for viability, strain degradation etc. is necessary.
- During recovery never thaw the frozen stocks.
- Each freeze-thaw cycle results in a 50% loss in cell viability.

Composition of LB medium (Luria-Bertani Medium) with or without antibiotic

Dissolve 10 g bacto-tryptone, 5 g bacto-yeast extract, and 10 g sodium chloride in 500 ml of distilled water. Maintain pH at 7.0. Autoclave and cool to 50–60°C. Add antibiotic, if required.

Preparation of overnight culture

Overnight cultures are required for fresh inoculation.

On solid medium

1. Scrape bacteria from glycerol stock or from an old agar tube/plate stored at 4°C.
2. Streak onto an agar plate with or without antibiotic.
3. Incubate the plate in inverted position overnight at 37°C.

In liquid broth

1. Pick a single colony from the agar plate using a sterile micropipette tip or a flamed loop.
2. Dip this in 2.5 ml of LB with or without antibiotics.
3. Place on a shaker for 4–8 hours at 37°C and at 200 rpm.
4. Take 0.5 ml of this culture and inoculate 100 ml of LB with antibiotics.
5. Keep on the shaker overnight at 37°C at 200 rpm.
6. Growth is indicated if the initial transparent culture turns cloudy after overnight incubation.
7. Overnight cultures can be stored at 4°C for several days.

Regular Periodic Subculture

Repeated culture of bacteria from stock culture is done on fresh medium to maintain its viability, but the method is not very reliable as there are very high chances of mutation leading to strain variation, contamination, etc. Medium, time of subculture, storage temperature, etc. vary with species. The time interval and type of medium chosen should support slow growth of organism. Frequency of subculture can be reduced if organism is grown on minimal medium which reduces the metabolism of the cells and hence growth is slower. e.g. egg saline slopes, semisolid nutrient agar slabs.

2. Overlaying with mineral oil

The slant with the organism growing in minimal medium is completely covered by half an inch thick layer of oil. This seals the surface, excludes air and prevents dehydration of medium. Culture can be then stored at room temperature or at 4°C. Advantage is that as and when required, a small amount of inoculum can be removed and rest remains preserved. Disadvantage is that change in strain character may occur due to mutations.

3. Preservation by drying

Bacteria can be stored in dried form on

- *Paper discs:* Discs are cut out of thick absorbent paper and sterilised. A thick suspension of bacteria is placed on the disc and dried over phosphorus pentoxide in a dessicator under vacuum.
- *Gelatin discs:* Nutrient gelatin and thick suspension of bacteria are mixed and placed on sterile waxed paper or plastic petri dishes and dried over phosphorus pentoxide in a dessicator under vacuum.
- *Cotton plugs:* Cotton plugs are saturated with peptone, starch or dextran and dried. Thick suspension of bacteria is placed on these pre dried plugs or cellophane paper and dried over phosphorus pentoxide in a dessicator under vacuum.
- *L-drying:* Bacterial suspension is filled in ampoules and dried using a vacuum pump and dessicant and a water bath at controlled temperature.

4. Storage in a refrigerator

Culture can be stored at 4–6°C in a refrigerator. The metabolic rate at this temperature is lowered to almost nil, but the cells remain viable. Storage at −20 to −30°C can be done in deep feezers or at −80 °C in ultra deep freezers. The disadvantage of these systems is that at low temperature cells may get damaged due to formation of ice crystals and concentration of electrolytes. The rate of cooling should be carefully monitored and the amount of electrolytes in the medium can be reduced.

This damage can be reduced by use of protective substances such as glycerol or dimethyl sulfoxide (DMSO), etc. Alternatively cell suspension is mixed with cryoprotective agent and then sealed in vials and frozen at controlled rates at −150°C and stored in liquid nitrogen container either by immersion or storage in gas phase above liquid nitrogen at −150°C.

5. Storage in dry ice

Culture can be stored in solid carbon dioxide at −70°C, but the amount of dry ice has to monitored carefully and maintained by quick replenishment. Reduction in amount may lead to loss of cultures.

6. Cryopreservation

Cultures can be stored in liquid nitrogen at –196°C.

7. Storage in sterile soil

Spore forming bacteria can be preserved by mixing the spore suspension with soil sterilised for 2–3 hours at an interval of 1–2 days. Soil is then stored in a refrigerator.

8. Storage in silica gel

Powdered, heat-sterilised, cooled silica gel is mixed with paste of cells and stored at low temperature.

9. Lypholization

It is also known as freeze drying. Material is dried rapidly in vacuum from the frozen state. Cell suspension is placed in a vial and frozen by immersing in dry ice or acetone or liquid nitrogen. Vials are removed, dried under vacuum, sealed and stored at low temperature (almost 30 years). Advantage of this method is that there is no change in strain character, there is long-term survival, and very little space is required for storage. But the procedure is complex and tedious and loss of vacuum leads to inactivation of cultures.

Important points to be remembered

- Always check the expiry date on the packet of dehydrated preformed commercially available media.
- Always enter the receiving data on the label.
- The remaining shelf life shall be minimum six month at the time of receiving.
- Media should always be stored as per the manufacturer's specifications.
- Media can be stored in a refrigerator if storage should be between –2°C and –8°C. Storage below –25°C has to be done in a deep freezer or an incubator with temperature maintained at –20°C or –25°C.
- Medium should be reconstituted as per the manufacturer's instructions.
- Sterilisation of media should be done at 121°C at 15 lbs pressure for the specified time as applicable or specified.
- After sterilisation of media always incubate the preferred media at 30–35°C for 24–48h.
- Confirm proper sterilisation of media by incubation at 30–35°C for 24 to 48 h. If growth occurs even in one sample during this time discard the complete lot.
- Only if prepared media tests positive for growth promotion test use the media for further experiments.
- All media after sterilisation should not be stored for more than 1 month except sterilising testing media, which should be stored for period of NMT 15 days after sterilisation.

7

LABORATORY EXPERIMENTS TO STUDY BACTERIA

1. STUDY OF MOTILITY BY HANGING DROP METHOD

Requirements: 24-hour broth culture of *Pseudomonas*, *Bacillus*, *Staphylococcus*, cavity slide, inoculation loop, spirit lamp, cover slip, microscope, petroleum jelly.

Method: Place a loopful suspension on a cover slip. Apply vaseline along the four edges of the cover slip. Place the cavity slide over the cover slip in such a manner that cavity is super-imposed on the bacterial suspension. Press to seal the coverslip and invert the slide. Observe under phase contrast microscope (See chap.4 for details).

Observations:

S.No	Name of organism	Flagellar motility	Brownian movement
1	*Pseudomonas* sp.		
2	*Bacillus* sp.		
3	*Staphylococcus* sp.		
4	Unknown		

Result:

2. CALIBRATION OF MICROMETER AND MEASUREMENT OF MICROBIAL SIZE

Requirements: 24-hour broth culture of bacteria; (*Pseudomonas*, *Bacillus*, *Staphylococcus* spp.) yeasts, *Paramecium*, fungal spores, sporangiophores, cyanobacterial colonies, heterocysts, etc. Ocular and stage micrometer, immersion oil, xylene, lens cleaning paper, microscope.

Method: Insert ocular micrometer reticle on the eyepiece diaphragm. Place stage micrometer on the stage and calculate the calibration factor as described in chapter 4. Replace the stage micrometer with slide bearing the sample. Measure the oculometer divisions covering the length and/width of the microorganism/object.

Take at least ten readings and then calculate their average. Multiply this by calibration factor to find out the size.

Observations:

S.No	No. of stage divisions	No. of oculometer divisions	Calibration factor		
			10X	40X	100X
1					
2					
3					

Calculations:

1. Calibration factor

1 division of stage micrometer = .01 mm = 10 µm
Therefore **Y** divisions of stage micrometer = **Y** × 10 µm
Now **X** divisions of ocular micrometer = **Y** division of stage micrometer.
Therefore 1 division of ocular micrometer = $\dfrac{Y \times 10}{X}$ µm

2. Size of microbe in µm (Length × Breadth) or Diameter

S.No	Name of organism	Length			Breadth		
		CF	N	(CF × N)	CF	N	(CF × N)
1							
2							
3							

CF= Calibration factor
N = No. of oculometer divisions

Result: Mention size in µ.

3. SIMPLE STAINING OF BACTERIA

All the methods for staining have been discussed in details in chapter 4.

Requirements: 24-hour broth culture of *Pseudomonas*, *Bacillus*, *Staphylococcus*, slides, spirit lamp, cover slip, microscope, blotting paper, needle or loop, coplin jars or staining rack, methylene blue, crystal violet, carbol fuchsin.

Method: Prepare a thin smear. Stain by the methods given in chapter 4. Observe.

Observations and Results:

S.No	Name of organism	Shape and arrangement	Cell colour and stain reaction
1	*Pseudomonas* sp.		
2	*Bacillus* sp.		
3	*Staphylococcus* sp.		
4	Unknown		

Results:

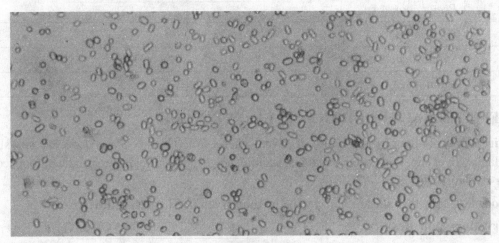

Rod shaped bacilli

4. NEGATIVE STAINING OF BACTERIA

All the methods for staining have been discussed in details in Chapter 4

Requirements: 24-hour broth culture of *Pseudomonas*, *Bacillus*, *Staphylococcus*, slides, spirit lamp, cover slip, microscope, blotting paper, needle or loop, coplin jars or staining rack, India ink or nigrosine.

Method: Prepare a thin smear. Stain by the methods given in chapter 4. Observe.

Observations and Results:

S.No	Name of organism	Shape and arrangement	Cell colour and stain reaction
1	*Pseudomonas* sp.		
2	*Bacillus* sp.		
3	*Staphylococcus* sp.		
4	Unknown		

5. DIFFERENTIAL STAINING: GRAM STAINING

Requirements: 24-hour broth culture of *Pseudomonas*, *Bacillus*, *Staphylococcus*, slides, spirit lamp, cover slip, microscope, Gram's stain, 95% ethyl alcohol or absolute alcohol, safranin, blotting paper, needle or loop, coplin jars or staining rack.

Method: Prepare a thin smear. Stain by the method given in chapter 4. Observe.

Observations and Results:

S.No	Name of organism	Shape and arrangement	Cell colour and stain reaction
1	*Pseudomonas* sp.		
2	*Bacillus* sp.		
3	*Staphylococcus* sp.		
4	Unknown		

6. DIFFERENTIAL STAINING: ACID FAST STAINING

Requirements: 24-hour broth culture of Mycobacteria, *Bacillus* sp., *Staphylococcus* sp., slides, spirit lamp, cover slip, microscope, acid fast staining reagents, blotting paper, needle or loop, coplin jars or staining rack.

Method: Prepare a thin smear. Stain by the method given in Chapter 4. Observe.

Observations and Results:

S.No	Name of organism	Shape and arrangement	Cell colour and stain reaction
1	*Pseudomonas* sp.		
2	*Bacillus* sp.		
3	*Staphylococcus* sp.		
4	Unknown		

Mycobacteria are usually found in milk and soil. *Mycobacterium tuberculosis* can be found in milk and *M. leprae* in soil samples.

7. CYTOLOGICAL STAINING: ENDOSPORE

Requirements: 24-hour to 72-hour broth culture of *Bacillus cereus*, *Clostridium* sp., slides, spirit lamp, cover slip, microscope, malachite green, safranin, blotting paper, needle or loop, coplin jars or staining rack.

Method: Prepare a thin smear. Stain by the method given in Chapter 4. Observe.

Observations:

S.No	Name of organism	Endospore Pr/Ab.	Shape and position
1	*Pseudomonas* sp.		
2	*Bacillus* sp.		
3	*Staphylococcus* sp.		
4	Unknown		

Results: Mention result as spore former or Non-spore former.

8. CYTOLOGICAL STAINING: CAPSULE

Requirements: 24-hour broth culture of *Leuconostoc mesenteroides*, *Enterobacter aerogenes*, unknown sample, slides, spirit lamp, cover slip, microscope, 1% aqueous crystal violet solution, 20% copper sulphate solution, blotting paper, needle or loop, coplin jars or staining rack.

Method: Prepare a thin smear. Stain by the method given in chapter no. 4. Observe.

Observations:

S.No	Name of organism	Cell shape and arrangement	Capsule Pr./Ab.
1	*L. mesenteroides*		
2	*E. aerogenes*		
3	Unknown		

Results: Mention result as capsulated or noncapsulated

9. CYTOLOGICAL STAINING: FLAGELLA

Requirements: 24-hour broth culture of *Pseudomonas*, *Bacillus*, *Staphylococcus*, slides, spirit lamp, cover slip, microscope, Gram's stain, 95% ethyl alcohol or absolute alcohol, safranin, blotting paper, needle or loop, coplin jars or staining rack.

Method: Prepare a thin smear. Stain by the method given in chapter no. 4. Observe

Observations:

S.No	Name of organism	Flagella Pr/Ab	Number and position
1	*Pseudomonas* sp.		
2	*Bacillus* sp.		
3	*Staphylococcus* sp.		
4	Unknown		

Results: Mention results as atrichous monotrichous..etc.

10. STAINING OF INTRACELLULAR METABOLITES

Requirements: 24-hour broth culture of *Pseudomonas*, *Bacillus*, *Staphylococcus*, slides, spirit lamp, cover slip, microscope, Gram's stain, 95% ethyl alcohol or absolute alcohol, safranin, blotting paper, needle or loop, coplin jars or staining rack.

Method: Prepare a thin smear. Stain by the method given in chapter 4. Observe.

Observations:

S.No	Name of organism	Lipids	Nuclear material	Metachromatic granules	Polysaccharides
1	*Pseudomonas* sp.				
2	*Bacillus* sp.				
3	*Staphylococcus* sp.				
4	Unknown				

Results:

11. CULTURE OF BACTERIA ON SOLID MEDIUM

Requirements: 24-hour broth culture of *Pseudomonas*, *Bacillus*, *Staphylococcus*, sterile culture media, inoculation loop, spirit lamp, petri dishes, bent glass rod, LAF bench, incubator.

- **Streak plate method:** Expose the culture medium, petri plates and inoculation instruments to 20 minutes of UV in laminar flow bench. Melt medium and pour in sterile petri plates. Allow the medium to solidify. Using an inoculation loop, pick a loopful of bacteria from stock culture/ dilution and inoculate the medium across the surface of agar by methods described in chapter 4 and incubate at 37°C for 48 hours.

- **Pour plate method:** Expose the culture medium, petri plates and inoculation instruments to 20 minutes of UV in laminar flow bench. Melt medium and mix one ml of inoculum from stock culture/ dilution in it. Pour in sterile petri plates and allow it to solidify. Inoculate the medium across the surface of agar by methods described in chapter 4 and incubate at 37 °C for 48 hours.

- **Spread plate method:** Expose the culture medium, petri plates and inoculation instruments to 20 minutes of UV in laminar flow bench. Melt medium and pour in sterile petri plates. Allow the medium to solidify. Place one ml of inoculum on one side of plate and spread it all over the agar surface with the help of a glass rod or a sterile cotton swab. Incubate at 37°C for 48 hours.

- **Agar slant:** Melt sterile medium and pour into culture tubes. Place the tubes at 45° angle and allow the medium to solidify. Expose the medium and inoculation instruments to 20 minutes of UV in laminar flow bench. With the help of sterile needle streak the inoculum on the surface of slant. Incubate at 37°C for 48 hours.

- **Stab culture:** Prepare agar slants as described earlier. Now inoculate the slants by stabbing the medium. The needle should penetrate the medium. Incubate at 37°C for 48 hours. This method helps to distinguish between aerobes (grow on surface) and anaerobes (grow inside the medium).

Bacillus sp. Colonies

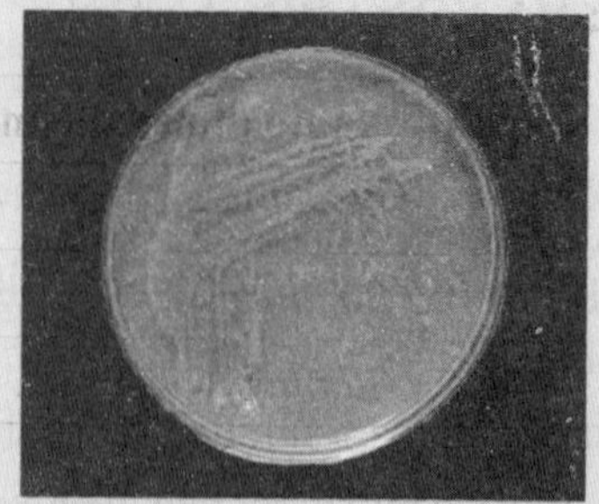

Pseudomonas sp. Colonies

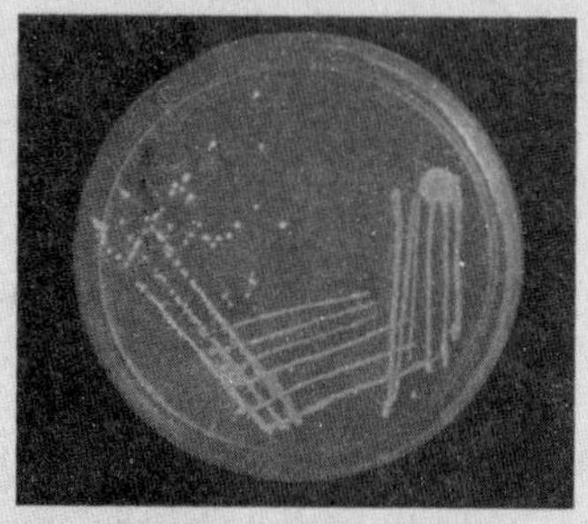

Staphylococcus sp. Colonies

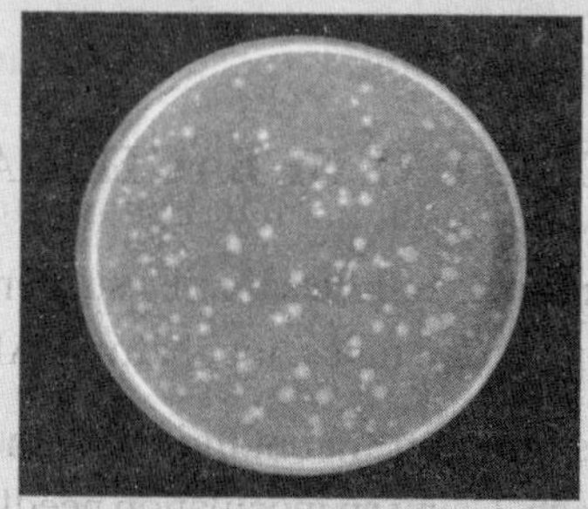

Proteus sp. Colonies

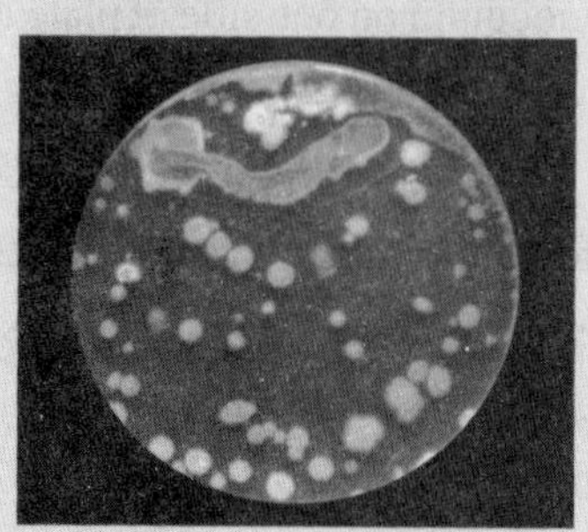

Enterobacter sp. Colonies

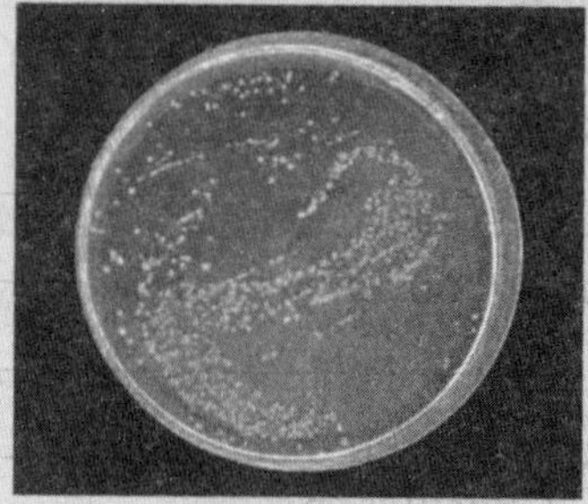

Streptococcus sp. Colonies

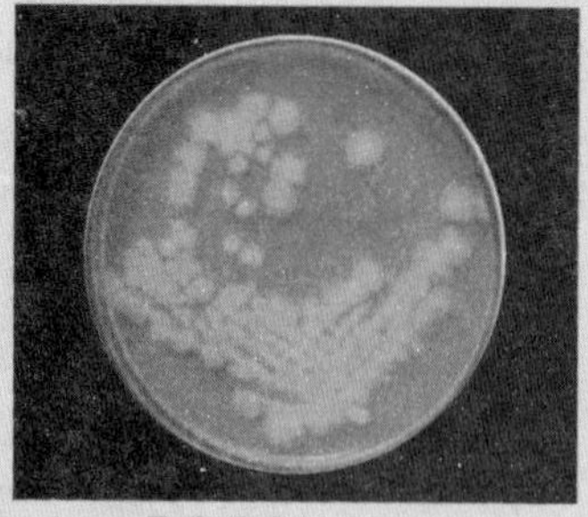

E. Coli Colonies

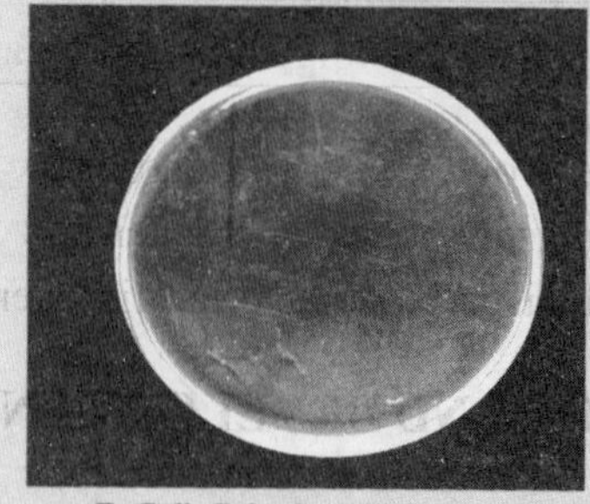

E. Coli Colonies (EMB Agar)

Bacterial colonies

Observations:

S.No	Name of organism	Growth (+/-)	Type of colony
1	*Pseudomonas* sp.		
2	*Bacillus* sp.		
3	*Staphylococcu*s sp.		
4	Unknown		

Result:

12. BROTH CULTURE OF BACTERIA

Requirements: Stock culture sample / suspension sterile nutrient broth (culture media), inoculation loop, spirit lamp, culture tubes, incubator, LAF bench.

Method: Expose the tubes containing broth and inoculation instruments to 20 minutes of UV in laminar flow bench. Using an inoculation needle, pick a loopful of bacteria from stock culture/ unknown sample suspension and inoculate the medium (method described in Chapter 4). Incubate culture tubes at 37°C for 48 h.

Observations:

S.No	Name of organism	Growth		
		Surface	Bottom	Dispersed
1	*Pseudomonas* sp.			
2	*Bacillus* sp.			
3	*Staphylococcu*s sp.			
4	Sample			

Result: Surface growth: Bacteria aerobic.
Bottom growth: Bacteria anaerobic
Growth dispersed: Bacteria Microaerophilic

13. ISOLATION AND DEVELOPMENT OF PURE CULTURES OF BACTERIA

Requirements: Soil/ water/ food/ blood / sample (containing unknown bacteria), 24-hour broth culture of any control organism such as *Pseudomonas*, *Bacillus*, *Staphylococcu*s, sterile nutrient broth and media, inoculation loop, spirit lamp, culture tubes, incubator, LAF bench.

Method:

Step 1

Prepare a homogenised suspension of sample in sterile water/normal saline. Inoculate a loopful into few culture tubes containing approximately 20 ml nutrient broth. Incubate to get 24-hour culture of the organism. Prepare serial dilutions (see chapter 4 for details) of the 24-hour cultures of both, sample as well as control (known organism).

Label the petri plates according to the dilutions made. Make two sets: one for control and one for unknown organism. Expose the culture medium, petri plates and inoculation instruments to 20 minutes of UV in laminar flow bench. Melt medium and pour in sterile petri plates. Allow the medium to solidify. Using an inoculation loop, pick a loopful of bacteria from specific dilutions of sample and inoculate the medium by streak plate method described in chapter 4. Repeat the procedure for control. Incubate at 37°C for 48 h. All colonies in control will be of some type. Different colonies will be seen in plates inoculated with inknown sample.

Step 2

Pick a well separated colony from one of the inoculated petri plates and inoculate a tube filled with nutrient broth. Note the dilution used. Incubate for 24 h and repeat step 1 again.

The whole procedure is repeated till identical colonies develop in the petri plates. Identify the bacteria by standard methods given in chapter 5.

14. DETERMINATION OF CELL NUMBERS BY TURBIDIMETRIC METHOD

Requirements: 24-hour broth culture of given bacteria, sterile liquid and solid nutrient media, petri plates, culture tubes, inoculation loop, spirit lamp, culture tubes, incubator, spectrophotometer.

Method:

Take 10 culture tubes and fill them with nine ml nutrient broth. Number them accordingly as 10^{-2}, 10^{-3},.....10^{-10}. Keep one tube as blank (only medium and no inoculum). Switch on the spectrophotometer one hour before starting the experiment. Set 100 % transmission by blank. Make serial dilutions of the 24-hour culture in the prepared test tubes by inoculating one ml of 24 h culture into 1st tube. Now take 1 ml from this tube and add to next tube continue till the last tube labelled 10^{-10} has been inoculated. Measure the optical density of undiluted culture and subsequent dilutions.

Determine plate counts of these dilutions by plating 1 ml inoculum at 600 nm in agar plates. Count number of colonies which develop. Multiply the number with dilution factor to get Standard Plate Count (SPC) or number of cells/ml.Plot a graph between OD observed and the corresponding bacterial count made according to standard plating method (See chap. 4 for details). Plot a standard curve with known concentrations of the bacteria. Compare the OD of unknown sample with this to get number of bacteria/ml.

Important: The OD and the SPC must be of the same dilutions.

Observations:

S.No	Dilutions used	OD	SPC	Cfu/ml
1	Undiluted suspension			
2	10^{-2}			
3	10^{-4}			
4	10^{-6}			
5	10^{-8}			
6	10^{-10}			

Express results in cfu/ml.
- OD = Optical density
- SPC = Standard plate count
- Cfu = Colony forming units

15. DETERMINATION OF CELL NUMBERS BY HEMOCYTOMETER

Requirements: 24-hour broth culture of *Pseudomonas* sp., *Bacillus* sp, *Staphylococcus* sp., unknown sample, spirit lamp, sterile culture tubes, incubator, Hemocytometer, microscope, Pasteur pipettes or transfer Pipettes, Balanced Salt Solution (BBS) or PBS, Trypan blue, 0.4% in BBS (or PBS).

Method:

- Dilute the sample in Trypan Blue dye exclusion medium in 1:20 ratio (5 µl of cell suspension in 95 µl Trypan blue).
- Place cover glass over hemocytometer chamber and with a Pasteur or transfer pipette, fill both chambers of the hemocytometer taking care that the liquid does not overflow.
- Do not disturb for 1 to 2 min. This allows deposition of cells on the counting plane. If the slide needs to be left undisturbed for longer period, use a humid chamber.
- Count the number of cells in each of 10 squares (1 mm² each) at 10X.
- If over 10% of the cells represent clumps, repeat entire sequence. If fewer than 200 or more than 500 cells are present in the 10 squares, repeat with a more suitable dilution factor.
- Ideally more than 200 and less than 500 cells per chamber should be counted.
- If cell suspension is way above 500 cells/square, count as many smallest squares you can and then calculate the mean.
- Mean multiplied by 0.9 will approximately be the number of million cells / ml.

Cells/ml = average count per square $\times 10^4$ (each square has a volume of 0.00001 or 10^4 cm³)

Total cells = cells per ml × any dilution factor × total volume of cell preparation from which the sample was taken.

Methods for counting cells:

- Count the number of cells in the 4 outer squares and calculate cell concentration as follows:

  Cell concentration per milliliter = Total cell count in 4 squares × 2500 × dilution factor

e.g. If one has counted 450 cells after diluting an aliquot of the cell suspension 1:10, the original cell concentration = 450 × 2500 × 10 = 11,250,000/ml

- Estimate cell concentration by counting 5 squares in the large middle square and calculate cell concentration as follows:

  Cell concentration per millilitre = Total cell count in 5 squares × 50,000 × dilution factor

e.g. If one has counted 45 cells after diluting an aliquot of the cell suspension 1:10, the original cell concentration = 45 × 50,000 × 10 = 22,500,000/ml

Observations:

Results:

16. DETERMINATION OF VIABLE COUNT

Requirements: 24-hour broth culture of *Pseudomonas* sp., *Bacillus* sp., *Staphylococcus* sp., unknown sample, sterile culture media, inoculation loop, spirit lamp, culture tubes, incubator, colony counter.

Method: Prepare serial dilutions of 24-hour broth culture. Using pour plate method, inoculate nutrient agar plates with 0.1 and one ml suspension into petri dishes labelled according to the dilutions used. Incubate the plates in inverted position at 37 °C for 48 h. After incubation period, count the number of colonies by colony counter or a manual counter. Discard plates having spreaders, more than 300 colonies (too numerous to count: TNTC), less than 30 colonies (too few to count) or colonies merging with each other. Count the colonies in plates having 30–300 colonies.

Calculate number of viable cells/ml by following formula:

$$\text{No. of cell/ml} = \frac{\text{No. of colonies}}{\text{Vol. of sample}} \times \text{dilution factor*}$$

e.g. No. of colonies = 40

Dilution factor = 1:1 × 10^6

Volume of sample taken = 1 ml

No. of cell/ml = 40 × 1,000,000 = 40,000,000 or 4 × 10^7 cells/ml

* Dilution factor is reciprocal of dilution ∴ is always positive.

17. ANAEROBIC CULTURE OF BACTERIA

1. Pyrogallic acid method

Requirements: 24-hour broth culture of anaerobic bacteria, sterile culture media, agar slants and plates, inoculation loop, spirit lamp, culture tubes, incubator, LAF bench, rubber stoppers, glass rod, sterile pipettes, 4% sodium hydroxide, pyrogallic acid.

Method: Prepare serial dilutions of 24-hour broth culture. Inoculate nutrient agar slants. Insert a cotton plug into the tube and burn it. Fill the space between the plug and the mouth of the tube with pyrogallic acid crystals. Add two ml of 4% NaOH and immediately close the mouth with the rubber stopper. Incubate the slants in inverted position at 37 °C for 48 h, 72 h.

2. In anaerobic jar

Requirements: 24-hour broth culture of anaerobic bacteria, sterile culture media, petri plates, inoculation loop, spirit lamp, culture tubes, incubator, rubber stoppers, glass rod, anaerobic jar, vacuum pump, nitrogen and carbon dioxide gas cylinders/gas pack.

Method: Prepare serial dilutions of 24-hour broth culture. Inoculate nutrient agar plates and place them in the jar. Place the gas pack also inside the jar. Insert a candle in the jar and ignite it. Incubate at 37°C for 48 h, 72 h.

Observations: Pr/ab of growth after 48h, 72 h.

Results : Given bacteria aerobic or anaerobic.

18. SAMPLING OF AIR FOR PRESENCE OF MICROBES

Places like culture chambers, growth chambers, laminar airflow bench, laboratory etc. should be free of microbial contamination. Most common reason for contamination is the presence of microbes in the air. Therefore, periodic sampling of air for microbial contamination is very essential. There are various ways by which air can be monitored for microbial presence:

1. Physical Methods:

Requirements: Sterile agar plates, agar medium strips, incubator, and slide holder.

- Leave a petri plate containing nutrient medium open in the area for a couple of hours. Incubate at 37°C for 24 h. Count the number of bacterial and fungal colonies. Calculate colony forming units (CFU). Pure culture of theses can be developed later and identification can be done by standard methods.
- Remove the wrapper from the agar medium strip (available commercially). Pull it out by holding by the edges. Mount it on the slide holder and leave it in the area for a couple of hours. Incubate at 37°C for 24 h. Count the number of bacterial and fungal colonies. Calculate colony forming units (CFU). Pure culture of these can be developed later and identification can be done by standard methods.

2. Mechanical Method:

Requirements: Air sampler system with air sampling probe and control box, agar medium strip and incubator.

Method: Remove the impeller cup from the handle of the probe. Sterilise at 121°C for 15 minutes. Disinfect the other parts of the stainless steel body with alcohol. Set the sampling time and connect the impeller handle to the power supply. Carefully remove the wrapper from the agar strip and pull it out by holding from the edges. Insert it into the slot of the metal cup carefully with the agar surface facing the impeller and about 2 cm of the strip tab hanging out. Hold the handle and switch on the control box to start the impeller. Both the power and timer "On" lights should light up when the impeller starts collecting the sample.

Sounding of the alarm indicates end of sampling time. Switch off the control box and remove the agar medium strip by carefully pulling it out. Place it back in its wrapper in such a manner that agar surface faces away from the sliding lid. Seal the wrapper by sliding the lid or by using an adhesive tape. Mark the wrapper and incubate for 24 hrs 37°C ± 1. Observe the different fungal and bacterial colonies that appear on agar strips. Calculate colony count.

Colony count:

$$CFU/lt = \text{Colonies on agar strip}/ 40 \times \text{sampling time (in minutes)}$$

$$CFU/m^3 = \frac{\text{Colonies on agar strip}}{\text{Sampling time (in minutes)}} \times 25$$

$$CFU/ft^3 = \frac{\text{Colonies on Agar Strip}}{\text{Sampling time (in minutes)}} \times 0.708$$

19. CULTURE OF BACTERIA TO SHOW UBIQUITOUS PRESENCE OF MICROORGANISMS

Requirements: Nutrient agar, petri plates, 70% alcohol, glass marker, LAF bench, bunsen burner or spirit lamp, inoculating needle or loop.

Inoculum: Scrapings from under the nails, tarter deposits in the mouth, saliva, soil, root surface, leaf surface, raw vegetables and fruits, curds etc.

Method: Wipe workbench as well as hands with cotton, soaked in 70 % absolute alcohol. Expose medium and instruments to UV for 20 mts. Melt agar and pour in petri plates. Allow to solidify. Label prepared agar plates according to the inoculum. Prepare the suspension of respective inoculum and streak the agar plates with a loopful. Leave few petri plates open in places like storage cupboards, refrigerator, incubator etc. for few hours. Incubate the plates at 37°C for 48 h. Observe for development of bacterial and fungal colonies.

Observation:

S.No	Inoculum	Fungal growth +/-	Bacterial growth +/-
1			
2			
3			
4			

Result:

20. ISOLATION OF BACTERIA BY USING DIFFERENTIAL AND SELECTIVE MEDIA

Requirement: Petri plates, spirit lamp, inoculating needle, LAF bench and glass marker.

Inoculum: Fresh 24/48 hour cultures of *E. coli, Enterobacter aerogenes, Streptococcus faecalis, Staphylococcus aureus, Salmonella typhimurium* and other coliforms.

Media

EMB agar: Allows identification of and differentiation between *E.coli* (blue black colonies with a metallic sheen) and *Enterobacter aerogenes* (thick mucoid colonies pink in colour). Lactose and eosin as well as methylene blue differentiate enteric bacteria into lactose and non-lactose fermenters (colourless colonies which appear purple due to colour of medium). Medium is also selective for gram-negative organisms.

MacConkey agar: Selects gram-negative organisms as crystal violet inhibits gram-positive organisms. Separates coliforms into lactose fermenters (acid producers) and non-lactose fermenters. Coliforms ferment lactose into acid due to which medium changes. Medium contains pH indicator neutral red and bile salts. Acid produced by lactose fermenting coliforms changes pH of medium, bile salts are precipitated, neutral red is absorbed and medium surrounding the colony becomes red.

Phenylethyl alcohol agar: A selective medium for gram-positive cocci.

Blood agar: Blood enriches the medium and supports the growth of fastidious streptococci. It is also used to differentiate bacteria into gamma hemolytic (do not lyse RBC, therefore no change in colour), alpha hemolytic (cells cause incomplete hemolysis of RBC and convert hemoglobin to methhemoglobib which imparts a greenish colour to medium surrounding the colony) and beta hemolytic (cells cause complete hemolysis resulting in removal of colour from medium surrounding the colony).

Mannitol salt agar: Due to its high salt concentration (7.5%), this medium allows the growth of staphylococci and inhibits other bacteria. It is also a differential medium as it differentiates mannitol fermenting bacteria on the basis of their ability to ferment mannitol into acid. Medium also contains phenol red, a pH indicator that changes colour. pH change is reflected by development of yellow halo around growth.

Method:

1. With the help of a glass marker, divide the plate into definite sections by marking the underside of the plate.
2. Label the section according to the organism to be inoculated.
3. Pour respective media into petri plates and allow it to solidify.
4. Inoculate each section with a different type of bacterial inoculum by making a single straight streak across the medium.
5. Incubate the plates in inverted position at 24–72 hors at 37°C.
6. Examine each plate after every 24 h and note time of appearance of growth, colony shape and colour, change in colour of medium, amount of growth as absent, poor, moderate or abundant.
7. Tabulate the results.

Observations:

Medium	Organism	Time taken for growth	Colony morphology	Change in medium
McConkey agar				
Blood agar				
Phenyl ethyl alcohol agar				
Mannitol salt agar				
Eosin methylene blue agar				

Result:

Indicate the selective and/or differential nature of each media used according to the growth.

STUDY OF BACTERIAL GROWTH

Growth means an increase in number/ mass/ size of cells. It is brought about by an increase in cell contents. In unicellular organisms this usually results in increase in number of cells, therefore measurement of cell number is frequently used to assess growth. In bacteria, the cell cycle does not have separate individual phases instead all processes such as macromolecular synthesis, cell elongation and DNA replication take place simultaneously. Reproduction in bacteria is mostly by binary fission, where each mother cell divides to form two daughter cells. The time taken by a population of bacterial cells to double its size (numbers) is called **doubling time** or **generation time**. This is the unit of measurement of growth rate. Growth rate, therefore is the time taken by all the cells to reproduce and depends on several physiological factors. Microbial growth has four sequential phases that reflect the standard growth curve.

1. **Lag Phase:** During this phase the number of cells does not increase. However, considerable metabolic activity is occurring as the cells prepare to grow. They absorb nutrients from the medium and increase in size.

2. **Logarithmic or exponential phase:** Now bacterial division begins and therefore cell number increases exponentially. In each generation time, the number of cells in a population increases by a factor of two. Number increases slowly at first, then extremely rapidly. Log growth is limited by depletion of nutrients, accumulation of metabolic wastes, oxygen depletion, pH change, etc.

3. **Stationary Phase:** In this phase there is no net increase in cell number. The cells undergo unbalanced growth, cell constituents are synthesised at unequal rates and growth rate is exactly equal to death rate. The metabolic rate decreases, changes in cells occur, they become smaller and synthesise cell components to help them survive longer periods without

growing (some may even produce endospores).The cells become resistant to environmental stresses. The signal to enter this phase may have to do with overcrowding (accumulation of metabolic byproducts, depletion of nutrients, etc.). It is also regulated by the activity of *katF* gene which synthesises protein *KatF* which in turn regulates the production of oxidising enzymes such as catalase, acid phosphatase, exonuclease, etc. The physiological changes are very significant and alter the characteristics of cells. Some cells undergo changes leading to formation of arthrospores, endospores, etc. which are morphologically different from vegetative cells.

4. **Death Phase:** The cells are no longer able to reproduce since cellular ATP reserves get depleted. Cell death also occurs exponentially. Death rate is higher than growth rate therefore cell number decreases.

Growth can be measured by following methods:

1. Enumeration of cell numbers i.e. change in number of cells/ ml/unit time.
2. Estimation of change in biomass/ unit time.
3. Assay of enzyme activity before and after growth.
4. Assay of total nitrogen content before and after growth.

1. ENUMERATION OF CELL NUMBERS

An increase in number of cells in a given amount of suspension in a unit time signifies growth. Change in cell number in a bacterial suspension is measured by incubating the suspension over a period of time and number of cells /ml is enumerated after every hour. Enumeration of cell numbers can be done by any of the following ways:

- **Turbidimetric method;** Cell concentration in bacterial suspension is measured as a function of light transmitted and result is expressed in terms of O.D. of solution.

- **Cell counting by hemocytometer:** Number of cells present/ml of bacterial suspension is counted / unit time with the help of a hemocytometer.

- **Standard plate count method:** 1 ml of bacterial suspension is streaked on nutrient medium and number of colonies developed after incubation are counted. The process is repeated for every hour sample. Change in cell number is expressed in terms of increase in cfu/ml/unit time. It also gives the viable count.

The results of each method are correlated and in totality are used to find the cell number in a given bacterial suspension.

Requirements: 24 hr cultures of bacterial isolate or unknown sample, culture tubes, petriplates, nutrient agar and broth, incubator, inoculation loop, spirit lamp, spectrophotometer, LAF Bench, hemocytometer, colony counter, Pasteur or transfer pipettes, balanced salt solution (BBS) or (PBS), trypan blue.

Method:

Serially dilute the bacterial suspension (see chapter 4 for details). Take 10 culture tubes and fill them with 5 ml nutrient broth. Number them accordingly as zero h, 1h, 2h, ... 12h. Keep one tube as blank (only medium and no inoculum). Inoculate the tubes with 1 ml inoculum from a specific dilution. **Estimate cell number from same tube and of every tube by all three methods simultaneously** so that the results can be correlated.

Number of cells in zero h tube are enumerated immediately. Rest of the tubes are incubated at 37°C. Cell number enumeration from these tubes is done after every successive hour.

Turbiditric method: Switch on the spectrophotometer one hour before starting the experiment. Set 100 % transmission with blank and immediately measure the optical density of the 1st tube i.e. zero h tube immediately at 600nm. Measure the optical density of subsequent tubes after every successive hour. As the number of cells increase the % transmission of light through broth decreases which is reflected as increase in O.D. Plot a graph between O.D and time.

Cell counting by hemocytometer: Number of cells in 1ml of zero h broth is counted immediately and then subsequently cells from the respective tubes are counted after every successive hour (see chapter 4 and chapter 7 for details).

Standard plate count method: 1 ml broth from zero h tube is immediately streaked on a sterile agar plate under sterile conditions. Similarly 1ml from subsequent tubes is streaked after every successive hour. All plates are incubated at 37°C for 24 and 48 hours. Number of colonies developed are counted and cfu / ml is calculated as shown in chapters 4 and 7.

Correlate the three results. O.D values give cell concentration/ml of the bacterial suspension. Counting by hemocytometer provides number of cells/ml in the same suspension and standard plate count provides the estimate of total viable cells present in the suspension. Repeat the procedure with other dilutions.

An increase in O.D, number of cells and cfu/ml signifies growth.

Important: O.D, number of cells and cfu/ml must be of same dilution.

Observations:

Tube .no	Number of cells	Optical Density	CFU/ml
Zero hour			
1 hour			
2 hour			

2. DETERMINATION OF VIABLE COUNT

Requirements: 24-hour broth culture of *Pseudomonas*, *Bacillus*, *Staphylococcus*, sterile culture media, inoculation loop, spirit lamp, culture tubes, incubator, colony counter.

Method:

Prepare serial dilutions of 24-hour broth culture. Using pour plate method, inoculate nutrient agar plates with 0.1 and 1ml suspension into petri dishes labelled according to the dilutions used. Incubate the plates in inverted position at 37 °C for 48 h. After incubation period, count the number of colonies by colony counter or a manual counter. Discard plates having spreaders, more than 300 colonies (too numerous to count: TNTC), less than 30 colonies (Too few to count) or colonies merging with each other. Count the colonies in plates having 30–300 colonies.

Calculate number of viable cells/ml by the following formula:

$$\text{No. of cells/ml} = \text{No. of colonies} \times \text{dilution factor}^*$$

e.g. No. of colonies = 40

Dilution factor = $1:1 \times 10^6$

Volume of sample taken = 1 ml

$$\text{No. of cells/ml} = 40 \times 1,000,000 = 40,000,000 \text{ or } 4 \times 10^7 \text{cells/ml}$$

* Dilution factor is reciprocal of dilution ∴ if dilution is 10^{-6}, dilution factor in 10^6.

3. TO PREPARE THE GROWTH CURVE OF BACTERIA

Requirements: 24-hour broth culture of given bacteria, sterile liquid and solid nutrient media, petri plates, culture tubes, inoculation loop, spirit lamp, culture tubes, incubator, spectrophotometer.

Method

Take 10 culture tubes and fill them with five ml nutrient broth. Number them accordingly. Keep one tube as blank (only medium and no inoculum). Switch on the spectrophotometer one hour before starting the experiment. Set 100 % transmission by blank. Inoculate the tubes with one ml of inoculum and immediately measure the optical density (OD) of the first tube at 600 nm. Incubate the rest at 37°C. Measure the optical density of the 2nd tube after one hour, 3rd tube after 2 h and so on.
Plot a graph between OD and time. This gives the growth curve. To find the number of cells, compare the obtained OD with reference curve drawn between OD and number of cells.

Observations:

Tube No.	Number of cells	Optical density	Incubation time

Results:

Important: The generation time taken by different bacteria varies, so at times one hour may not be a sufficient time or it may be too long. In case of known bacteria, generation time is known, therefore, incubation period can be adjusted accordingly.

Preparation of reference curve: Prepare several aliquots of nutrient medium containing known number of cells (this can be calculated by standard plate count or by hemocytometer). Take OD of these aliquots and plot a graph between number of cells and OD. This graph now indicates the OD obtained due to presence of specific number of cells in a suspension and can be used to find out the number of cells present at time T in unknown sample.

4. TO COMPARE GROWTH OF BACTERIA ON DIFFERENT MEDIA

Requirements: Twenty-four hour broth culture, inorganic synthetic broth, lactose broth, glucose salt broth, yeast extract broth, petri plates, culture tubes, inoculation loop, spirit lamp, culture tubes, incubator, spectrophotometer.

Method:

Take culture tubes and label them according to the name of medium. Fill them with 5 ml of respective nutrient medium. Keep one tube as blank (only medium and no inoculum) for each. Inoculate the tubes with 1 ml inoculum and incubate at 37°C for 24–48 hours. Switch on the spectrophotometer one hour before starting the experiment. Set 100 % transmission with blank. Blank will be different for each medium. Measure the optical density of each tube at 600 nm corresponding to respective blanks.

Observations:

S. No	Name of Organism	Synthetic broth	Lactose broth	Glucose salt broth	Yeast extract broth
1.					
2.					
3.					

Result:

Express result in terms of medium best suited for growth and comparative suitability in increasing order for each bacterium in the following manner:

Name of organism: Synthetic broth < lactose broth < glucose salt broth < yeast extract broth.

STUDY OF EFFECT OF ENVIRONMENT
ON BACTERIAL GROWTH

Although bacteria are ubiquitous, but specific species have very specific and optimal environmental preferences. Any change in the environment is reflected as either positive or negative effect on growth. Every environmental condition has a range in which there is a minimum, maximum and an optimum value.

FACTORS THAT EFFECT GROWTH

Environmental influences considered important enough to effect bacterial growth are:

Physical: Temperature, pH and osmotic balance of medium, light, gases (oxygen/carbon dioxide).

Chemical: Nutrient concentration, type, response to toxin and other additives in medium.

Biological: Interaction with other species.

Physical Factors

1. pH: On the basis of preferred medium of pH, bacteria can be classified as:

- **Acidophiles or acid-loving:** which grow best at a pH of 1–5.4; e.g. *Lactobacillus* (ferments milk).
- **Neutrophiles:** which grow from a pH of 5.4 – 8.5; include most human pathogens.
- **Alkaliphiles or base loving:** which grow best from pH 7.0 to 11.5; e.g. *Vibrio cholerae* (causes cholera).

2. Temperature: On the basis of preferred temperature, bacteria can be classified as:

- **Psychrophiles or cold loving:** which grow best between 15°C and 20°C; some can grow at 0°C.
- **Mesophiles:** which grow best between 25°C and 40°C; human body temperature. is 37 °C.
- **Thermophiles or heat loving:** which grow best between 50°C and 60°C, found in compost heaps and in boiling hot springs.

3. Moisture: Almost all microbes require moist and damp environment. Only the spores can exist in a dormant state in a dry environment.

4. Osmotic pressure (hypotonic, hypertonic, isotonic medium): On the basis of preferred medium bacteria can be classified as:

- **Halophiles or salt lovers:** inhabit the oceans, require NaCl for growth.
- **Osmotolerant:** can grow at high solute concentrations.
- **Osmophiles:** require high solute concentrations for growth.

5. Radiation: UV rays and gamma rays can cause mutations in DNA and even kill microorganisms. Some bacteria have enzyme systems that can repair some mutations.

6. Oxygen Requirements: Based on O_2 requirements, bacteria can be classified as:

- **Strict or obligate anaerobes:** Cannot tolerate oxygen, e.g. *Clostridium tetani.*
- **Strict or obligate aerobes:** Require oxygen, lack of oxygen kills the bacteria; e.g. *Pseudomonas.*
- **Facultative anaerobes:** Can exist as anaerobes if oxygen is absent or aerobes if oxygen is present. e.g. *E. coli, Staphylococcus.*
- **Aero tolerant:** Do not use oxygen, but oxygen doesn't harm them; e.g. *Lactobacillus.*
- **Microaerophiles:** Require very low oxygen concentrations and higher carbon dioxide concentrations. e.g. *Campylobacter.*

7. Nutritional or Biochemical Factors: Microorganisms require the following nutrients for growth:
- **Carbon:** as an energy source and for building blocks.
- **Nitrogen:** for amino acids and nucleotides; some can synthesise all the 20 amino acids; others need to be provided some in their medium.
- **Sulphur:** needed for amino acids, coenzymes.
- **Phosphorus:** required for synthesis of ATP, phospholipids, and nucleotides.
- **Vitamins:** required in small amounts and used as a coenzyme. Some bacteria can synthesise their own, but some require vitamins as supplements in the medium. Symbionts of human intestine manufacture vitamin K, needed for blood clotting and B vitamins.
- **Trace elements:** Copper, iron, zinc, sodium, chloride, potassium, calcium, etc. serve as cofactors in enzymatic reactions.

Deficiency or shortage of any of nutrient limits the growth. The amount of nutrient essential for growth can be found out by using **assay medium**. This is a **defined medium** deficient in a specified nutrient. Small amounts of the essential nutrient is added gradually in the medium and growth response as well as acid production is observed. The amount of nutrient can be calculated by titration of the acid with a base till neutral end point. The amount of acid produced is directly proportional to the concentration of nutrient.

Nutritional classification of microbes depends on:

1. Method of ATP and reducing power generation.
2. The source of carbon atoms used to make precursor metabolites.

Microbes are classified into the following nutritional groups:

- **Autotrophs:** obtain carbon from carbon dioxide.
- **Heterotrophs:** obtain carbon from organic compounds like carbohydrates, proteins, lipids, etc.
- **Chemotrophs:** obtain carbon from oxidation of inorganic compounds like sulphur nitrite, ammonia, iron etc.
- **Photoautotrophs:** use solar energy. Growth is directly related to the amount of nutrients present in the medium.

Photosynthetic bacteria use sunlight energy to reduce carbon dioxide to carbohydrate. They do not use water as a source of electrons. Instead they use hydrogen sulphide to supply the electrons needed to synthesise NADPH and ATP. In the process, they produce elemental sulphur.

$$2H_2S + CO_2 \rightarrow (CH_2O) + H_2O + 2S$$

Photosynthetic bacteria contain bacteriochlorophylls and carotenoids but phycobilins are absent. The absorption spectrum of bacteriochlorophylls lies mostly in the infrared region of the spectrum so that they can trap energy missed by the green plants. **They lack Photosystem II**, which is the reason why they cannot use water as a source of electrons. Most photosynthetic bacteria are obligate anaerobes. They are found in anaerobic habitats such as in the bottom of shallow ponds and estuaries.

They are of the following types:

- Purple sulphur bacteria or Thiorhodaceae e.g. *Chromatium, Thiospirillum*, and *Thiopedia*.
- Purple non-sulphur bacteria or Athiorhodaceae e.g. *Rhodospirillum, Rhodobacter* and *Rhodomicrobium*.
- Green sulphur bacteria or Chlorobacteriaceae e.g. *Chlorobium, Prosthecoi* and *Pelodictyon*.
- Green non-sulphur bacterium or Achlorobacteriaceae e.g. *Chloroflexus*.
- Cyanobacteria: e.g. *Oscillatoria, Nostoc* etc.

Sulphur bacteria are colourless **chemoautotrophic** bacteria. Instead of solar energy, they utilise chemical energy liberated by the oxidation of reduced substances present in their environment to reduce carbon dioxide to carbohydrate. e.g. Sulphur bacteria present in the water of sulphur springs, oxidise H_2S in their surroundings to produce energy:

$$2H_2S + O_2 \rightarrow 2S + 2H_2O; \Delta G = -100 \, kcal$$

$$2H_2S + CO_2 \rightarrow (CH_2O) + H_2O + 2S$$

They are of following types:

- **Iron bacteria:** Oxidise iron compounds, responsible for the brownish scale that form inside the tanks of flush toilets.
- **Nitrosomonas:** Oxidise NH_3 to nitrites (NO_2^-).

1. STUDY OF EFFECT OF TEMPERATURE ON GROWTH

Requirements: Twenty-four hour cultures of *E. coli, Bacillus* sp. *Pseudomonas* sp. glucose broth, culture tubes, incubator and inoculation loop.

Method:

- Fill culture tubes with glucose broth.
- Prepare a set of five tubes for each organism and label them with the name of the organism and the temperature of incubation.
- Inoculate the tubes with a loopful of respective microbe.
- Incubate one tube from each set at 5, 20, 37, 45 and 55°C.
- Observe for growth after two and seven days.

Observation

S. No.	Name of organism	Growth at specific temperatures (in °C)				
		5	20	37	45	55

Result: Express result as mesophile, thesmophile or psychrophile.

2. STUDY OF LETHAL EFFECT OF TEMPERATURE ON GROWTH

Requirements: Twenty-four hour cultures of *E. coli, Bacillus* sp. *Pseudomonas* sp. *Saccharomyces cerevisiae, Aspergillus niger,* glucose broth, culture tubes, incubator, inoculation loop and water bath.

Method:

- Fill culture tubes with glucose broth.
- Prepare three sets of five tubes and label them with the name of the organism and the lethal temperature.
- Inoculate the tubes with respective organism and heat the tubes in water bath at 20°, 30°, 40°, 50°, 70° and 80°C for 10 min each.
- Remove from water bath and cool by immersing in cold water and immediately incubate at 37°C.
- In another set repeat the same procedure but do not heat the tubes. Observe both for growth after two days.

Observation:

Name of organism	Growth in unheated tubes (in °C)			Growth in heated tubes (in °C)		
	50	70	80	50	70	80

Result: Express result as lethal temperature.

3. STUDY OF HEAT RESISTANCE OF VEGETATIVE CELLS AND SPORES

Requirements: Spores of *Lactobacillus* sp., *Streptococcus* sp., *Saccharomyces cerevisiae*, *Aspergillus niger*, twenty four hour broth cultures of *Pseudomonas* sp. *Saccharomyces cerevisiae*, *Aspergillus niger*, soil suspension (may contain spores and bacteria) glucose broth, culture tubes, incubator, inoculation loop, water bath.

Method:

Fill culture tubes with glucose broth.

- Prepare two sets of culture tubes and label them with the name of organism, vegetative cell/ spore.
- Inoculate the tubes with the samples i.e. soil suspension, bacterial cultures and spores, respectively, and heat the tubes in water bath at 80°C for 10 min each.
- Remove from water bath and cool by immersing in cold water and immediately incubate at 37°C.
- In another set, inoculate the tubes but do not heat the tubes.

- Maintain a control for each set (uninoculated medium).
- Incubate both at 37°C.
- Observe all tubes for growth after two days.
- From these tubes inoculate agar slopes, incubate and observe growth after four days.
- Make gram and spore stains of tubes showing growth.

Observation:

Name of organism	Growth of vegetative cells			Growth of spores		
	Unheated	Heated	Control	Unheated	Heated	Control

Result: Express result as heat resistant spores/cells.

4. STUDY OF EFFECT OF OSMOTIC PRESSURE ON GROWTH

Requirements: Twenty-four hour cultures of *E.coli*, *Bacillus* sp. *Pseudomonas* sp. *Streptococcus* sp. glucose broth or nutrient broth, culture tubes, incubator, inoculation loop, NaCl and sucrose.

Method:

Fill culture tubes with glucose broth containing 0.5, 5, 15 and 25% solutions of NaCl.

- Similarly prepare tubes containing broth and 0.5, 15, 30 and 60% solutions of sucrose.
- Pour the media into plates and label them specifying the concentration of NaCl or sucrose.
- Divide the plate into four quadrants by marking the bottom of the plate with a glass marker.
- Inoculate each quadrant with a different organism and label the quadrant accordingly.
- Inoculate with highly diluted suspension and in very small area, to prevent overcrowding of colonies.
- Incubate at 37°C. Observe for growth after two and seven days.

Observation:

Name of organism	Growth in NaCl (in %)				Growth in sucrose (in %)			
	0.5	5	15	25	0.5	5	15	25

Result: Express result as osmotolerant, ralophilic or osmiophilic.

5. STUDY OF EFFECT OF MEDIUM pH ON GROWTH

Requirements: Twenty-four hour cultures of *E.coli*, *Bacillus* sp. *Pseudomonas* sp. *Streptococcus* sp. glucose agar or nutrient agar, culture tubes, incubator, inoculation loop, NaOH and HCl.

Method:

Readjust pH of broth to 3, 5, 7 and 9 with NaOH and HCl.

- Pour the media into respective plates and label them specifying the pH.
- Divide the plate into four quadrants by marking the bottom of the plate with a glass marker.
- Inoculate each quadrant with a different organism and label the quadrant accordingly.
- Inoculate with highly diluted suspension and in very small area, to prevent overcrowding of colonies.
- Incubate at 37 °C. Observe for growth after two and seven days.

Observation:

Organism	Growth at pH 3	Growth at pH 5	Growth at pH 7	Growth at pH 9

Result: Express result as acidophile, neutrophile or alkaliphile.

6. STUDY OF ENERGY SOURCE PREFERENCE ON GROWTH

Requirements: Twenty-four hour cultures of *Streptococcus* sp. *Bacillus*, *Aspergillus* sp. *Saccharomyces cerevisiae* culture tubes, incubator, inoculation loop and four different types of media containing different sugars: Glucose broth, Lactose broth, PDA and Sugar solution.

Method:

Label the tubes according to the name of the medium and the organism.

- Pour the respective media and inoculate with a loopful of suspension of specified bacteria.
- Incubate at 37°C. Observe for growth after two days.
- Turbidity indicates growth.
- Observation can be done visually or with the help of a **turbidimeter** or by measuring OD by a spectrophotometer.

Observation:

Name of organism	Growth on medium 1	Growth on medium 2	Growth on medium 3	Growth on medium 4

Result: Express result in terms of increasing or decreasing order of preference.

7. STUDY OF EFFECT OF ENERGY SOURCE AND ROLE OF BUFFERS ON GROWTH

Growth is mainly dependent on the availability of carbon and energy. The organic substances are broken down by extra cellular enzymes into energy, acids, gas, water and other substances. These breakdown products accumulate in the medium causing a change in pH. To overcome this problem usually buffers are added in the medium.

Requirements: Twenty-four hour cultures of *Streptococcus* sp. culture tubes, pH meter, inoculation loop, incubator, pH indicator strips and four different types of media:

1. Medium without energy source: 1% tryptone and 1% yeast extract.
2. Medium with energy source: 1% tryptone, 1% yeast extract and 1% glucose.
3. Medium with energy source and buffer: 1% tryptone, 1% yeast extract, 1% glucose and 0.5% K_2HPO_4.
4. Medium with reduced amount of energy source and buffer: 1% tryptone, 1% yeast extract, 0.1% glucose and 0.5% K_2HPO_4.

Method:

Pour the media into well-labelled culture tubes.

- Inoculate each tube with a loopful of respective bacterial suspension.
- Incubate at 37°C.
- Observe for growth after two days.
- Turbidity indicates growth.
- Observation can be done visually or with the help of a turbidimeter or by measuring OD by a spectrophotometer.
- Measure the pH of each tube by a pH indicator strip or pH meter.

Observation:

Type of medium	pH	Growth
Medium no. 1		
Medium no. 2		
Medium no. 3		
Medium no. 4		

Result: Express result in terms of growth in presence or absence of energy source and effect of buffer on medium pH.

8. MICROBIOLOGICAL ASSAY FOR NUTRIENT REQUIREMENT

Requirements: Culture tubes, inoculation loop, incubator, pH indicator strips, pipette, titration assembly, 0.1 N NaOH, indicator (bromothymol blue), spectrophotometer and Niacin requiring strain of *Lactobacillus* sp., Niacin assay medium of different niacin concentrations:

1. Medium with 0.0 µg of niacin/10 ml of medium.
2. Medium with 0.025 µg of niacin/10 ml of medium.
3. Medium with 0.05 µg of niacin/10 ml of medium.
4. Medium with 0.1 µg of niacin/10 ml of medium.
5. Medium with 0.15 µg of niacin/10 ml of medium.
6. Medium with 0.2 µg of niacin/10 ml of medium.
7. Medium with 0.3 µg of niacin/10 ml of medium.
8. Medium with 0.5 µg of niacin/10 ml of medium.

Method:

Pour the respective medium into well-labelled culture tubes.

- Inoculate each tube with a loopful of respective bacterial suspension.
- Maintain a control (uninoculated niacin assay medium).
- Incubate at 37°C.
- Observe for growth after 24 h.
- Turbidity indicates growth.
- Measure OD by a spectrophotometer at 650 nm (use control as blank).
- Plot a graph between OD and niacin concentration of medium.
- Once again incubate these tubes at 37°C for 24 h. After 24 h, add a drop of bromothymol blue to each tube.
- Titrate with 0.1 N NaOH till the colour changes to greenish blue indicating end point at pH7.
- Record the amount of NaOH used/titration.
- Plot a graph between niacin concentration of the medium and the amount of NaOH used/titration.

Observations:

Niacin concentration in medium	OD	Amount of NaOH used/titration
0.0 µg of niacin/10 ml medium		
0.025 µg of niacin/10 ml medium		
0.05 µg of niacin/10 ml medium		
0.1 µg of niacin/10 ml medium		
0.15 µg of niacin/10 ml medium		
0.2 µg of niacin/10 ml medium		
0.3 µg of niacin/10 ml medium		
0.5 µg of niacin/10 ml medium		

Result:

9. CULTURE OF PHOTOSYNTHETIC BACTERIA

Requirements: Sediment sample (for isolation), cultures of photosynthetic bacteria, culture tubes, incubator, inoculation loop, 250 ml sterile flasks, Lascelle's medium, light source and aluminum foil.

Method:

- Fill two flasks labelled as aerobic growth with shallow layer of medium (approximately one cm depth).
- Fill the other two with same medium up to the base of the neck and label as anaerobic.
- Inoculate each with 0.1 ml of bacterial culture.
- Completely cover one aerobic and one anaerobic flask with aluminum foil to prevent exposure of contents to light.
- Incubate all the four at 30 °C under light for 2–7 days till growth occurs.
- Remove foil from flasks and compare the growth and pigmentation.

To isolate photosynthetic bacteria from sediment sample, repeat the same procedure but use one ml of soil suspension as inoculum. Once growth takes place, plating on solid medium can separate bacteria. Pure cultures can be identified by standard methods.

Observations:

S.No.	Growth in light		Growth in dark	
	Aerobic medium	Anaerobic medium	Aerobic medium	Anaerobic medium

Result:

10. STUDY OF SEQUENTIAL DEPENDENCY OF BACTERIA BY WINOGRADSKY COLUMN

Requirements: Sediment sample (for isolation), cylindrical glass jar, sodium sulphate, sodium carbonate, shredded newspaper or leaves, twigs etc. and water.

Method:

- Layer the bottom of glass jar with sediment, paper or leaves, twigs etc. mixed with little sodium sulphate and sodium carbonate.
- Add another layer of mud. Repeat the process.
- The column should be tightly packed.
- Pour little water (enough to cover the layers but should not disturb them).
- Incubate in light at room temperature for few days.
- Observe for formation of different coloured layers.

Observations:

- Bottom zone will be black in colour due to H_2S formation. Activities of cellulose degrading bacteria such as *Clostridium* will result in the formation of fermentation products. *Desulfovibrio* uses these and sulphate to form H_2S.
- Next will be olive green and/or purple zone where photoautotrophs like *Chlorobium* and *Chromatium* which use H_2S and CO_2 from sodium carbonate.
- Third zone will be rust coloured due to the presence of photoheterotrophs like *Rhodospirillum* and *Rhodopseudomonas*.
- Next light brown coloured zone will be formed due to the presence of algae and sulphide oxidising microbes like *Beggiatoa*, *Thiobacillus* and *Thiothrix*.
- The uppermost watery layer will support the growth of diatoms and cyanobacteria.

11. STUDY OF OXYGEN REQUIREMENTS OF BACTERIA

Oxygen is the main hydrogen acceptor in cell respiration, therefore, it plays an important role in oxidation–reduction reactions of the cell. The presence or absence of oxygen alters the oxidation–reduction potential of the cell. Bacteria are classified into the following groups based on their response to oxygen:

1. **Aerobic:** Presence of oxygen is essential for continued growth and existence.
2. **Anaerobic:** Cannot tolerate gaseous oxygen. The presence of oxygen is lethal for them.
3. **Facultative anaerobes:** Prefer presence of oxygen, but can grow without it.
4. **Microaerophiles:** Require very low concentration of oxygen. Both absence and higher concentrations are lethal for them.
5. **Aerotolerant:** Anaerobes, which can tolerate oxygen pressures less than atmospheric pressures.

Requirements: Unknown sample (for isolation), cultures of *E.coli*, *Clostridium perfringes*, *Streptococcus feacalis*, *Lactobacillus* sp. culture tubes, incubator, inoculation loop, spirit lamp, yeast extract/tryptone agar and broth.

Method:

Fill culture tubes with broth and medium respectively.

- Boil the medium at 100°C for 10 min to expel all air.
- Cool to 42–45°C.
- Inoculate one ml of bacterial culture in both the broth and the medium (one organism/tube).
- Gently mix the inoculum and incubate at 37°C for two days.
- Observe the location and appearance of growth.

Observation:

Result:

10

STUDY OF CONTROL OF MICROBIAL GROWTH BY ANTIMICROBIAL AGENTS

EDUCATIONAL OBJECTIVES

In this chapter you will learn about:

- Various antimicrobial agents.
- Experiments to test the efficacy of antimicrobial agents.
- Experiments to study effect of antimicrobial agents.

Several methods are used to prevent microbial growth. Sterilisation reduces the number of pathogens to a level at which they pose no danger of disease. Disinfectants kill microbes present on inanimate objects but most disinfectants do not kill spores. Antiseptics are applied to living tissue and kill microbes or inhibit their growth on skin or other living tissue. Sanitisers are mainly detergents used to reduce bacterial numbers on food-handling equipment and eating utensils. The various antimicrobial agents therefore can be grouped into the following categories:

PHYSICAL AGENTS OF CONTROLS

- **Heat:** Heat denatures proteins. Refrigeration at low temperature is microbiostatic. It does not sterilise but slows down enzymes. Freezing kills most bacteria, but survivors can remain alive for long periods in the frozen state.
- **UV (ultraviolet) Light & Ionizing Radiation:** denatures DNA.

- **Membrane Filtration:** physically removes cellular organisms.
- **Osmotic Strength:** high concentrations of salt or sugar deprives cells of water and causes crenation or shrinkage.

CHEMICAL AGENTS OF CONTROLS

Several chemicals exhibit or kill microbes due to their effect on cell proteins, membranes, cell wall formation, nucleic acid structure and cell metabolism. Such chemicals are called germicides. Germicides are of the following types:

1. **Surfactants:** compounds with hydrophilic & hydrophobic parts, which break oily substances and coat the droplets to form an emulsion, which can be rinsed away. Surfactants are not germicidal by themselves.
2. **Quaternary ammonium salts:** kill microbes by disrupting membranes.
3. **Phenol and phenolics:** denature cell proteins and disrupt cell membranes.
4. **Alcohols:** disrupt lipids in cell membranes & denature proteins.
5. **Halogens:** inactivate enzymes by oxidation.
6. **Hydrogen peroxide:** denatures proteins.
7. **Heavy metals:** react with the sulphydryl groups of proteins and cause denaturation.
8. **Alkylating agents:** attach short chains of carbon atoms to proteins and nucleic acids.
9. **Dyes:** block cell wall synthesis. They effectively inhibit growth of G (+) bacteria in cultures and in skin infections. Dyes can be used to treat yeast infections.

The effectiveness of a chemical anitmicrobial agent is affected by time, temperature, pH, and concentration. Germicides can be tested in 3 ways:

1. **Phenol coefficients:** by comparing their effectiveness to phenol, a traditional germicide. The ratio of the effective dilution of the chemical agent to the dilution of phenol that has the same effect is the **phenol coefficient.**
2. **Paper disc method** – paper discs are saturated with the chemical agent and placed on the surface of an agar plate inoculated with a test organism. Clear "zones of inhibition" appear around the discs if the chemical agent is effective.
3. **Use-dilution test** – The test microbe is added to dilutions of the chemical agent. The highest dilution that remains clear after incubation indicates a germicide's effectiveness.

1. STUDY OF EFFECT OF ANTISEPTIC AND DISINFECTANT ACTION ON BACTERIA

Requirements: cultures of *E.coli, Clostridium perfringes, Streptococcus feacalis, Lactobacillus* sp. etc. petri plates, incubator, inoculation loop, spirit lamp, nutrient agar, filter paper discs, various antiseptics, disinfectants.

Method:

Mix molten agar with bacterial inoculum, and pour in a sterile petri plate (one organism / plate).

- Soak filter paper discs in antiseptic/disinfectant and place in the centre of an inoculated plate.
- Use an unsoaked disc as control.
- Repeat for all chemicals that need to be tested.
- Incubate at 37°C for 2 days.
- Observe the presence, time taken for formation and size of zone of inhibition.

Observations:

Name of organism	Antimicrobial ⚠ Agent 1	Antimicrobial ⚠ Agent 2	Antimicrobial ⚠ Agent 3	Antimicrobial ⚠ Agent 4
E.coli				
Clostridium perfringes				
Streptococcus feacalis				
Lactobacillus sp.				

⚠ Inhibition zone size/time taken

Result: Express the result in terms of effectiveness of disinfectants.

2. DETERMINATION OF PHENOL COEFFICIENT

The ratio of the effective dilution of the chemical agent to the dilution of phenol that has the same effect is the phenol coefficient. This method is used to compare the effectiveness of germicides by comparing their action with phenol (carbolic acid), the most commonly used germicide.

Requirements: Cultures of *Salmonella typhi, Staphylococcus aureus, E.coli, Clostridium perfringes, Streptococcus feacalis, Lactobacillus* sp. culture tubes, incubator, inoculation loop, spirit lamp, nutrient broth, dilutions of phenol, dilutions of chemical agent.

Method:

- Prepare several dilutions of a chemical agent as well as phenol and label with the time of inoculation, incubation and dilution.
- Inoculate these tubes with one drop of bacterial cultures and incubate the tubes at 37°C.
- Periodically (such as every five minutes) remove one loopful of liquid from each tube and introduce into tubes containing nutrient broth. Label the tubes accordingly.
- Incubate the tubes at 37°C and check for cloudiness in the tubes at periodic intervals.

- Phenol coefficient is determined by dividing the highest dilution of chemical agent being tested that destroyed microorganisms in 10 minutes with highest dilution of phenol being tested that destroyed microorganisms in 10 minutes. Both dilutions do not destroy microbes after five minutes.
- Note the dilution of disinfectant that did not kill bacteria in 5 minutes but killed them in 10 minutes.
- Similarly note the highest dilution of phenol that did not kill bacteria in 5 minutes but killed them in 10 minutes.
- Calculate phenol coefficient (PC) by following formula:

$$\frac{\text{Reciprocal of test chemical dilution noted}}{\text{Reciprocal of phenol dilution noted}}$$

For example test chemical dilution noted = 1:450 and phenol dilution noted = 1:100, then

$$PC = \frac{1/450}{1/100} = \frac{450}{100} = 4.5$$

- A disinfectant with a phenol coefficient of 1.0 has the same effectiveness as phenol. Less than 1.0 means it's less effective. Greater than 1.0 means it's more effective.

Observation:

Disinfectant	Growth in minutes			
	Dilution	**5 minutes**	**10 minutes**	**15 minutes**
Phenol	1 : 10 – 1 : 70			
	1 : 80			
	1 : 100			
Test Disinfectant	1 : 400			
	1 : 450			
	1 : 500			

Result: Express the result in terms of comparative effectiveness.

3. STUDY OF LETHAL ACTION OF UV LIGHT AND PHOTO REACTIVATION ON BACTERIA

Bactericidal effect of UV is due to its absorption by DNA resulting in production of pyrimidine dimers. Due to this bacterial DNA cannot replicate and unless DNA repair takes place, bacteria will die. DNA repair is known as photo reactivation and takes place if cells are immediately exposed to 365–450 nm wavelengths of visible light. This process activates an enzyme, which cleaves the dimmers and resumes replication.

Requirements: 24-hr cultures of common bacteria like *Salmonella typhi, Staphylococcus aureus, E.coli, Clostridium perfringes, Streptococcus feacalis,* culture tubes, petri plates, trypticase soy agar, incubator, inoculation loop, spirit lamp.

Method:

- Prepare serial dilutions of bacterial suspension. Use 10^{-6}, 10^{-8} and 10^{-10} dilutions to inoculate sterile plates containing trypticase soy agar. Incubate at 37°C. Count number of colonies and determine number of cells/ml.

- Place 2 ml suspension from 10^{-6} in four petri plates and expose these to germicidal UV light for 5, 10, 15 and 20 sec. respectively, without the cover. Take 0.1ml of bacterial suspension each from these plates and inoculate into new agar plates. Label them as irradiated for 5, 10, 15 and 20 sec. respectively.

- In another set, inoculate four agar slants with respective irradiated samples and then expose these to light from a 500 W bulb for 30 minutes. Make sure that temperature of tubes does not rise. Keep the medium cool by using a fan or placing the tubes in a beaker full of ice. Seed four plates with these photo-reactivated samples and label accordingly as photo reactivated along with the dilution used and the time of irradiation.

- Incubate all plates and tubes at 37°C for 48 hrs. Count the number of colonies in each plate. Compare the number of colonies in original sample and in irradiated and photo reactivated plates. Calculate % survivors and plot them against time of irradiation.

Observation:

Name of Organism	Effect of radiation			
	UV exposure ⚠ for 5 sec.	UV exposure ⚠ for 10 sec.	UV exposure ⚠ for 15 sec.	UV exposure ⚠ for 20 sec.

Name of Organism	Effect of reactivation			
	UV exposure ⚠ for 5 sec.	UV exposure ⚠ for 10 sec.	UV exposure ⚠ for 15 sec.	UV exposure ⚠ for 20 sec.

⚠ Number of colonies

Result: Express the result in terms of lethal exposure.

4. STUDY OF ACTION OF ANTIMETABOLITES ON BACTERIA

Antimetabolites are inhibitory chemicals which act as analogues to essential metabolites and thus interfere with either the synthesis or utilisation of these metabolites. e.g. Sulfa drugs.

   Requirements: 24-hr cultures of common bacteria like *Salmonella typhi*, *Staphylococcus aureus*, *E.coli*, *Clostridium perfringes*, *Streptococcus feacalis*, culture tubes, petri plates, minimal broth, incubator, inoculation loop, spirit lamp, and following solutions: 0.02 M sulphanilamide, 2×10^{-8} para-aminobenzoic acid, 2×10^{-7} para-aminobenzoic acid, 7×10^{-5} folic acid, 3×10^{-3} L-methionine, 2×10^{-4} Thymine, 1×10^{-4} Serine and 7×10^{-5} Xanthine.

Method:

Prepare media with the following composition and label tubes from 1-6:

1. Only minimal broth (control).
2. Minimal broth + 0.02 M sulphanilamide (antimetabolite).
3. Minimal broth + 0.02 M sulphanilamide + 2×10^{-8} para-aminobenzoic acid (metabolite required for synthesis of folic acid).
4. Minimal broth + 0.02 M sulphanilamide + 2×10^{-7} para-aminobenzoic acid.
5. Minimal broth + 0.02 M sulphanilamide + 7×10^{-5} folic acid (Vitamin for which para-aminobenzoic acid is essential).
6. Minimal broth + 0.02 M sulphanilamide + 3×10^{-3} L-methionine + 2×10^{-4} Thymine + 1×10^{-4} Serine + 7×10^{-5} Xanthine (metabolites synthesis of which requires folic acid).

Inoculate these media with 0.1ml of bacterial suspension. Then incubate for 2 days at 37°C. Observe the growth visually as well by measuring the OD.

Result and Observations:

Name of organism	Tube 1: Control	Tube 2	Tube 3	Tube 4	Tube 5	Tube 6

Tip: Addition of antimetabolite inhibits incorporation of para- aminobenzoic acid for synthesis of folic acid. Addition of para- aminobenzoic acid compensates for the inhibition. Addition of folic acid reverses inhibition. Addition of metabolites whose synthesis requires folic acid also reverses inhibition.

5. ISOLATION OF ANTIBIOTIC PRODUCING STRAIN OR SPECIES

Most antibiotics are produced by soil inhabiting Actinomycetous bacteria like *Streptomyces*, *Actinomyces* etc. and fungi like *Penicillium* etc.

Requirements: Soil sample. 24-hr cultures of common bacteria like *Salmonella typhi*, *Staphylococcus aureus, E.coli, Clostridium perfringes, Streptococcus feacalis,* culture tubes, petri plates, glucose salt agar (selective medium for *Streptomyces*), nutrient agar, incubator, inoculation loop, and spirit lamp.

Method:

- Pour glucose salts agar in petri plates and allow to solidify.
- Prepare serial dilutions of soil sample.
- Inoculate the plates with different dilutions of sample.
- Incubate for 2 days at 37°C.
- In a culture tube mix molten nutrient agar and any of the bacterial suspensions provided.
- Mix the medium with inoculum and pour it on the plate showing *Streptomyces* growth.
- Incubate at 37°C.
- Observe the plates for appearance of clear zones around some *Streptomyces* colonies.
- Isolate bacteria from these colonies on glucose/yeast extract agar plate.
- Test for comparative growth inhibition against other bacteria.
- Observe which bacteria are inhibited and extent of inhibition as well as bacteria, which show resistance to antibiotic.

Observations:

Result:

11

STUDY OF BACTERIAL GENETICS

Bacteria consist of a single chromosome present as a closed circle of double-stranded DNA with no associated histones. They mostly reproduce by binary fission thus we assume that in a pure culture all cells are clones of a single cell. But, any small change in culture conditions generates few cells, which are different from the parent cells either genotypically and/or phenotypically. Such cells are called **variants or mutants.** These changes may be **temporary or permanent.** The change may be reflected in their physiological behavior like failure to degrade a particular substance or inability to synthesize a specific growth factor or toxin, conversion of nonpathogenic stains into pathogenic and vice versa, development of antibiotic resistance etc.

Permanent variations take place due to change in genotype. Such changes are inherited and passed on to the progeny of the mutant. This leads to development of new strains and species. Changes in the genotype may also occur due to genetic recombination by conjugation, transformation and transduction processes. Fortunately, rate of mutation and genetic recombination is very slow and is seen in a very few species. Temporary change takes place in response to change in environment and is not transferred to the progeny. If the factor is removed, the cells revert back to their original behaviour or phenotype.

1. ISOLATION OF TEMPERATURE VARIANTS

Requirements: 24-hr cultures of *Serratia mascerans* and common bacteria like *Salmonella typhi, Staphylococcus aureus, E.coli, Clostridium perfringes, Streptococcus feacalis,* culture tubes, petri plates, nutrient agar, incubator, inoculation loop and spirit lamp.

Important: The temperature preference of bacteria should be known.

Method:

- Pour nutrient agar in petri plates and allow it to solidify.
- Inoculate two plates with each bacterial sample.
- Incubate one plate at normal incubation temperature and other at a temperature different from the normal range.
- Observe the effect of temperature on growth, colony characters such as colour, size etc.
- Isolate the colony showing variation and use it to inoculate fresh agar plates.
- Incubate the plates at normal temperature to see whether the change was temporary or permanent.

Observations:

Result:

2. ISOLATION OF ANTIBIOTIC RESISTANT VARIANTS

Requirements: 24-hour cultures of common pathogens like *Salmonella typhi*, *Staphylococcus aureus*, *E.coli*, *Clostridium perfringens*, *Streptococcus feacalis*, sterile culture tubes, petri plates, nutrient agar, trypticase soy agar (TSA), trypticase soy broth (TSB), incubator, inoculation loop, spirit lamp, glass rod, antibiotic (either a specific one or a broad range one).

Method:

- Pour molten TSA in the sterile petri plate and tilt it a little by inserting a glass rod or a stick under one side.
- Agar should cover the entire surface of the petri plate.
- When the agar solidifies remove the rod.
- This forms a gradient agar plate in which the depth of medium is more on one side and less on the other.
- In a tube filled with molten TSA, add 0.1 ml of antibiotic solution.
- Pour this on the surface of gradient agar plate.
- The antibiotic agar forms an uneven layer on gradient surface in which the concentration of antibiotic is proportional to the depth or thickness of agar.
- Now inoculate the plate with 0.3 ml of bacterial sample by spread plate technique. Incubate at 37°C. Isolate the resistant colonies growing on highest concentration of agar and use them to inoculate TSB.

- Incubate at 37°C and after 24 h use this suspension to inoculate tubes containing TSB mixed with 0.01, 0.05, 0.1 mg of antibiotic respectively.
- Observe for growth.

Observations:

Result:

3. ISOLATION OF NUTRITIONAL MUTANTS

Requirements: 24-hour cultures of *E.coli*, sterile culture tubes, petri plates, nutrient agar, complete nutritient broth and medium for *E.coli*, deficient medium (medium lacking one essential growth factor), incubator, inoculation loop, spirit lamp and glass rod.

Method:

- Prepare sample by first culturing bacteria in complete nutrient broth.
- Centrifuge and wash the pellet several times with sterile water to remove traces of nutrients.
- Pour deficient medium into sterile petri plates.
- Inoculate plates with bacterial sample by spread plate technique.
- Maintain a control with complete medium.
- Incubate at 37°C. Isolate the colonies growing on deficient medium and prepare a suspension in deficient broth.
- Incubate at 37°C and after 24 hours, use this suspension to once again inoculate deficient agar plates.
- Repeat the procedure 2–3 times to get a pure culture of the mutant strain.

Observations:

Result:

4. ISOLATION OF CONJUGATION MUTANTS

Conjugation in bacteria results in exchange of genetic material between two different strains leading to development of recombinants showing characteristics of both cells. It can be assumed that the recombinant cell is similar to wild type from which the mutant strains must have arisen due to changed nutritional conditions and growth requirements. For e.g. Strain C–600A is a F⁻ strain, which cannot synthesize threonine and leucine, is unable to ferment lactose and is resistant to streptomycin;

(T⁻L⁻Lac⁻Sʳ). The second strain Hfr-235B is an Hfr strain able to synthesise threonine and leucine, can ferment lactose and is susceptible to streptomycin; (T⁺L⁺Lac⁺Sˢ). Conjugation between these two will result in a wild type recombinant.

Requirements: Cultures of two different strains C–600A and Hfr-235B of *E.coli* differing in growth factor utilisation and antibiotic susceptibility. The strains can be procured from microbial culture collection centres or strains isolated in the laboratory by methods described earlier can be used, sterile culture tubes, petri plates, nutrient agar, complete nutrient broth and medium for *E.coli*, minimal medium (lacking one or other essential growth factor), Eosin Methylene Blue Agar (EMB agar), incubator, inoculation loop, spirit lamp, glass rod, streptomycin solution.

Method:

- Prepare sample by first culturing bacteria in complete nutrient broth.
- Centrifuge and wash the pellet several times with sterile water to remove traces of nutrients.
- Concentrate in sterile saline to 1/20 of the original volume.
- Spread 0.1 ml of C-600 A on minimal + streptomycin agar plate.
- Label the plate accordingly and incubate.
- Similarly, spread 0.1 ml of Hfr-235B on a second minimal + streptomycin agar plate. Label the plate accordingly and incubate.
- Add 1ml suspension of C-600 A and 0.1ml of Hfr-235B to a sterile wasserman tube. Gently rotate the tube and then agitate the solution by drawing the liquid through the pipette and letting it flow out once only in order to mix the cells.
- Let the liquid stand for 30–60 minutes.
- Now inoculate a third minimal + streptomycin agar plate with 0.1 ml of this mixture and label accordingly.
- Inoculate the plates in inverted position at 37°C for 2 days.
- Maintain control plates of minimal agar medium to check for presence or absence of chemical markers: leucine and threonine requirement and streptomycin resistance or sensitivity.
- With EMB agar plates check the lactose marker.
- Lactose fermenter will be darker and non-lactose fermenter will be pinkish or lighter coloured.

Observations:

Result:

5. ISOLATION OF TRANSFORMATION MUTANTS

Transformation is a process in which genetic recombination takes place between a naked DNA and a competent cell.

Requirements: Tryptophan requiring histidine synthesizing strain of *Bacillus subtilis*, auxotrophic strain of *B. subtilis* (unable to synthesise histidine), transformation medium (minimal medium containing tryptophan and histidine), minimal medium containing tryptophan, tryptose blood agar, saline-citrate solution, 4M NaCl solution, lysozyme, membrane filter, refrigerator, centrifuge, shaker, bent glass rod, sterile culture tubes, petri plates, incubator, inoculation loop, spirit lamp and glass rod.

Method:

DNA extraction:

- Centrifuge tryptophan requiring histidine synthesising strain of *B. subtilis*.
- Under aseptic conditions separate the supernatant and cell pellet.
- Suspend cell pellet in five ml of sterile saline–citrate solution.
- Add 2 ml of 2mg/ml solution of lysozyme, mix well and shake gently for 15 minutes at room temperature.
- Clearing of the solution indicates cell lysis.
- If required add another 0.25 ml of lysozyme and once again shake for 15 minutes.
- To the cleared mixture add five ml of 4M NaCl solution and mix well.
- Filter the solution through membrane filter of 0.45 μm pore size.
- Collect the filtrate containing the extracted DNA and refrigerate it.
- Sterility of the solution can be tested by streaking on tryptose blood agar.

Preparation of competent cells:

- Add 1 ml of 5-hour culture of auxotrophic strain of *B. subtilis* into 9ml of transformation medium.
- Aerate by keeping it on shaker for one hour.

Transformation:

- Take three tubes and label them 1, 2 and 3.
- In first tube add 0.9ml of competent cell suspension, and 0.1ml of DNA solution.
- In second tube add 0.1ml of 2mg/ml DNAse in addition to 0.9ml of competent cell suspension, and 0.1ml of DNA solution.
- In the third tube add 0.9ml of competent cell suspension and 0.1ml of sterile saline-citrate solution (this tube acts as DNA less control).
- Incubate all tubes on a shaker for one hour.

- Pour minimal + tryptophan agar into three petri plates and label them also as, 1, 2 and 3. Inoculate the plates with 0.1ml suspension from respectively numbered tubes by spread plate technique.
- Make 1/10 and 1/100 dilutions of suspension in tube 1 and use these to inoculate minimal + tryptophan agar plates labelled as 1–10 and 1–100 by spread plate method. Incubate at 37°C for 2–4 days.
- Count the number of transformants.

Observations:

Result:

12

WATER MICROBIOLOGY

A large part of our earth is covered with water and microbes are an integral part of aqueous biology. They play an important role in production of food as constituents of phytoplanktons (cyanobacteria), as well as degrading of organic matter as part of detritus food chain (heterotrophs). Usually fresh water bodies have lesser number of heterotrophs as they are nutrient-deficient but if organic matter enters these bodies from sewers, industries, etc. several types of microbes develop in it. Highly polluted water bodies become anaerobic and favour the growth of anaerobes. Anaerobic decomposition results in foul smell. These water bodies also harbour several pathogenic bacteria such as coliforms. Microbiology of water is studied in order to estimate the number of live heterotrophic bacteria in any sample and to monitor water quality for the presence of coliforms in water. The number of bacteria present in a sample is an indication of degree of contamination.

Sample collection: See Chapter 4.

Sample analysis: Analyse the sample as soon as possible after collection to minimise changes in bacterial population. Maximum gap between collection and examination of samples should not exceed more than 8 h. If it is not possible to analyse immediately, refrigerate the sample at or below 4°C. This will minimise any changes in microbial populations. Never freeze the sample. If time between collection and analysis exceeds 24 h discard the sample.

Maximum storage time for drinking water/swimming pool water is 48 h, for mineral water, distilled water and deionised water is 72 h and for water samples from natural water bodies or industrial effluent is 2–7 days.

Enumeration of bacteria in water is done by the following methods:
1. Standard plate count
2. Membrane filtration technique (MF)
3. Biochemical Oxygen Demand (BOD)

1. STANDARD PLATE COUNT (HETEROTROPHIC PLATE COUNT: HPC)

Standard plate count can be done by three methods:
- Spread plate
- Pour plate
- Membrane filtration

Spread Plate Technique

Requirements: General purpose nutrient agar or Standard Methods Agar (SMA) also called Standard Plate Count Agar (SPCA), water sample, petri plates, 0.1% peptone water and colony counter.

Method:

Pour molten agar in a petri plate and spread 1 ml of inoculum on its surface. Incubate at 35°C for 48h. Record the number of colonies that develop. See chapter 3 for details.

Observations:

Count plates with SPC counts between 20 and 300 colonies.

   Spreaders: Spreading colonies are called spreaders. If spreaders are observed on the agar plate, count bacterial colonies on representative portions only if they are well distributed in spreader-free areas and the area covered by the spreader(s) does not exceed one-half the plate area.

   Counting of spreaders: When spreading colonies must be counted, each of the following types is counted as one:
- A chain of colonies that appears to be caused by disintegration of a bacterial clump as agar and sample were mixed
- A spreader that develops as a film of growth between the agar and bottom of petri dish.
- A colony that forms in a film of water at the edge or over the agar surface
- Similar-appearing colonies growing in close proximity but not touching, provided the distance between them is at least equal to the diameter of the smallest colony. These are counted as individual colonies
- Impinging colonies that differ in appearance, such as morphology or colour, are counted as individual colonies

Calculation of cfu/ml

Multiply the total number of colonies or average number (if counting duplicate plates of the same dilution) per plate by dilution factor.

E.g. If 39 colonies are counted on the pour plate which was inoculated with 1.0 ml of the 10^{-2} dilution, the result is reported as:

$$\text{cfu/ml at } 35°C/48 \text{ h:} = \frac{3.9}{1 \times 10^{-2}} = 3.9 \times 10^2$$

When colonies on duplicate plates and/or consecutive dilutions are counted and results are averaged before being recorded, counts are rounded to two significant figures only when converting to colony-forming units.

Calculated standard plate count is reported as cfu/ml along with information about the incubation time, temperature, method and type of media used as follows:

- Standard Plate Count: (cfu/ml) at 35°C/48 h
- Method used : streak plate
- Medium used : SPAC
- Fungal growth : Absent
- Spreader or non–spreader : Non spreader

Sometimes fungi are present when SPC (HPC) is done. If there are a few isolated colonies of mold on the agar plate, which do not interfere with the reading of the colonies, the number of mold are not reported, but their presence is noted.

E.g. If 40 colonies were counted (on 0.1 ml of sample tested), the SPC (HPC) result would be:

- Standard Plate Count: (cfu/ml) at 35°C/48 h: 4.0×10^2
- Method used: Spread plate method
- Medium used: Standard Methods Agar
- Fungi also present
- Spreader

In cases where the mold growth on the plate interferes with counting bacterial colonies the count is reported with a comment as follows:

- Standard Plate Count: (cfu/ml) at 35°C/48 hours: 1.5×10^2
- Method used: Spread plate method
- Medium used: Standard Methods Agar
- Spreader

Note: Fungal growth in the sample may result in an incorrect estimate of the actual heterotrophic plate count.

If **only** fungi are growing on the plate (0.1 ml of sample tested), i.e. if there is no visible growth of bacteria, HPC is reported as:

- Standard Plate Count: (cfu/ml) at 35°C/48 h: Less than 10.
- Method used: Spread plate method

- Medium used: Standard Methods Agar
- Spreader
- Note: Fungi present

2. MEMBRANE FILTRATION TECHNIQUE (MF)

The membrane filtration technique can be used to test large volumes of sample.
- Prepare sample dilutions in the same manner as for pour plate and spread plate techniques
- Use sterilised forceps to remove the membrane filter from the package
- Place the filter on the holder base, grid side up
- Place the filter funnel onto the assembly and secure it
- Pour the sample into the funnel of the filtration unit, with minimal splashing on the funnel walls
- Vacuum the water from the surface of the filter
- Rinse the funnel with 0.1% of peptone water, and vacuum off
- Remove the membrane from the holder and place it on to the media surface. Ensure that there are no air bubbles between the filter and the media
- Incubate at 35°C ± 0.5°C for 48–72 ± 2h
- Count the number of colonies and report as cfu/ml
- Colonies can be counted by a colony counter or a stereoscopic microscope set at a 45° angle to the agar plate with the light source adjusted vertical to the colonies.
- Calculate cfu/ml as in SPC method

Indicator organisms or microbes which indicate faecal contamination are:

- **Presumptive coliforms:** usually found in sewage and faeces, but since these are also found in environment free of faeces and sewage therefore they are not true indicators.
- **Faecal coliforms:** are thermotolerant bacteria and their presence is a clear indication of contamination.
- *E.coli:* is an essential indicator of contamination.
- **Faecal streptococci:** are gram-positive and catalase negative. They mostly inhabit human intestines but may be found in food and water also.These indicate contamination by human and ruminant especially horse faeces. e.g. *Enterococcus faecalis, E. faecum, Streptococcus bovis* and *S. equis.*
- **Clostridia or members of clostridium group:** reduce sulphite to sulphide. e.g. *Clostridium perfringens.*
- *Pseudomonas aeruginosa:* is not a true indicator but since it is responsible for food spoilage therefore it is an important organism.
- **Phages:** infecting *E.coli* are a clear cut indicator of presence of these bacteria.

3. STANDARD ANALYSIS OF WATER FOR THE PRESENCE OF COLIFORMS

Requirements: Water sample, single strength lactose broth, double strength lactose broth, Endo agar or eosin–methylene blue agar, EMB agar, nutrient agar and durham tubes.

This test is performed in three steps:

Presumptive test

- Take 10 ml of sample and inoculate 10 ml of double strength lactose broth.
- Inoculate one ml and 0.1ml of sample in single strength lactose broth, respectively.
- Incubate at 37°C for 48h.
- Insert durham tubes in all culture tubes. Press them down so as to expel all air. The tubes should become filled with the medium.
- Observe at 24 and 48h for the presence of gas. This can be observed by appearance of air bubble in the Durham tubes.
- Appearance of gas in the first 24–hour period is a positive presumptive test.
- Appearance of gas in the next 24–hour period is a doubtful test.
- Absence of gas after the 48–hour period is a negative test. This indicates that sample does not have coliform bacteria.

Confirmed test

Test is performed for all samples giving positive or doubtful results.

- Streak an Endo agar or EMB agar plate with the smallest inoculum from the tubes testing positive or doubtful for presumptive test.
- Incubate at 37°C for 48h.
- If colonies are not well spread, prepare dilutions and then use for streaking.
- Observe the colonies. If colonies show typical characters of coliforms (given in Chapter 4) test is positive.

Completed test

- Pick colonies from Endo agar or EMB agar and transfer to tubes containing lactose broth and nutrient agar slant.
- Incubate at 37°C for 48h.
- Perform Gram and endospore staining.

The presence of darkish colonies exhibiting greenish metallic sheen, formation of gas in lactose broth, non-spore-forming gram-negative rods confirms the presence of coliforms. This is an indication that the water is polluted.

Tip: High concentrations of the general bacterial population may hinder the recovery of coliforms.

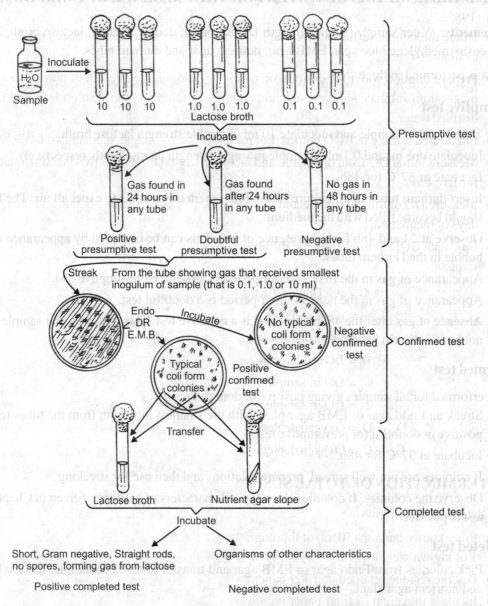

Fig. 12.1. Standard analysis of Water for the presence of Coliforms

4. DETERMINATION OF BIOCHEMICAL OXYGEN DEMAND (BOD)

Biochemical oxygen demand is the rate of consumption of oxygen by microorganisms to degrade the organic matter present in the effluent. It is a measure of organic pollution and is one of the standard parameters routinely checked for quality of potable water. BOD is measured by the method given in APHA (1985).

- Dilute the sample sufficiently so that the demand of oxygen does not exceed the amount of available oxygen.
- Prepare dilution water for dilution of sample by adding one ml each of phosphate buffer, magnesium sulphate solution, calcium chloride solution and ferric chloride in one litre of distilled water.
- Fill diluted sample in BOD bottles. For each dilution of sample, prepare two BOD bottles.
- Measure initial DO from one bottle and incubate second bottle in a BOD incubator at 20°C for five days.
- Measure DO of diluted sample and dilution water blank before incubation and then after 5 days of incubation.
- Calculate BOD in mg/l from the difference between initial and final DO by the following formula:

$$BOD = A \times \text{Dilution Factor}$$
$$A = (C_0 - C_1) - (D_0 - D_1)$$

D_0 = DO in dilution water

D_1 = DO in dilution water, after incubation

C_0 = DO in sample

C_1 = DO in sample after incubation

$D_0 - D_1$ = Oxygen depletion in dilution water

$C_0 - C_1$ = Oxygen depletion due to biomass in diluted original sample

5. DETERMINATION OF TOTAL SUSPENDED SOLIDS (TSS)

TSS is determined as follows:

- Filter known amount (50 ml) of thoroughly stirred sample through Whatman filter paper no. 1 of known weight.
- Dry the residue retained on filter paper at 103°C–105°C.
- Increase in weight of filter paper represents the amount of total suspended solids present.

$$TSS \text{ in mg/l} = \frac{(A-B)}{V} \times 10^6$$

where

A = Final weight of the dish

B = Initial weight of the dish

V = Volume of sample taken

6. DETERMINATION OF DISSOLVED OXYGEN (DO) BY AZIDE MODIFICATION METHOD

- Fix the dissolved oxygen on the spot by adding one ml each of manganous sulphate reagent and alkali–iodide–azide reagent to known amount (250 ml) of sample. Manganous sulphate reacts with potassium hydroxide and forms a white precipitate of manganous hydroxide. Oxygen present in the sample oxidises this manganous hydroxide to basic manganic oxide, which is brown in colour. Thus the manganic oxide formed is equivalent to the amount of dissolved oxygen present in the sample.

- Add two ml of concentrated sulphuric acid to the sample. In acidic medium, basic manganic oxide reacts with sulphuric acid and forms manganic sulphate. The reaction between manganic sulphate and potassium iodide liberates iodine. The amount of free iodine liberated is equivalent to the basic manganic oxide, which in turn is equivalent to the dissolved oxygen present in the sample.

- Titrate 50 ml of this sample with 0.025 N sodium thiosulphate ($Na_2S_2O_3$) using few drops of starch as an indicator. Starch combines with free iodine to form a blue coloured complex. Addition of hypo breaks this coloured complex into colourless sodium per sulphate and sodium iodide. Disappearance of blue colour signifies the end point. Dissolved oxygen is calculated by the following formula:

$$D.O. \text{ in mg/l} = \frac{8 \times 1000 \times N}{\text{Volume of sample taken}} \times V$$

where

N = Normality of titrant (N/40)
V = Volume of titrant used

7. CHEMICAL OXYGEN DEMAND (COD) BY OPEN REFLUX METHOD

- Add one gram of mercuric sulphate, several glass beads, and five ml of sulphuric acid reagent to 50 ml of sample in a refluxing flask.

- After cooling the above solution, add 25 ml of 0.0417 M potassium chromate, followed by 70 ml of sulphuric acid reagent and reflux for two hours.

- Refluxing of the sample in a strong acid solution oxidises the organic matter present in it. Dilute the refluxed solution to twice its volume with distilled water, cool and titrate with ferrous ammonium sulphate (FAS) using ferroin indicator.

- Change in colour from blue–green to reddish–brown, indicates the end point.

- Blank containing reagents and distilled water is also refluxed and titrated simultaneously.

- Titration of the refluxed solution with FAS denotes the amount of unreduced potassium dichromate left.

- Amount of potassium dichromate consumed is equivalent to the oxidisable organic matter.
- It is calculated by the following formula and the results are expressed in mg/l.

$$\text{COD in mg/l} = \frac{(A-B) \times M \times 8000}{\text{Vol. of sample taken}}$$

where

A = Volume of FAS used for blank

B = Volume of FAS used for sample

M = Molarity of FAS

8. DETERMINATION OF DISSOLVED ORGANIC MATTER AND DISSOLVED INORGANIC MATTER (DOM AND DIM) (Four hour permanganate test)

DOM and DIM indicates the amount of oxidisable organic and inorganic matter present in the sample.

- Take 250 ml each of well agitated sample in two-stoppered glass bottle.
- Add 10 ml of 25% sulphuric acid solution followed by 10 ml of N/80 potassium permanganate solution to each bottle.
- If potassium permanganate decolourises, 20–30 ml may again be added, to ensure that some potassium permanganate remains in the sample after incubation.
- Prepare a blank in other set of two bottles using distilled water instead of the sample.
- Same amount of reagents are added in all the bottles.
- Incubate both the sets of bottles i.e. two sample bottles and two blank bottles in dark at 27°C.
- Take out one set of bottles exactly after three minutes and titrate with N/80 sodium thiosulphate using potassium iodide as indicator.
- Take out second set after four hours and titrate in the same manner.

When sulphuric acid and potassium permanganate are added to the sample, oxygen is produced which is allowed to react with the organic and inorganic matter present in the sample. Excess oxygen is used to oxidise the iodide to iodine, which is then titrated against sodium thiosulphate. Inorganic salts present in the sample also utilise the oxygen produced. Therefore, one set is titrated after three minutes of incubation, which gives the amount of oxygen used immediately by inorganic salts. Whereas the other set which is titrated after four hours of incubation, gives the amount of oxygen used by both inorganic as well as organic matter. Thus the difference in values obtained by four hour sample and three minutes sample gives a measure of the organic matter present.

Calculate the amount of oxygen absorbed by the sample by the following formula and express in mg/l.

$$O_1 = \frac{A_1 - B_1}{\text{Vol. of sample taken}} \times 1000 \times 1 \times 1.6$$

$$O_2 = \frac{A_2 - B_2}{\text{Vol. of sample taken}} \times 1000 \times 1 \times 1.6$$

where

A_1 = Volume of N/80 hypo used for blank in three minutes

A_2 = Volume of N/80 hypo used for blank in four hours

B_1 = Volume of N/80 hypo used for sample in three minutes

B_2 = Volume of N/80 hypo used for sample in four hours

O_1 = mg/l oxygen absorbed in three minutes

O_2 = mg/l oxygen absorbed in four hours

Thus DIM and DOM can be calculated by the following relations:

DIM = O_1 (Amount of Oxygen absorbed in three minutes)

DOM = $O_2 - O_1$

9. IDENTIFICATION OF ENTERIC BACILLI BY IMViC TEST

Enteric bacteria are a group of bacteria found in the intestinal tract of humans and most animals of the order mammalia. These are short gram-negative rods responsible for several intestinal diseases. Some of them also constitute the normal flora of intestines and some are pathogenic. e.g. *Salmonella* and *Shigella*, which are pathogenic, *Proteus* and *Klebsiella*, which are occasional pathogens and *E.coli* and *Enterobacter* which are symbionts of intestine. Mutant strains of these can cause diseases like dysentery, diarrhoea etc. Enteric bacteria are further divided into lactose fermenters and non-lactose fermenters. *E.coli*, *Enterobacter aerogenes* and *Klebsiella pneumoniae* are lactose fermenters. *Salmonella typhimurium*, *Shigella dysenteriae*, *Proteus vulgaris*, *Pseudomonas aeruginosa* and *Alcaligenes faecalis* are non–lactose fermenters.

IMViC test is a series of tests done to identify these bacteria on the basis of their biochemical properties and enzymatic reactions. IMViC stands for:

I : Indole test

M : Methyl red test

V : Voges–Proskauer test

C : Citrate utilisation test

Method:

See Chapter 5 for methods (Identification of Bacteria).

10. DETERMINATION OF MOST PROBABLE NUMBER (MPN)

This method of counting the number of bacteria in a sample is used when the sample contains very few bacteria, which cannot be counted by standard plate count method, or the bacteria do not grow on agar. It measures the number of viable cells and is a statistical method based on probability. A series of progressive dilutions of the sample are made. At certain higher dilutions, there will be either no organisms or a single cell will be left in the sample. Observing the turbidity or colour changes in the tubes can confirm the presence or absence of bacteria. The MPN is then calculated by comparing the observed results with the table of statistical probable values. Values in MPN table are based on the assumption that cells in original table have 95% chances of being in that dilution. The number of tubes showing growth are counted and pattern of negative and positive growth is used to estimate bacterial concentration in the sample. More is the number of tubes showing positive growth at higher dilutions more is the number of cells present in the sample.

Since known amount of water and dilutions of water are added to a number of tubes arranged in a series and containing liquid indicator growth medium, this test is also known as **Multiple tube test**.

The common indicator medium used is MacConkey broth containing bromocresol purple. It indicates acid formation by development of yellow colour. Insertion of Durham tube helps to find out if gas formation is taking place or not.

Method:

- Make serial dilutions of the sample (bacterial suspension).
- Prepare five sets of three tubes each containing 10, 1 and 0.1 ml of sample of any one dilution and the liquid indicator broth.
- Insert a Durham tube in each tube and incubate at 37°C for 48 hrs.
- Observe the tubes for turbidity or gas formation after incubation.
- Compare the results with MPN table to find the number of cells in the original sample.
- MPN index/100 ml is the number of bacteria present in the sample.

Table 12.1 MPN table for combinations of positive and negative results when five tubes are used per dilution (five tubes each of 10, 1 and 0.1 ml)

			Number of tubes with positive results				
10 ml	1 ml	0.1 ml	MPN index /100 ml	10 ml	1 ml	0.1 ml	MPN index /100 ml
0	0	0	Less than 2	4	3	1	Thirty three
0	0	1	Two	4	4	0	Thirty four
0	1	0	Two	5	0	0	Twenty three
0	2	0	Four	5	0	1	Thirty
1	0	0	Two	5	0	2	Forty
1	0	1	Four	5	1	0	Thirty
1	1	0	Four	5	1	1	Fifty
1	1	0	Six	5	1	2	Sixty
1	2	1	Six	5	2	0	Fifty
2	0	0	Four	5	2	1	Seventy
2	0	1	Seven	5	2	2	Ninety
2	1	0	Seven	5	3	0	Eighty
2	1	0	Nine	5	3	1	One hundred and ten
2	2	0	Nine	5	3	2	One hundred and forty
2	3	1	Twelve	5	3	3	One hundred and seventy
3	0	0	Eight	5	4	0	One hundred and thirty
3	0	1	Eleven	5	4	1	One hundred and seventy
3	1	0	Eleven	5	4	2	Two hundred and twenty
3	1	1	Fourteen	5	4	3	Two hundred and eighty
3	2	0	Fourteen	5	4	4	Three hundred and fifty
3	2	1	Seventeen	5	5	0	Two hundred and forty
4	0	0	Thirteen	5	5	1	Three hundred
4	0	1	Seventeen	5	5	2	Five hundred
4	1	0	Seventeen	5	5	3	Nine hundred
4	1	1	Twenty one	5	5	4	Sixteen hundred
4	1	2	Twenty six	5	5	5	More than Three hundred
4	2	0	Twenty two				
4	2	1	Twenty six				
4	3	0	Twenty seven				

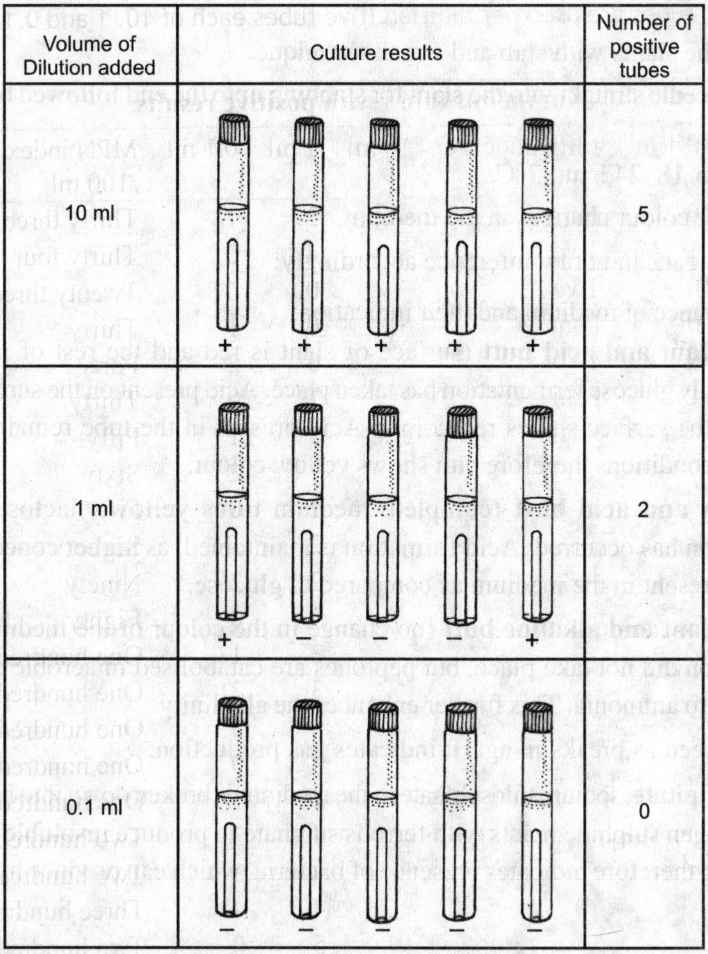

Volume of Dilution added	Culture results	Number of positive tubes
10 ml		5
1 ml		2
0.1 ml		0

Fig. 12.2. Test for Most Probable Number (MPN). Tubes are labeled (+) if there is gas accumulation in Durham tubes. These tubes contain bacteria.

11. SCREENING OF INTESTINAL BACTERIA BY TSI AGAR

Triple sugar iron (TSI) agar medium is used to differentiate between various enteric bacteria from other gram-negative intestinal bacilli on the basis of their ability to ferment different sugars into acid or acid and gas. Medium contains 1% lactose and sucrose and 0.1% glucose as well as the acid base indicator phenol red. Change in the colour of the medium indicates the preference for a particular sugar and the path of its metabolism. Red colour indicates alkaline nature and yellow shows presence of acid. Black precipitate indicates formation of hydrogen sulphide.

Requirements: 18 to 24-cultures of intestinal bacteria, TSI slants, inoculating needle and loop, spirit lamp and LAF bench.

Method:

1. Inoculate the slants with stab and streak technique.
2. Insert the needle straight into the slant for stabbing upto the end followed by streaking of the surface.
3. Incubate for 18–24 h at 37°C.
4. Observe for colour changes in the medium.
5. Record the data and draw inference accordingly:

Possible appearance of medium and their indication:

- **Alkaline slant and acid butt** (surface of slant is red and the rest of medium becomes yellow): Only glucose fermentation has taken place. Acid present on the surface gets oxidised, therefore, the surface shows red colour. Acid present in the tube remains as such due to anaerobic conditions therefore butt shows yellow colour.

- **Acid slant and acid butt** (complete medium turns yellow): lactose and/or sucrose fermentation has occurred. Acid formation is maintained, as higher concentration of these sugars is present in the medium as compared to glucose.

- **Alkaline slant and alkaline butt** (no change in the colour of the medium) carbohydrate fermentation did not take place, but peptones are catabolised in aerobic and or anaerobic conditions to ammonia. This further enhances the alkalinity.

- **Bubbles** (seen as breaks in agar): indicates gas production.

- **Black precipitate**; sodium thiosulphate in the medium is broken down into hydrogen sulphide. This hydrogen sulphide reacts with ferrous sulphate to produce insoluble ferrous sulphide. Blackening therefore indicates presence of bacteria, which can produce hydrogen sulphide.

Observation:

Organism	Colour and reaction of slant	Color and reaction of Butt	Sugar fermented	Hydrogen sulphide production

Results:

- Entire agar slant red coloured: (no change in colour) *Pseudomonas, Acinetobacter*, and *Alcaligenes*.
- Entire agar slant yellow coloured without black precipitate: *Escherichia, Enterobacter, Klebsiella*.
- Entire agar slant yellow coloured with black precipitate: *Citrobacter, Arizona* and some *Proteus* sp.
- Agar slant surface red and rest of the medium yellow coloured without black precipitate: *Shigella* and some *Proteus* sp.
- Agar slant surface red and rest of the medium yellow coloured with black precipitate: Mostly *Salmonella, Citrobacter* and *Arizona*.

12. SELECTIVE CULTURE OF E. COLI

E. coli can be selectively cultured on bromothymol blue agar containing lactose. *Shigella* is unable to utilise lactose and therefore only *E. coli* will grow on this medium.

Requirements: Bromothymol blue agar, water sample, test tubes, petri plates, spirit lamp and inoculation loop.

Method:

- Prepare serial dilutions of the sample.
- Melt agar and pour in petri plates.
- Allow it to solidify.
- Inoculate highest dilution on agar plate.

Observe for colony development.

Confirm by performing other identifying tests mentioned in Chapter 4.

13. ENRICHMENT CULTURE OF *PSEUDOMONAS spp.*

Pseudomonas can be isolated from soil, water and other habitats by enrichment culture. Before isolation, the sample should be kept at 20°C for 24 h preferably in phosphate buffer to enhance the numbers. *Pseudomonas spp.* produce pyocyanin, a water soluble pigment and fluorescein, a pigment which fluoresces in UV light on *Pseudomonas* P and F agar, respectively.

Requirements: *Pseudomonas* F and P agar, Novobiocin–penicillin–cycloheximide agar, sample, test tubes, petri plates, spirit lamp and inoculation loop.

Method:

- Prepare serial dilutions of the sample.
- Melt agar and mix in Novobiocin–penicillin–cycloheximide mixture (See Annexure III) to prepare Novobiocin–penicillin–cycloheximide *Pseudomonas* F and P agar.

- Pour into petri plates and allow it to solidify.
- Prepare sample by adding soil/water/food source to 10 ml of distilled water.
- Prepare serial dilutions of the sample.
- Inoculate highest dilution on *Pseudomonas* F and P agar.
- Observe for pigment production and colony characters.

Confirm by performing other identifying tests mentioned in Chapter 4.

If organisms produce greenish colour, which turns the medium green, it is *P. aeruginosa*.
If it produces fluorescent colours and is gelatinase positive, it is *P. fluorescens*.

If it produces fluorescent colours but is gelatinase negative, it is *P. putida*.

P. aeruginosa can be isolated on cetrimide agar also.

14. ENRICHMENT CULTURE OF *SHIGELLA*

Shigella are gram-negative, rod shaped non-motile bacteria which cause dysentery.

Requirements: MacConkey agar, EMB agar, Selenite-F broth, sewage sample, test tubes, petri plates, spirit lamp and inoculation loop.

Method:

- Prepare sample by adding one gram of sewage sample to 10 ml of distilled water.
- Melt MacConkey agar or EMB agar and pour into petri plates and allow it to solidify.
- Prepare serial dilutions of the sample.
- Inoculate 0.1 ml of MacConkey agar or EMB agar plates.
- Incubate for 48 h at 37°C.
- Develop pure cultures.
- Inoculate in Selenite-F broth.
- Incubate for 48 h at 37°C.
- Inoculate 0.1 ml on MacConkey agar.

Confirm by performing other identifying tests mentioned in Chapter 4.

15. ENRICHMENT CULTURE OF *PROTEUS*

Proteus is a gram-negative, rod-shaped motile bacterium found in soil, sewage, food products, water etc.

Requirements: Deoxycholate citrate agar, nutrient agar, sewage sample, test tubes, petri plates, spirit lamp and inoculation loop.

Method:

- Prepare sample by adding one gram of sewage sample to 10 ml of distilled water.

- Prepare serial dilutions of the sample.
- Inoculate on nutrient agar plates with 0.1 ml sample.
- Melt deoxycholate agar and pour into petri plates and allow it to solidify.
- Select swarming colony and inoculate deoxycholate citrate agar plates with it.
- Incubate for 48 h at 37°C.
- Observe for inhibition of swarming.
- Inoculate once again on nutrient agar.
- Swarming should resume.

Confirm by performing other identifying tests mentioned in Chapter 4.

16. ENRICHMENT CULTURE OF *SALMONELLA*

Salmonella is usually observed in water contaminated with feaces and causes enteric fever. It is a gram-negative, rod-shaped motile bacterium found in soil, sewage, food products, water etc.

Requirements: MacConkey agar, deoxycholate citrate agar, Wilson and Blair's medium, SS agar, sewage sample, test tubes, petri plates, spirit lamp and inoculation loop.

Method:

- Prepare sample by adding one gram of sewage sample to 10 ml of distilled water.
- Melt MacConkey and deoxycholate agar and pour separately into petri plates and allow to solidify.
- Prepare serial dilutions of the sample.
- Inoculate 0.1 ml on agar plates.
- Incubate for 48 h at 37°C.
- Observe for colourless, translucent, non-lactose fermenting colonies.
- Inoculate once again on Wilson and Blair's medium and SS agar.
- On Wilson and Blair's medium, *E.coli* are inhibited and green colonies with dark centre of *Salmonella* develop.
- On SS agar, black smooth colonies of *Salmonella* develop which gradually become rough.

Confirm by performing other identifying tests mentioned in Chapter 4.

17. ENRICHMENT CULTURE OF *VIBRIO*

Gram-negative, comma shaped, motile bacteria are known as *Vibrio*. Most important species of this group is *V. cholerae* which causes cholera.

Requirements: TCBS agar (do not autoclave medium), Modified Wilson and Blair's medium, sample, test tubes, petri plates, spirit lamp and inoculation loop.

Method:

- Prepare sample by adding one gram of sample to 10 ml of distilled water.
- Melt TCBS agar, pour into petri plates and allow to solidify.
- Prepare serial dilutions of the sample.
- Inoculate 0.1 ml on agar plates.
- Incubate for 48 h at 37°C.
- Observe for colonies.
- Inoculate once again on modified Wilson and Blair's medium.
- On Wilson and Blair's medium, non-agglutinable *Vibrio* are inhibited.

Confirm by performing other identifying tests mentioned in Chapter 4.

18. TOTAL COLIFORM TEST

Coliform bacteria can be isolated from sewage water, water contaminated with faeces etc.

Requirements: Membrane filteration assembly. M-Endo broth, sample, test tubes, petri plates, spirit lamp and inoculation loop.

Method:

- Filter sample.
- Melt agar, pour into petri plates and allow it to solidify.
- Insert a sterilised absorbent pad in the petri plates and cover it with molten agar.
- Remove the membrane from the filtration assembly with a sterile forceps and place on the pad.
- Press a little to ensure that no air is trapped in between.
- Incubate plates in inverted position 37°C.
- Observe for colonies which are pinkish with a golden metallic sheen.
- Calculate number of organisms/100 ml by the formula:

Organisms per 100 ml = 100 × No. of colonies counted/amount of sample taken

Confirm by performing other identifying tests mentioned in Chapter 4.

13

SOIL MICROBIOLOGY

Soil contains several types of pathogenic as well as non-pathogenic bacteria. Most important are the bacteria involved in biogeochemical cycles such as nitrifying and denitrifying bacteria, ammonifying bacteria etc. Apart from these, actinomycetes, the antibiotic producing bacteria, plant pathogenic fungi and bacteria, protozoa etc. are also found in soil. Soil also contains the spores of bacteria like *Clostridium*, *Bacillus* etc.

1. ISOLATION AND ENUMERATION OF SOIL BACTERIA

Requirements: Soil samples from different places such as rich garden soil, sandy soil etc. trypticase–glucose agar, inoculation loop, culture tubes, petri plates, incubator, colony counter, Gram-stain and endospore-stain.

Method:

- Add one gram of soil to 99ml of sterile water.
- Mix well and let it stand for 30 min.
- Prepare serial dilutions of this suspension and streak them on trypticase–glucose agar plates.
- Incubate at 30°C for four days.

- Count the number of colonies and calculate number of bacteria/g of soil. Make gram-stained and endospore-stained slides of soil suspension as well as bacteria taken from individual colony and observe under microscope.

- Pure cultures of these can be further developed by methods given in Chapter 4.

- Compare the results of all soil samples taken.

Observations for each sample:

Type and number of colonies:

Number of cell/g:

Gram reaction:

Endospore stain:

Result:

2. ISOLATION OF AMMONIFYING BACTERIA

Requirements: Soil samples from different places such as rich garden soil, sandy soil etc. 4% peptone solution, Nessler's reagent, trypticase–glucose agar, inoculation loop, culture tubes, petri plates, incubator, colony counter, cultures of *Bacillus cereus, Pseudomonas fluorescence, Proteus vulgaris* cultures.

Method:

- Take five test tubes and fill 15–20 ml of 4% peptone solution in them.

- To the first tube add a loopful of soil suspension prepared earlier.

- In the second tube add a loopful of *Bacillus cereus*.

- Inoculate the third and fourth with a loopful of *Pseudomonas fluorescence* and *Proteus vulgaris*, respectively. These bacteria act as positive and negative control.

- Leave the fifth as control. Incubate at 30°C for 48h.

- Observe for growth and test for the presence of ammonia with Nessler's reagent.

- Place a drop of culture on a slide and add a drop of the reagent.

- Development of yellow colour indicates presence of ammonia.

- If soil suspension tests positive to ammonia, isolate bacteria by streaking serial dilutions of soil suspension on peptone agar and develop pure cultures.

- Inoculate cells from pure cultures in 4% peptone water, incubate and test for the presence of ammonia.

- Develop cultures of only those showing positive test.

Observations:

So.N.	Name of sample	Reaction with Nessler's reagent
1	Soil	
2	*B.cereus*	
3	*P.fluorescens*	
4	*P.vulgaris*	
5	Control	

Result:

3. ISOLATION OF NITRIFYING (NITRITE FORMING) BACTERIA

Requirements: Soil samples from different places such as rich garden soil, sandy soil etc. nitrite formation medium, HCl, NaOH, Trommsdorf solution, Sulphuric acid, Nessler's reagent, Gram-stain, inoculation loop, culture tubes, glass rod, petri plates, incubator, colony counter and endospore-stain.

Method:

- Add 0.1 g of soil sample to nitrite formation medium (adjust pH to neutral of slightly alkaline) and incubate at 25°C for seven days.
- Test the culture for presence of nitrites by adding three drops of Trommsdorf solution with one drop of sulphuric acid to a drop of culture.
- Positive test is indicated by the appearance of intense blue–black colour.
- Test for ammonia with Nessler's reagent.
- Negative result for ammonia indicates complete oxidation of ammonia to nitrites. Isolate and develop pure cultures.
- Make a gram-stain preparation to observe morphology.

Observations:

S.No.	Name of sample	Reaction with Trommsdorf solution	Reaction with Nessler's reagent
1			
2			
3			
4			
5			

Result:

4. ISOLATION OF NITRIFYING (NITRATE FORMING) BACTERIA

Requirements: Soil samples from different places such as rich garden soil, sandy soil etc. nitrate formation medium, diphenylamine solution, Trommsdorf solution, Sulphuric acid, Nessler's reagent, Gram-stain, inoculation loop, culture tubes, glass rod, petri plates, incubator, colony counter and endospore-stain.

Method:

- Add 0.1 g of soil sample to nitrate formation medium and incubate at 30°C till test for nitrites with Trommsdorf solution turns negative.
- On a slide add one drop of diphenylamine solution and two drops of concentrated sulphuric acid to a drop of culture.
- Positive test is indicated by the appearance of intense blue–black colour. Negative result for nitrites is essential as diphenylamine tests positive for both nitrate as well as nitrites. Isolate and develop pure cultures.
- Make a gram-stain preparation to observe morphology.

Observations:

S.No.	Name of sample	Reaction with Trommsdorf solution	Reaction with diphenylamine reagent
1			
2			
3			
4			
5			

Result:

5. ISOLATION OF DENITRIFYING BACTERIA

Requirements: Soil samples from different places such as rich garden soil, sandy soil etc. culture of *Pseudomonas aeruginosa*, nitrate broth, nitrate–free broth, diphenylamine solution, Trommsdorf solution, Sulphuric acid, Nessler's reagent, Gram-stain, Durham tubes, inoculation loop, culture tubes, glass rod, petri plates, incubator, colony counter and endospore-stain.

Method:

- Add 0.1 g of soil sample to nitrate broth in a culture tube.
- Inoculate another tube of nitrate broth with *Pseudomonas aeruginosa*.
- Repeat the procedure with nitrate-free broth in another set of two tubes.
- Insert a Durham tube in each. Incubate at 30°C for seven days.
- Observe for gas formation indicated by rising up of durham tube.
- Test for nitrites and ammonia.
- Presence of gas indicates complete denitrification.
- Presence of ammonia or nitrites shows incomplete denitrification. Isolate and develop pure cultures.
- Make a gram-stain preparation to observe morphology.

Observations:

S.No.	Name of sample	Reaction with Trommsdorf solution	Reaction with Nessler's reagent
1	Soil 1		
2	Soil 2		
3	Soil 3		
4	Soil 4		
5	*P.aeruginosa*		

Result:

6. ISOLATION OF RHIZOBIA FROM ROOT NODULES

Rhizobial bacteria can be isolated from infected root and confirmed by growing on selective indicator media. Typical rhizobial colonies show little or no congo red absorption when incubated in dark. Bacterial growth changes the colour of Bromothymol blue to yellow showing acid production or blue indicating alkali production. No growth or poor growth is obtained on peptone glucose agar plates.

Requirements: Nodulated roots of leguminous plants, $CaCl_2$, silica gel, cotton, 95% ethanol, isopropanol, 3% (w/v) solution of sodium hypochlorite, 0.1% of mercuric chloride, sterile water, forceps, alcohol, spirit lamp, sterile petri plates, watch glass, yeast extract mannitol agar, Gram-stain, cavity slide, cover slip, vaseline, congo red, bromothymol blue and peptone glucose agar.

Method:

1. Perform all the steps under sterile conditions in a laminar flow bench.
2. Wash the roots and remove nodules leaving a small piece of root attached to it.
3. Place the nodules with desiccant such as anhydrous $CaCl_2$ or silica gel in screw cap vials for some time followed by immersing in sterile cool water overnight to rehydrate.
4. Sterilise the rehydrated nodules by treating intact and undamaged nodules in 95% ethanol or isopropanol for 1–2 min.
5. Surface sterilise the nodules by soaking for 2–4 minutes in 3% (w/v) solution of sodium hypochlorite or 0.1% mercuric chloride.
6. Wash repeatedly four–five times in sterile water.
7. Now, crush the surface sterilised nodules in a drop of water in a sterile petri dish or watch glass with a pair of heat sterilised blunt forceps.
8. Streak a loopful of nodule suspension on Yeast Extract Mannitol Agar (YEMA) medium and incubate at 25–30°C for 2-5 days.
9. Observe colony characters, cell morphology and motility by gram-staining and hanging drop method.
10. Streak isolated colonies from the primary plate on YEMA containing congo red, YEMA containing bromothymol blue and peptone glucose agar.
11. Incubate at 25–30°C for 4–10 days.
12. Perform antibiotic resistance test by spreading one ml of test suspension evenly on YEMA medium.
13. Place antibiotic discs on the medium and incubate at 28 ± 1°C for 72 h.
14. Measure the zone of inhibition.
15. Sensitivity of bacterial isolates towards a specific antibiotic is determined on the basis of zone size interpretative chart (Baver, 1966: PSADST, 1984).

Observations:

Colony characters: colour, shape, size etc.:_____

Gram's reaction: Gram+/–, cell shape, size: _____

Reaction on YEMA containing bromothymol blue and peptone glucose agar: _____

7. DIFFERENTIATION OF RHIZOBIA

Variuos species of rhizhobia can be differentiated and separated from each other by their response to thiamine and pentothenate (0.01g in YEMA) and by their response to biuret utilisation (2% in YEMA). *Rhizobium* can be differentiated from *Agrobacterium* by adding congo red to YEMA as both give same reactions for all identifying tests.

Requirements: YEMA, YEMA with congo red/thiamine/pentothenate soil sample, test tubes, petri plates, spirit lamp and inoculation loop.

Method:

- Take soil from field where legumes are grown.
- Prepare soil suspension by dissolving one gram of soil in 10 ml of distilled water.
- Prepare serial dilutions of the soil suspension.
- Melt YEMA.
- Pour into petri plates and allow it to solidify .
- Inoculate highest dilution on YEMA plates.
- Observe for colony characters.
- Perform gram-staining and motility test.
- If large white colonies testing negative for gram-stain and showing flagellar motility are seen inoculate on YEMA plates and YEMA with congo red/thiamine Pentothenate/biuret additives.

Agrobacterium is indicated if there is development of pink or red colonies on YEMA supplemented with congo red.

R. melliloti tests positive for pentothenate and negative for thiamine and biuret utilisation.

R. phaseoli tests positive for both pentothenate and biuret utilisation and negative for thiamine.

R. trifoli tests positive for thiamine and negative for pentothenate utilisation.

Confirm by performing other identifying tests mentioned in Chapter 4.

8. ENRICHMENT CULTURE OF *BACILLUS*

Bacillus is a gram-negative, motile, spore-forming bacteria found in soil.

Requirements: Nutrient broth, nutrient agar, soil extract agar, soil, test tubes, petri plates, spirit lamp and inoculation loop.

Method:

- Prepare sample by adding one gram of soil to 10 ml of distilled water.
- Inoculate nutrient broth.
- Melt agar, pour into petri plates and allow it to solidify.
- Prepare serial dilutions of the broth.
- Inoculate 0.1 ml on agar plates.
- Incubate for 72 h at 30°C.
- Observe for colonies.
- For selective isolation inoculate soil extract agar.

Confirm by performing other identifying tests mentioned in Chapter 4.

9. SELECTIVE CULTURE OF *BACILLUS*

Gram-positive, non-motile, non-spore forming *bacilli* which form the normal microbial flora of human gastro-intestinal tract. It is also one of the most commercially used genus.

Requirements: SL medium, curd, test tubes, petri plates, spirit lamp and inoculation loop.

Method:

- Prepare sample by adding five ml of curd to five ml of distilled water.
- Melt agar, pour into petri plates and allow it to solidify.
- Prepare serial dilutions of sample.
- Inoculate 0.1 ml on agar plates.
- Incubate for 48 h at 37°C.
- Observe for colonies.

Confirm by performing other identifying tests mentioned in Chapter 4.

10. DETECTION OF SIDEROPHORE-PRODUCTION BY SOIL BACTERIA

Siderophores are low molecular weight molecules secreted by certain microorganisms. They have a high affinity for iron. Siderophore production by bacterial isolates can be detected by Chrome Azurol S (CAS) assay. This assay is based on the iron chelating property of siderophores. The iron dye complex is blue in colour but when siderophores bind with iron, dye is released and the free dye is yellow in colour.

Requirements: 24 hr cultures of bacterial isolate, culture tubes, petriplates, YEM agar, CAS reagent, 0.7% agar, incubator, inoculation loop, spirit lamp.

Method:

- Make a uniform lawn of bacterial isolate by spreading it over YEM agar.
- Incubate for 48 h at30°C.
- Spread a thin layer of CAS reagent in 0.7% agar on the bacterial growth and incubate once again for 24 h at 30°C.
- Formation of yellow-orange halo around the colonies indicates siderophore production.

Result:

11. DETECTION OF PSOSPHATE SOLUBILIZING ABILITY OF SOIL BACTERIA.

A large number of microbes heterotrophic microbes have the capacity to solubilize inorganic phosphate through their metabolic activities. These phosphate solubilising microbes are usually found in the rhizosphere and increase phosphate availability to plants. Phosphate solubilising ability of soil bacteria can be detected by culturing the organism on pikovskaya medium.

Requirements: 24 hr cultures of bacterial isolate, culture tubes, petriplates, pikovskaya medium, nutrient agar, incubator, inoculation loop, spirit lamp.

Method:

- Streak a loop ful of bacterial suspension on nutient agar plates and incubate for 24 h at 37°C.
- Pick up an isolated colony and spot inoculate on pikovskaya medium.
- Incubate for 48 h at 37°C.
- Observe for formation of clear zones/ halo.
- Measure the halo and colony diameter.
- Calculate solubilization efficiency by following formula:

$$SE = \frac{\text{Solubilisation diameter (Size of halo)}}{\text{Growth diameter (Colony diameter)}} \times 100$$

Result: Express result in terms of solubilizing effeciency.

14

FOOD MICROBIOLOGY

SAMPLE COLLECTION AND PREPARATION

When collecting samples of food substances few things are very important: adequate amount, condition of the sample or specimen, statistically significant number of units that comprise a representative sample, all samples should be from same lot, composition, nature, homogeneity and uniformity of the total sample mass. Proper statistical sampling procedure must be used.

1. DIRECT MICROSCOPIC EXAMINATION OF FOODS (EXCEPT EGGS)

Cooked, uncooked and preserved foods are analysed to make sure that they are not contaminated. It is also done to find out the cause of food poisoning.

Large numbers of gram-positive cocci indicate the presence of staphylococcal enterotoxin, which is not destroyed by heat treatments that destroy enterotoxigenic *Staphylococcus aureus* strains. Detection of spore forming, gram-positive rods in frozen food specimen indicates the presence of *Clostridium perfringens*, an organism that is sensitive to low temperatures. Other gram-positive, spore forming rods such as *C. botulinum* or *Bacillus cereus* may also be present in the food.

Requirements: Glass slides, wire loop, Gram-stain reagents and Microscope.

Method:

- Homogenise the sample with sterile water and prepare a thin film of 10^{-1} dilution.
- Air-dry the film and heat fix or air-dry film and fix with methanol for 1–2 min and then drain excess methanol.

- Cool to room temperature before staining.
- De-fat films of food with high fat content by immersing films in xylene for 1–2 min.
- Drain, wash in methanol, drain, and dry.
- Stain film by Gram-stain.
- Observe microscopically with oil immersion objective (95–100X) and 10X ocular.
- Examine at least 10 fields of each film, noting predominant types of organisms, especially clostridial forms, gram-positive cocci, and gram-negative bacilli.

2. DIRECT MICROSCOPIC EXAMINATION OF FROZEN OR LIQUID EGG PRODUCTS

Requirements: Microscope, slides, pipette or metal syringe, North aniline (oil)–methylene blue stain, 0.1 N lithium hydroxide, 0.85% sterile physiological saline solution, Butterfield's phosphate-buffered dilution water, xylene and 95% Ethanol.

Method:

1. Place 0.01 ml of undiluted egg material on clean, dry microscope slide and spread it evenly over a 2 cm² area.
2. Add a drop of water for uniform spreading.
3. Air dry on level surface at 35–40°C.
4. Immerse the slide in xylene for one minute followed by 95% ethanol for one minute.
5. Stain film with North aniline (oil)–methylene blue stain for 10–20 min.
6. Wash repeatedly in a beaker filled with water and thoroughly air-dry before examining (do not blot).
7. Count microorganisms observed in 10–60 fields.
8. Multiply average number per field by microscopic factor and since 2 cm² used area.
9. Express final results as number of bacteria (or clumps) per gram of egg material.

3. DIRECT MICROSCOPIC EXAMINATION OF DRIED EGG PRODUCTS

Requirements: Eggs, 0.1N lithium hyroxide, sterile physiological salt solution, sterile glass beads, test tubes, glass slides, microscope.

1. Thoroughly mix the sample and prepare 1:10 dilution by adding 11 g of material and 99 ml of diluent (0.1 N lithium hydroxide) or sterile physiological salt solution and one tablespoon of sterile glass beads.
2. Rapidly shake the tube to thoroughly agitate the diluted sample.
3. Let the bubbles escape. Now place 0.01 ml of 1:10 or 1:100 dilution of the solution on a

clean microscope slide and spread evenly over 2 cm² and follow the procedure for direct microscopic examination of eggs.

4. Multiply average number of microorganisms per field by twice the microscopic factor (since 2 cm² area was used) and multiply by 10 or 100, depending on whether film was prepared from 1:10 or 1:100 dilution.

5. Express result as number of bacteria (or clumps) per gram of egg material.

4. MICROBIOLOGY OF CHEESE

Requirements: Different types of cheese samples, Gram-stain, glass slides, test tubes, sterile water, brom-cresol-purple agar and lactose agar.

Method:

- Emulsify the cheese sample.
- Prepare serial dilutions.
- Inoculate brom-cresol-purple agar and/or lactose agar with different dilutions.
- Incubate for two days at 37°C.
- Count the number of colonies and types of colonies. Develop pure cultures.
- Prepare gram-stained slides and perform identifying tests.
- Note the aroma, texture and flavour of the different cheese samples.

Result and observations:

5. PREPARATION OF SAUERKRAUT

Requirements: Cabbage, 1% phenolphthalein, 0.1 N sodium hydroxide, methylene blue stain, uniodised table salt, wide mouthed jar, wooden board, heavy weights, cheese cloth, glass slide, burette, flask cover slip, pH paper or pH meter and spirit lamp.

Sauerkraut is prepared by making alternate layers of shredded cabbage leaves and salt. For every 100 kg of cabbage used 3 kg of salt is required i.e. (3% by weight). The layers are compressed till juice in the leaves is squeezed out. The container is covered by a soft muslin cloth and incubated at 30°C. Salt extracts the water from cabbage leaves. Acid-forming bacteria thrive in the resultant brine. Fermenting activity of these bacteria converts sugar into acids which act as preservatives.

Sauerkraut is the result of fermenting activities of *Leuconostoc mesenteroides*, *Lactobacillus plantarum*, *L. brevis* and *Enterococcus faecalis*. It is a series of combined reactions. *L. mesenteroides* initiates production of lactic acid and *L. plantarum* sustains it. When acid concentration reaches

0.7–1% *L. plantarum* takes over completely along with *L. brevis* and *E. faecalis*. Total acidity of final product is 1.5–2% of which 1–1.5% is due to lactic acid.

Method:

- Wash the cabbage leaves.
- Remove the outer leaves and shred the cabbage head.
- Note the weight of the shredded leaves.
- Take table salt equal to 3% of total weight of cabbage leaves.
- Place the cabbage and salt in alternate layers in the jar.
- Press with the wooden board to squeeze out the juice.
- Put the weight on the board and cover the jar with cheese cloth.
- Incubate the jar for 14 days at 30°C.

Periodically observe the kraut for pH, taste and smell. The kraut is fermented till the typical smell and taste develop. To check for bacteria, a methylene blue-stained slide can be prepared.

Note:

1. Odour: Acidic, earthy, spicy or putrid.
2. Colour: Brown, pink, straw yellow, pale yellow or colourless.
3. Texture: Soft (fermenting organism is *L. plantarum*), slimy (fermenting organism *L. cucumeris* is growing rapidly), rotted (spoiled by bacterial, yeast or mold action).
4. pH should be in the range of 3.1–3.7.
5. Express total acidity as % lactic acid.

Calculation of % lactic acid:

- Boil 10 ml of fermented juice with 10 ml of DW to remove carbon dioxide.
- Titrate the cooled solution with 0.1 N sodium hydroxide adding five drops of 1% phenolphthalein as indicator till pink colour appears.
- Calculate % lactic acid by the following formula:

$$\% \text{ Lactic acid} = \frac{\text{amount of NaOH} \times \text{normality of NaOH} \times 9}{\text{Weight of sample in g}}$$

$$1 \text{ ml} = 1g$$

6. PREPARATION OF BUTTER MILK

Butter milk is actually soured milk or milk in which fermentation of lactose by lactic acid bacteria leads to formation of acid and gas and coagulation of milk protein into curds. The characteristic taste and flavour depends on the starter culture used.

Several starter cultures are available commercially. e.g. *Streptococcus lactis, Leuconostoc dextranicum, L. citrovorum, etc.*

Method:

- Warm the pasteurised milk slightly and add 1% volume of starter culture.
- Mix the solution by rolling the tube or agitating the flask.
- Incubate at 21 °C till milk curdles.
- Agitate the milk to break the curd.
- Note the smell and flavour.
- Prepare a stained slide for observation of bacteria.

7. PREPARATION OF ALCOHOL FROM FRUIT JUICES

Fruit juices especially grape juice contain a very high amount of sugar. At 20–30°C, this sugar is fermented by yeasts into alcohol.

Method:

- Inoculate tubes containing apple juice, grape juice, orange juice, sugarcane juice, carrot juice etc. with 0.1 ml broth of *Saccharomyces cerevisiae var ellipsoideus.*
- Incubate at 25°C.
- Check for the presence or absence of alcoholic aroma and time taken by each type of juice. Test for alcohol by rapid methods of alcohol estimation.

Observation:

Result:

8. PRODUCTION OF WINE

Requirements: Grape juice, 48-h grape juice broth culture of *Saccharomyces cerevisiae var ellipsoideus,* 1% phenolphthalein solution, 0.1 N sodium hydroxide, sucrose, Erlenmeyer flask and stopper containing a glass tube plugged with cotton.

Method:

- In a one litre flask add 500 ml of grape juice, 20 g of sugar and 50 ml of broth culture.
- Close the flask and close the vent with cotton plug.
- Incubate for two days and then add 20 g of sucrose.
- Incubate for 21 days at 25°C.
- Titrate 10 ml of the fermented solution with 0.1 N sodium hydroxide using phenolphthalein as indicator.
- Calculate total acidity as % tartaric acid by the following formula:

$$\% \text{ Tartaric acid} = \frac{\text{Amount of sodium hydroxide used} \times \text{Normality of sodium hydroxide} \times 7.5}{\text{Weight of sample in grams}}$$

$$1 \text{ ml} = 1 \text{ g}$$

Calculate volatile acidity as % acetic acid by following formula:

$$\% \text{ Acetic acid} = \frac{\text{Amount of sodium hydroxide used} \times \text{Normality of sodium hydroxide} \times 6.0}{\text{Weight of sample in grams}}$$

$$1 \text{ ml} = 1 \text{ g}$$

9. EXAMINATION OF FOOD FOR PRESENCE OF FOOD POISONING BACTERIA

Requirements: Food samples, 15 peptone water, selenite broth, deoxycholate agar, brilliant green MacConkey agar, culture tubes, petri plates, blood agar or mannitol agar, gentamycin, Robertson's cooked meat broth, homogenizer, incubator.

1. ***Salmonella:*** Uncooked meat, eggs, poultry, cream and processed food made from these are likely to be contaminated with *Salmonella*. To test for its presence:

- Take 100 ml of 1% peptone water. Add 50 g food sample to it and homogenize completely.
- Add 50ml of this mixture to 50 ml selenite broth in two different tubes respectively.
- Incubate one tube for 24 h at 37°C and the other for 24 h at 43°C.
- Plate the cultures developed on deoxycholate agar and brilliant green MacConkey agar.
- Incubate at 37°C for 24 and 48 h.
- Observe for colonies.
- Perform identifying tests given in chapter 5.

2. **Staphylococcus aureus:** This organism contaminates synthetic creams, custards, trifles and salted foods. To test for its presence:

- Take 10g sample and homogenize it with 0.1% peptone water.
- Plate on blood agar or mannitol salt agar containing egg yolk emulsion.
- Incubate aerobically the blood agar plates for 24 h at 37°C and mannitol agar plates for 36 h at 35°C/72 h at 32°C.
- Perform identifying tests given in chapter 5.

3. **Clostridium perfringens:** Meat, poultry are especially susceptible to contamination by this bacterium. To test for its presence:

- Take 10g sample and homogenize it with 100 ml of 0.1% peptone water (10% sample).
- Prepare 10 fold dilutions.
- Melt blood agar containing 5 mg gentamycin/l.
- Pour into plates. Maintain two sets of plates for each dilution.
- Inoculate with 0.1 ml of different dilutions of sample.
- Incubate one set of plates at 37°C for 16–24 h.
- Simultaneously inoculate 2–3 ml sample in two tubes containing Robertson's cooked meat broth containing 5 mg gentamycin/l.
- Heat one tube for one hour at 100°C*.
- Incubate both heated and unheated tubes overnight at 37°C.
- Plate cultures developed in these tubes on two blood agar plates containing gentamycin.
- Incubate one plate aerobically and other anaerobically overnight at 37°C.
- Identify by performing gram-stain, colony appearance and Nagler test as given in chapter 5.

* heat sensitive strains will not grow in heated tubes.

15

MILK MICROBIOLOGY

Standard norms of purity requirements of dairy products are specified by of Indian or/and International standardisation bodies such as: ISI (Indian Standards Institution, IDF (International Dairy Federation), ISO (International Organisation for Standardisation), AOAC (Official Analytical Chemists). In order to confirm purity, the following checks need to be performed:

- Measurement of microorganisms by standard plate count at 30°C.
- Measurement of somatic cells in milk.
- Determination of phosphatase activity in milk.
- Determination of reductase activity in milk.
- Determination of peroxidase activity in milk.
- Detection of pathogenic micro-organisms such as *Staphylococcus aureus, Salmonella* spp, *Listeria monocytogenes* and *Escherichia coli.*
- Measurement of coliforms at 30°C; Enumeration of presumptive *E. Coli* content by the most probable number technique.
- Colony count technique at 44°C using membranes.

1. MEASUREMENT OF COLIFORMS IN MILK

This is done by the following methods:

- Quantitative plating method.

- Direct plating on desoxycholate, which is both a selective as well as a differential medium for coliforms. It promotes growth of coliforms only. The lactose fermenting coliforms produce red colonies and non-lactose colonies are whitish.

- Standard plate count is more accurate but is dependent on the concentration of bacteria present. Fewer the number, the more is the accuracy.

- Direct microscopic examination is simplest and less time consuming but is also less accurate. It is used when the samples contain large number of bacteria.

Method:

- Prepare serial dilutions of both raw and pasteurised milk as described in Chapter 4.
- Plate the samples containing higher concentration of bacteria on Standard plate count (1/1000, 1/10,000) agar and lesser ones on desoxycholate agar (1/10, 1/100).
- Incubate at 37°C.
- Observe number of colonies and report as cfu/ml.

Direct microscopic method:

- Take a glass slide and mark a one cm² area on it. Alternatively use a guide card for spreading milk in the specified area.
- Guide card has several one cm² areas demarcated on it. Holding it beneath the slide highlights the area in which milk should be spread.
- Take a breed pipette and draw milk sample up to 0.1 ml mark.
- Place the tip of pipette on the slide and blow out the sample.
- With the help of a sterile needle spread the sample in the demarcated area.
- Make two smears of same sample on same slide.
- Dry the film.
- Stain by methylene blue solution for two mins.
- Wash, drain and dry the slides.
- Observe under oil immersion.
- Count number of bacterial cells/microscopic field.
- Counting should be done in at least 100 fields for high-grade milk and 10 fields for low-grade milk.
- From these results calculate the number of bacteria/ml of milk.
- Clusters, clumps and chains are counted as one.

Calculation:

Diameter of microscopic field = 0.16 mm

r (radius) = 0.08 mm, therefore $\pi r^2 = 0.0064$ mm²

Area of microscopic field = $\pi r^2 = 3.1416 \times 0.0064 = 0.02$ mm^2

Area of one microscopic field = $0.02 \times 0.01 = 0.0002$ cm^2

1 divide by 0.0002 = 5000 fields in one cm^2

Since 1/100 ml of milk was spread in one cm^2, each field will cover $1/100 \times 1/5000 = 1/500,000$ ml of milk and therefore each cell in a field = 500,000/ml of milk.

0.1 ml or 1/100 ml of milk was spread in one cm^2 in a microscopic field of 0.16 mm.

Diameter, number of bacterial cells/field multiplied by 500,000 = number of bacteria/ml of milk.

Total number of cells in one/two/ten/fifty fields $\times$ 500,000 = cells/ml

2. DETERMINATION OF PHOSPHATASE ACTIVITY (PHOSPHATASE TEST)

This test was developed by Kay and Graham (1933). Phosphatase test is performed to determine whether pasteurisation was effectively done or not and to detect the possibility of addition of raw milk into pasteurised milk. It is based upon the inactivation of alkaline phosphatase, an enzyme present in raw milk by pasteurisation. The presence of phosphatase enzyme, which should have been destroyed by pasteurisation, indicates that proper heat was not used during pasteurisation and hence pathogens may be present or raw/unpasteurised milk has been added to the sample.

It can be determined by two methods:

1. Scharer method

Principle: Alkaline phosphatase enzyme in raw milk liberates phenol from a disodium phenyl phosphate substrate. Phenol reacts with 2,6-dichloroquinone-4-chloroimide (CQC) to form indophenol blue, intensity of which can be measured calorimetrically at 620 nm. The amount of phenol or phenolphthalein liberated from the substrate is proportional to the activity of the enzyme. The level of enzyme activity is quantified by the use of a standard curve.

2. Rutgers method

Principle: Alkaline phosphatase enzyme in raw milk liberates phenolphthalein from a phenolphthalein monophosphate, a very stable substrate. The amount of phenolphthalein liberated is proportional to the activity of the enzyme present. Phenolphthalein is detected by adding of sodium hydroxide. Phenolphthalein monophosphate is easily hydrolysed by alkaline phosphatase to yield free phenolphthalein.

Important: Run both positive and negative controls in either of the methods.

Negative control: Heating the sample to 90°C for one minute followed by rapid cooling. Development of colour when a test is run on the control indicates contamination of reagents or presence of interfering colouring materials or both.

Positive control: Add 0.2 ml of fresh, raw milk sample containing milk of different animals to 100 ml of milk which has been heated at 90°C for one minute, followed by rapid cooling to room temperature. This test should give a positive result.

Reagent Blank: Run a single reagent blank for each series of sample units tested.

Place 10 ml of freshly prepared buffered substrate in a 25×150 mm test tube, and proceed with the analysis by incubating the blank along with the samples.

3. Test to distinguish residual alkaline phosphatase from microbial alkaline phosphatase:

Phosphatases released by microbes are more heat resistant than alkaline milk phosphatase. Differentiation between the two can be done by pasteurisation and retesting of the sample. If there is no significant difference in the results of the test, it can be concluded that the original positive result was due to microbial phosphatase.

4. Test to distinguish residual alkaline phosphatase from reactivated phosphatase:

Reactivated phosphatase is differentiated from residual phosphatase by its enhanced reactivation when exposed to magnesium salts. The enzymatic activity of the diluted sample (1:6) is then compared to the enzymatic activity of a portion of the undiluted sample.

Interfering substances: Substances reacting directly with the 2,6-Dichloroquinone-4-chloroimide (CQC) reagent produce a background colouration in the 620 nm range (e.g. free phenol, vanillic acid) thus giving false positive results. False negative results also appear when substances such as flavouring or colouring agents inhibit the phosphatase activity (e.g. chocolate or cocoa).

Requirements: Anhydrous sodium carbonate, anhydrous sodium bicarbonate, phenol free phenyl phosphate disodium salt, trichloroacetic acid, hydrochloric acid, Calgon (Sodium hexametaphosphate, common synonyms: Metaphosphoric acid, Hexasodium salt; Glassy sodium metaphosphate; SHMP), 2,6-Dichloroquinone-4-chloroimide (CQC), phenol, ethyl alcohol, magnesium chloride $6H_2OH$, spectrophotometer or colorimeter for measurements at 620 nm, glass rods, 11 cm Whatman No. 42 paper, volumetric pipettes of 1, 2 and 10 ml, volumetric flasks and water baths capable of maintaining $34°C$, $37°C$, $63.3°C$, $90–95°C$.

Note: Do not wash glassware with detergent containing phenolic substances. Ensure that the temperature of the incubators or water baths are maintained at the recommended temperatures with $\pm 1.0\,°C$ variations.

5. Sample preparation

Ice milk and ice cream: Melt a suitable amount of product, let it stand for one hour to release trapped air, (or overnight in the refrigerator) remove any fruit and nuts, then proceed as for milk.

Milk and other fluid dairy products: Pipette one ml of sample into 25×150 mm test tubes.

Powdered whole, partly skimmed or skimmed milk: Reconstitute by dispersing 10 ± 0.1 g of dry milk powder in 90 ml of distilled water at room temperature and mix in a blender for 90 s. Let the mixture stand at room temperature for five minutes before using.

Cheese: Make a paste of 0.5 g of cheese and remove 0.5 inch below the freshly exposed surface with 0.5 ml of distilled water. In the case of a goat cheese, take 1.5 g and add 1.5 ml of distilled water. Take precautions to avoid contamination of the sample by surface phosphatases in case of soft or semi-soft ripened cheeses.

Butter: Place 1.0 g amount directly into test tubes.

6. Phenol calibration line

Place 1–10 ml of working phenol solution into separate test tubes, and dilute to 10 ml with carbonate buffer. To this, add one ml of the Calgon solution, one ml of the CQC reagent and mix. Maintain a "blank" containing only 10 ml of the carbonate buffer solution treated similarly. Place all tubes in a water bath at 37°C for 15 min to completely develop the colour. Measure the resulting blue colour with respect to "blank" at 620 nm. Calculate the slope of the phenol calibration line using absorbance values (optical density), i.e. 2-log%T divided by the μg of phenol present over the linear range and use this figure to calculate the phenol content of unknown sample units.

7. Determination of initial alkaline phosphatase activity

Include a negative control, a positive control and a reagent blank for each sample. Add 10 ml of the buffered substrate to each tube containing one ml of test solution. Mix thoroughly and incubate for one hour in a 37°C water bath with occasional swirling. After one hour incubation, add one ml of trichloroacetic-HCl reagent slowly down the side of each tube. Mix and filter through 11 cm Whatman No. 42 paper. Pipette five ml of clear filtrate into a test tube, and add one ml of Calgon solution, five ml of sodium carbonate and one ml of 2,6-dichloroquinone chloroimide (CQC) reagent. Place all tubes in a 37°C water bath for 15 min. Set spectrophotometer on zero at 620 nm, first with distilled water then with reagent blank. Measure the resulting optical density (or %T) of the negative control.

If the negative control shows a result of < or = 5 μg phenol/g for cheese or < or = 2μg phenol/ml of milk, set spectrophotometer at zero for negative control and then measure optical density (or %T). If the negative control shows a result of >5 μg phenol/g for cheese or >2 μg phenol/ml of milk, determine interfering substances.

For each determination multiply the observed absorbency (2-log %T) by a factor of 1.2 and divide this value by the slope of the phenol calibration line. This gives the phosphatase value in terms of μg of phenol/0.25 g of solid or per 0.5 ml of fluid product. Convert the final result in units of μg of phenol/g or ml of product.

If the results are negative then report them as ≤ 5 μg phenol/g for cheese or ≤ 2 μg phenol/ml of milk.

If the results are positive (indication of incomplete pasteurisation) then perform confirmation tests for interfering substances, microbial phosphatase and reactivated phosphatase.

8. Interfering substances control

Replace buffered substrate with same amount of carbonate buffer solution to which no disodium phenylphosphate has been added and repeat the phosphatase test using the same amount of test material. Appearance of blue colour in this control indicates presence of interfering substances. If phenol equivalent of the actual test using buffered substrate is greater than the phenol equivalent of the control, under-pasteurisation, contamination with raw product, or presence of microbial phosphatase is indicated .

9. Microbial phosphatase control

Repasteurise the sample at 63.3°C for 30 min, stirring frequently and then cool the sample. Now once again analyse the repasteurised sample as well as the original sample and the negative control. If there is no significant reduction in phenol equivalent of repasteurised sample then the initial result was due to the presence of heat resistant microbial phosphatase, therefore, was false positive and should be reported as negative.

Reagents used:

- **Carbonate Buffer:** Dissolve 23.0 g of anhydrous Na_2CO_3 and 20.3 g of anhydrous $NaHCO_3$ in distilled water and dilute to 2000 ml. pH = 9.80. Test carbonate buffer for deterioration by heating to 85° ± 1°C and holding at that temperature for two minutes. Add two drops of freshly prepared CQC to 10 ml of buffer and incubate at room temperature for five minutes. If any colour develops, discard buffer.

- **Buffered substrate:** Dissolve 270 ± 3 mg of disodium phenyl-phosphate (phenol-free) in 250 ml of carbonate buffer. Prepare just before use.

- **Sodium carbonate 8% solution (w/v):** Dissolve 80 g of anhydrous Na_2CO_3 in 1000 ml of distilled water.

- **Trichloroacetic-hydrochloric acid solution:** Dissolve 125 ± 1 g of TCA in distilled water, dilute to 250 ml. Just before use add 250 ml of HCl (ca 37% HCl) and mix thoroughly. Prepare fresh solution on the day of analysis.

- **Calgon solution:** Dissolve 10 g of sodium hexametaphosphate in distilled water and dilute to 100 ml. pH = approximately 6.3.

- **2,6-Dichloroquinonechloroimide (CQC) (0.02% solution (w/v)):** Dissolve 10 mg of CQC in 25 ml of absolute ethyl alcohol and add 25 ml of distilled water. Store solution in a refrigerator (in the dark), can be kept for a maximum of two days.

- **Stock phenol solution (1000 μg/ml):** Dissolve 100 ± 1 mg of dry phenol crystals in distilled water and dilute to 100 ml.

- **Working phenol solution (2 μg/ml):** Add one ml of stock phenol solution to 500 ml with carbonate buffer (9.1).
- **Magnesium chloride solution:** Prepare a solution of $MgCl_2$ by dissolving 100 g of $MgCl_2 \cdot 6H_2O$ in 25 ml of distilled water. Warm slightly and pour into a 100 ml volumetric flask. Rinse the original container several times with five ml portions of distilled water, and add the rinse to the volumetric flask. Allow the solution to cool and make up to 100 ml. This solution contains 0.1196 g of mg per ml.

3. ORGANOLEPTIC TESTS

These tests permit rapid testing and grading of poor quality milk. The test does not require any equipment but a good sense of sight, smell and taste is a must.

Procedure: Smell the milk, observe the appearance, and if required taste the milk, but do not swallow it. Look at the can lid and the milk can to check cleanliness.

Abnormal smell and taste such as barn or cow odour may be caused by: atmospheric factors, physiological factors such as hormonal imbalance, late lactation, spontaneous rancidity, presence of bacteria, chemicals or advanced acidification (pH < 6.4).

4. CLOT ON BOILING (C.O.B) TEST

Boil very little amount of milk and observe for clotting, coagulation or precipitation. If a sample fails the test, milk contains many acid or rennet producing microorganisms or the milk has an abnormal high percentage of proteins like colostral milk. Such milk cannot stand the heat treatment in milk processing and must therefore be rejected.

5. THE ALCOHOL TEST

Mix equal amounts of milk and 68% of ethanol solution in a small bottle or test tube (68 ml 96% (absolute alcohol and 28 ml distilled water). Observe for coagulation, clotting or precipitation. The test is based on instability of the proteins when the levels of acid and/or rennet are increased and acted upon by the alcohol. Also increased levels of albumin (colostrum milk) and salt concentrates (mastitis) results in a positive test.

6. THE ALCOHOL–ALIZARIN TEST

The procedure is the same as for alcohol test. Alizarin is an indicator which changes colour according to the acidity. Ready-made Alcohol–Alizarin solution is available or it can be prepared by adding 0.4 g of alizarin powder to one litre of 61% alcohol solution.

Results:

Parameter	Normal milk	Slightly acidic milk	Acidic milk	Alkaline milk
pH	6.6–6.7	6.4–6.6	6.3 or lower	6.8 or higher
Colour	Red brown	Yellowish–brown	Yellowish	Lilac
Appearance	No coagulation/lumps	No coagulation	Coagulation*	No Coagulation**

7. ACIDITY TEST

Amount of 0.1 N sodium hydroxide used to neutralise the lactic acid produced by bacteria present in raw milk is measured and percentage of lactic acid present is calculated from this. Natural acidity of milk is 0.16–0.18%. Figures higher than this signify acidity due to the action of bacteria on milk sugar.

   Requirements: A porcelain dish or small conical flask pipette (10 ml, 1 ml), burette with 0.1 ml graduations, glass rod for stirring the milk, phenolphthalein indicator solution (0.5% in 50% alcohol), 1 N sodium hydroxide solution.

Method:

Mix nine ml of milk and one ml of phenolphthalein in a dish/flask/test tube and then slowly add 0.1 N sodium hydroxide with a burette. Mix continuously till a faint pink colour appears. The amount of sodium hydroxide solution used divided by 10 expresses the percentage of lactic acid.

8. RESAZURIN TEST

Resazurin, an indicator dye is dissolved in distilled boiled water and used to test the microbial activity in a given milk sample. This is the most widely used performed test to check for hygiene and the potential keeping quality of raw milk. It can be done as 10 min (Table) test or a one hour test or a three hour test. Ten minute test is rapid and used at the milk collection centres. One hour and three hour tests are more accurate take long time and are carried out in the laboratory.

Requirements: Resazurin tablets, test tubes, one ml pipette or dispenser, water bath, Lovibond comparator with Resazurin disc 4/9.

Method:

Mix 10 ml of milk with one ml of Resazurin solution. Close the tube with a sterile stopper and mix gently. Mark the tube before incubation in a water bath. Place the test tube in a Lovibond comparator with Resazurin disk and compare it colorimetrically with blank (test tube containing 10 ml of same sample without the dye).

Readings and results: Ten min resazurin test. (See Table)

Resazurin disc No.	Colour	Grade of milk	Action
6	Blue	Excellent	Accept
5	Light blue	Very good	Accept
4	Purple	Good	Accept
3	Purple pink	Fair	Separate
2	Light pink	Poor	Separate
1	Pink	Bad	Reject
0	White	Very bad	Reject

Preparation of Resazurin solution: Add one tablet to 50 ml of distilled sterile water and store in a cool, dry and dark place as it must not be exposed to sunlight. Its shelf life is very short and therefore it should not be used for more than eight hours as it loses its strength.

9. METHYLENE BLUE REDUCTASE TEST

This test is a rapid way of determining quality of milk. The time taken by dehydrogenases produced by bacteria, to reduce methylene blue dye and decolorise it is an indicater of degree of bacterial contamination.

Requirements: Samples of raw and pasteurised milk, methylene blue solution (1:250,000), sterile screw cap test tubes, sterile pipettes, water bath, spirit lamp and marker.

Method:

- Take 10 ml of milk sample in a test tube and add one ml of methylene blue solution.
- Cap the tubes.
- Invert the tubes 25 times to mix the two solutions.
- Place the tubes in water bath at 37°C and note the time of incubation.
- After five minutes, remove the tubes, invert them once more and replace in the water bath.
- Observe the samples every 30 min for methylene blue reduction for 3–6 h.
- Note the time taken for colour change from blue to white, which indicates reduction.
- Report result as milk quality as good, fair, poor or very poor.
- Maintain two controls. One of the controls is 10 ml milk which has been held at 100°C for 3 min with 1 ml methylene blue. Heating inactivates the reducing system. The other control

is 10ml milk plus 1 ml tap water.

Observations:

Milk sample	Reduction time	Milk quality
Raw milk		
Pasteurised milk		

16

STUDY OF VIRUSES

Virus means poison in Latin. They are also called "a piece of bad news wrapped in a protein", or "Hereditary material in search of cell". Virus are obligate intracellular parasites i.e. they can reproduce/ replicate only inside a host cell. They lack cell wall or cell membrane, cytoplasm and nucleus. They contain nucleic acids (DNA or RNA) enveloped in a protein coat called capsid. They enter into a host cell and use its metabolic machinery to replicate. They can infect all cellular organisms but are species specific.

Animate characteristics of virus: Replication, presence of nucleic acids and enzymes (in a few viruses e.g. Retro viruses), adaptation to changing environments and mutation, pathogenicity and species specificity. They can evolve, contain very few macromolecules which direct their own reproduction.

Inanimate characteristics of virus: Absence of cytoplasm, a cell membrane, organelles, ribosomes, or a nucleus, independent metabolism and inability to produce ATP. Viruses can be crystallised.

Size: Viruses are submicroscopic and their size is measured in nanometers. It ranges from about 1/10th to 1/3rd the size of a small bacterial cell.

Structure: Viral particle is made up of a nucleic acid surrounded by a protein capsid. In some viruses, an envelope derived from host's membrane is found outside of the capsid. Viruses that lack envelopes are called naked viruses. A complete viral particle (= capsid plus nucleic acid plus envelope if it is present) is called a virion. Viruses have DNA or RNA which may be double stranded (ds) or single stranded (ss). RNA viruses can be (–) sense or (+) sense types. (+) sense RNA acts like mRNA and can be translated into proteins by the host cell's ribosomes. (–) sense RNA is so called because it does not make sense to the host cell's ribosomes and when it enters the host cell, a complementary (+) sense strand is made from its (−) sense strand. Only (+) sense strand RNA can be read by the host cell's ribosomes. Capsid or protein coat that surrounds the nucleic acid is made up of capsomeres.

Initially viruses were cultivated by infecting plants and animals. By 1930 scientists began to use chick embryo to cultivate virus. By 1950's cell culture and tissue culture methods were developed. Now viruses are cultured in continuous cell lines derived from cancer cells/tissues. Most commonly used are Hela cell lines (Hela comes from the name of Helen lock).

Viral envelope: A lipoproteinaceous membrane surrounds some viruses. Such viruses are called enveloped viruses. Viruses without envelope are called naked viruses. This membrane is derived from host membrane during budding. It protects viruses from host antibodies and also helps in fusion of viral particle with host cells.

Symmetry of viral particles: Virus exists in three basic shapes depending on the arrangement of capsomeres:

- **Helical:** Capsomeres are arranged in a spiral form as a rod-shaped structure.
- **Polyhedral:** Capsomeres are arranged in equilateral triangles that fit together to form a geodesic dome-shaped structure, which appears almost spherical.
- **Complex:** This is a combination of both types of symmetry where the particle has a helical portion (tail) attached to a polyhedral portion (head); e.g. many bacteriophages along with appendages such as tail sheath plate, pins, and tail fibres.

Viruses that infect bacteria are called **bacteriophage**, which meaning "**bacteria eating**". Viruses replicate by either **Lytic Cycle** in which they multiply within the cell and are released by lysis of cell, or by **Lysogenic Cycle** (Lysogeny or Temperance), in which their genome is inserted into the host genome. A temperate phage is a phage capable of integrating with the host genome. The integrated viral DNA is called a prophage. In this case no new viral components are synthesised and the host cell is not harmed. The virus may remain latent for long periods of time before initiating a lytic cycle. Viral nucleic acid replicates along with the host cell's chromosome and is passed to daughter cells during cell division. Expression of viral genes changes the host cell's phenotype (e.g. diphtheria is caused by lysogenic strains of *Corynebacterium diphtheriae* because the disease-causing toxin is encoded in the prophage of the infecting virus). Any environmental shock can result in the release of viral genome, which can revert to lytic or lysogenic form.

Viral nomenclature: Viral names usually indicate the host or disease or its symptoms. Family names end in *viridae* and genus names end with virus. Species names are English words. E.g. Retroviridae: Human immunodeficiency virus. Virus are also classified on the basis of structure, size, nucleic acid, capsid structure, capsomeres type and number, presence and absence of envelope and host range.

Viroids are circular molecules of ssRNA without a capsid. They cause several economically important plant diseases but are not known to infect animals. They are 1/10 the size of the smallest plant virus. E.g. Potato spindle tuber disease.

Prions are infectious agents composed only of protein. They affect the central nervous system. They cause Scrapie disease of sheep, Creutzfeldt–Jakob disease (CJD) of humans and mad cow disease.

Viruses are responsible for several plant and human diseases, cancers and terato, but they are extremely host-specific. Their specificity extends to host cells also. The specificity resides in receptor sites on host cell surface and specific attachment sites on virus surface.

EXAMPLES OF SOME COMMON VIRAL PATHOGENS

- **Herpes simplex 1:** Causes herpes.
- **Varicella Zoster:** Causes chickenpox
- **HIV** (Human Immunodefficiency Virus): ssRNA virus, Retrovirus (*Retro* means backward), causes AIDS (Acquired Immune Deficiency Syndrome) and infects T cells. Capsid contains two copies of the same (+) sense RNA molecule, hence called a diploid virus, and the enzyme reverse transcriptase. The enzyme **reverse transcriptase** makes DNA from RNA, which can be integrated into the host cell's chromosome. DNA is transcribed into mRNA and translated into viral proteins to assemble new viruses for release. The provirus can stay in a latent stage in which it replicates along with host cell DNA, causing the host cell no damage but the person becomes a carrier.
- **Rubella** virus causes Rubella or German measles.
- **Enterovirus** causes poliomyelitis.
- **Rhinovirus** causes common cold.
- **Hepatovirus** causes Hepatitis A.
- **Influenza Viruses:** These are of 3 types: A, B and C of which A is the most common. It infects animals and humans. It is also responsible worldwide epidemics. B and C infect only humans and do not cause pandemics.
- **Rhabdovirus** causes rabies.
- **Paramyxoviruses** cause mumps, measles, viral pneumonia and bronchitis.
- *Hantavirus* causes "4 corners disease".
- **Ebola virus** is one of the most dangerous viruses and is responsible for a large number of deaths in African countries.
- **Rotavirus** causes diarrhoea in infants and young children under the age of two.
- **Adenoviruses** are responsible for human respiratory diseases and also causes diarrhoea in babies and young children.
- **Simplex virus** causes Herpes simplex 1 (oral) and 2 (genital and neonatal).
- **Varicellovirus** causes Varicella zoster—chicken pox and shingles.

- **Roseolovirus** causes Roseola infantum—roseola in infants (rash and fever).
- **Lymphocryptovirus**—Epstein Barr virus—causes infectious mononucleosis and Burkitt's lymphoma; also linked to Hodgkin's disease.
- **Orthopoxvirus** causes small pox and cowpox.
- **Papillomaviruses** cause warts (some associated with cervical cancer).
- **Canine Parvovirus** causes severe and sometimes fatal gastroenteritis in dogs.
- **Retrovirus** (B19) causes 5th disease (*Erythema infectiosum*): development of deep red rash on children's cheeks and ears and both a rash and arthritis in adults; can cross placenta and damage foetus.

Examples of Viral Cancers:

- Human T-cell leukaemia (blood cancer).
- *Epstein–Barr virus* causes Burkitt's lymphoma.
- *Hepatitis B virus* causes hepatocellular carcinoma (liver cancer).
- *Human papillomavirus* causes skin and cervical cancers.
- Kaposi's sarcoma is thought to be associated with *Herpes virus*.

Examples of Teratogenic viruses:

Teratogenesis means induction of defects during embryonic development and **Teratogen** is a drug or other agent that induces such defects. *Cytomegalovirus* (CMV), *Herpes Simplex virus* (HSV), and *Rubella virus* account for a large number of teratogenic effects.

1. ISOLATION OF BACTERIOPHAGES FROM RAW SEWAGE BY PLAQUE CULTURE

Requirements: Raw sewage, deca broth (contains nutrients at ten times higher concentration than normal), 24-h culture of *E.coli*, petri plates, nutrient agar, soft nutrient agar (0.7% agar), incubator, inoculation loop, spirit lamp, sterile culture tubes, centrifuge and membrane filter.

Method:

Enrichment culture: Inoculate 5 ml of deca broth with 45 ml of raw sewage sample and 5 ml of 24-hour culture of *E.coli* and incubate at 37°C. This increases the number of phages.

Phage separation: Centrifuge 10 ml of sewage culture at 2500 rpm for 10 min. Filter this liquid through a membrane filter with pore size of 0.45 mm. Clear liquid contains phages.

Plaque development: Melt normal nutrient agar and bring to 45°C. Pour it in three petri plates and allow it to solidify. Melt tubes containing 3 ml of soft nutrient agar and bring to 45 °C. Add 0.1 ml of *E.coli* culture to each tube and label as 1, 2 and 3. To first tube add a drop of sewage culture filtrate, mix and spread on one of the agar plates prepared earlier. Repeat the procedure with second tube using five drops of sewage culture filtrate. Spread the contents of third tube on agar plate without mixing the sewage culture filtrate. This plate serves as control. Once the soft agar solidifies, invert the plate and incubate at 37°C till plaques (clear areas where viruses have infected bacterial cells) are clearly visible.

To develop pure culture of phages, pick phages from the centre of a single plaque and transfer it to young turbid culture of *E.coli*. Incubate at 37°C for 24 h till liquid becomes clear. Incubate at 37°C.

Observations:

Result:

2. ENUMERATION OF BACTERIOPHAGES

Requirements: Twenty-four hour cultures of *E. coli*, B and T2 phages, tryptone agar plates, tyrptone soft agar, tryptone broth, petri plates, incubator, inoculation loop, spirit lamp, sterile culture tubes, centrifuge, membrane filter, water bath, one-ml sterile pipettes, Pasteur pipettes and glass marker.

Method:

- Take nine tubes and fill them with nine ml of tryptone broth and label them from 10^{-1} to 10^{-9}. Dilute the phage culture by serial dilution to obtain ten fold dilution.
- Fill two ml of tryptone soft agar in five tubes and label them from 10^{-5} to 10^{-9}.
- Prepare five tryptone hard agar plates and label them from 10^{-5} to 10^{-9}.
- Heat the soft agar tubes in a water bath at 100°C keeping the water level slightly above the agar level and then maintain the molten agar at 45°C.
- Mix two drops of *E. coli* culture and 0.1 ml of 10^{-4} tryptone broth dilution thoroughly and pour contents on hard agar plate labelled 10^{-5}. This results in a double-layered plate culture.
- Using different Pasteur pipettes repeat the whole procedure for dilution broth ranging from 10^{-5} to 10^{-8} to produce 10^{-6}–10^{-8} double-layered plate cultures.
- Incubate plates in inverted position at 37°C for 24 h.
- Observe for plaque formation and count number of plaques formed. Each plaque is designated as a plaque forming units (PFU).
- Multiply by dilution factor and report results as number of plaque forming units (PFU)/ml of phage solution.
- Plates showing more than 300 PFUs are reported as TNTC or too numerous to count.
- Less than 30 plaques are reported as TFTC or too few to count.

Observations:

Result:

Number of plaques	Dilution factor	PFU/ml

3. CULTURE OF VIRUSES IN MONOLAYERS

Cell Culture and Preparation of Monolayer:

Requirements: Single cell suspension of known concentration, growth medium (5–20% serum containing stimulants necessary for cell division, serum free medium with added stimulants, balanced salt solution, amino acids, vitamins, glucose for e.g. Eagles's minimal medium) antibiotics: penicillin/streptomycin, buffer and phenol red indicator, flasks and petri plates.

Method:

- Inoculate flasks containing sterile growth medium with cell suspension and incubate at 37°C overnight with trypsin or versene solution to separate into single cells.
- Seed petri plates containing sterile growth medium with single cell suspension of known concentration and incubate at 37°C overnight.
- Cells adhere to surface and replicate to form a monolayer.
- Inoculate the monolayer with viral inoculum.
- Cover with a thin layer of agar to immobilise viruses and restrict them to single cells.
- Host cells are infected at low multiplicity of infection (m.o.i) i.e. 0.1–0.01 infectious units per cell.
- For viral purification high m.o.i is used (10 units/cell).
- Incubate at 37°C and observe for plaque formation.
- Flood petri plate with neutral red or trypan blue to distinguish living cells from dead ones.
- To harvest virus from cells, cells are lysed by freezing and thawing or by ultrasonic treatment.
- Quantification can be done by infectivity assay.

4. CULTURE OF VIRUSES IN CHICK EMBRYO

Requirements: Fertilised eggs, iodine–alcohol disinfectant, 18-gauge injection needle with syringe, scotch tape, incubator and New castle disease virus suspension.

Method:

Surface sterilise an embryonated egg. Locate the position of air sac by holding the long axis of egg horizontally in front of a light source. Mark the position. Sterilise the needle by dipping in alcohol and then flaming it. Use this needle to make a small hole or puncture the shell over the air sac. The membrane at base should not be punctured. Inject 0.2 ml of dilute viral suspension into the allantoic cavity. Seal the hole in the shell with scotch tape. Maintain an uninoculated control by injecting 0.2 ml of sterile saline. Incubate at 37°C. Observe the egg periodically in front of light source for embryo death (usually takes 3–4 days). This is indicated by cessation of movement and disappearance of veins from eggshell. Once embryo death has been confirmed, crack the shell and collect the contents in a petri plate. Repeat the procedure with control egg. Compare the two embryos. The infected embryo will have lesions on it. Virus can be separated from the infected tissue and purified.

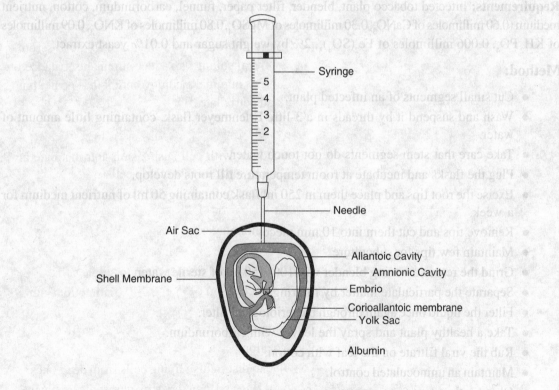

Fig. 16.1 Culture of virus in chick embryo

5. CULTURE OF PLANT VIRUSES

Requirements: Infected plant material, blender, filter paper, funnel, carborundum and cotton.

Method:

- Grind the infected plant material in a waring blender with100–150 ml sterile water.
- Separate the particulate matter by filtering.
- Filter the liquid once again through bacteriological filter.
- Take a healthy plant and spray the leaves with carborundum.
- Rub the viral filtrate on the leaf with cotton.
- Maintain an uninoculated control.
- Observe the plant for appearance of symptoms.
- Virus can be separated from the infected tissue and purified.

6. CULTURE OF TMV IN ROOT TIPS

Requirements: Infected tobacco plant, blender, filter paper, funnel, carborundum, cotton, nutrient medium (0.60 millimoles of $CaNO_3$, 0.30 millimoles of $MgSO_4$, 0.80 millimoles of KNO_3, 0.09 millimoles of KH_2PO_4, 0.006 millimoles of $Fe_2(SO_4)_3$, 2% by weight sugar and 0.01% yeast extract.

Method:

- Cut small segments of an infected plant.
- Wash and suspend it by threads in a 3-litre Erlenmeyer flask, containing little amount of water.
- Take care that stem segments do not touch water.
- Plug the flasks and incubate at room temperature till roots develop.
- Excise the root tips and place them in 250 ml flask containing 50 ml of nutrient medium for a week.
- Remove tips and cut them into 10 mm pieces.
- Maintain few tips for subculture.
- Grind the rest in a waring blender with100–150 ml of sterile water.
- Separate the particulate matter by filtering.
- Filter the liquid once again through bacteriological filter.
- Take a healthy plant and spray the leaves with carborundum.
- Rub the viral filtrate on the leaf with cotton.
- Maintain an uninoculated control.
- Observe the plant for appearance of symptoms.
- Virus can be separated from the infected tissue and purified.
- Several plant viruses can be cultured and maintained in insect vectors.
- Isolated protoplasts can also be used for viral culture.

17

STUDY OF FUNGI

Fungi (GK; mykes: mushroom) are non-motile, eukaryotic, thalloid organisms. Most fungi are microscopic but fruiting bodies of higher fungi form conspicuously visible macroscopic structures. Lower fungi do not show division of labour and structural organisation, but fruiting bodies of higher fungi show both characters e.g. mushrooms. Fungi show heterotrophic mode of nutrition and derive their energy by breaking up highly complex substances. Both parasitic and saprophytic forms are known. Fungal cell wall is made up of chitin and glucans and they have sterols in their plasma membrane.

Thallus organisation differs from unicellular to filamentous type and based on this fungi can be divided into two basic morphological forms, unicellular or yeast forms and hyphal or mycelial forms (mass of hyphal elements is termed as mycelium). Filaments may be coenocytic i.e. lack cross walls or septate depending on the species. In higher fungi, clamp connections are present at the septa which connect the hyphal cells. Few unicellular forms exhibit dimorphism.

Unicellular fungi reproduce asexually by **blastoconidia** formation (budding) or fission. Multicellular fungi reproduce asexually and/or sexually. Asexual reproduction is brought by production of conidia or spores. Relatively large and complex conidia are termed **macroconidia** while the smaller and more simple conidia are termed **microconidia**. Conidia may be attached either singly or in chains at the tip of conidiophores or may be enclosed in a sac-like structure called sporangium. The presence/absence of conidia and their size, shape and location are major features used in the laboratory for fungal identification.

Earlier, fungi were classified within plant kingdom as class Mycota, but now they constitute the fifth kingdom of Whittaker's classification and mycota has been renamed as kingdom **Mycetae**.

This kingdom includes several ecologically and economically important organisms. Fungi decompose dead organic material and therefore play an important role in biogeochemical cycling. They are the source of numerous antibiotics such as penicillin, griseofulvin etc. Mushrooms, truffles and morels are eaten and are rich in proteins. Some mushrooms are poisonous: *Amanita* (Fly and Deadly *Agaricus*), whereas others, such as *Agaricus, Cantharellus*, etc., are among the best edible varieties. Yeast are used in making bread and champagne. They convert sugar into alcohol therefore are used to make beer and other alcoholic beverages. Yeasts are also used for studying genetics and molecular biology.

Fungi also cause a number of diseases in humans such as ringworm, athlete's foot, mycoses, aspergilloses, oral thrush etc. Plant diseases caused by fungi include rusts, smuts, and leaf, root, and stem rots, which cause severe damage to crops. Ingestion of poisonous chemicals such as Ergot alkaloids produced by *Claviceps purpurea*, psychotropic agents like psilocybin, psilocin and lysergic acid diethylamide (LSD), aflatoxins (carcinogens) produced by *Aspergillus flavus* is harmful for humans and animals.

Fungi can be studied inside the host tissue by treatment of tissue preparations with 10% potassium hydroxide. They exhibit positive staining with Lactophenol cotton blue, Grocott silver stain, Haematoxylin and Eosin and negative staining with India ink. Under ultraviolet light fungi show fluorescence. Culture of fungi can be done at 25°C and 37°C on Sabouraud's agar (favours fungal growth because of low pH), Mycosel agar (selective for pathogenic fungi because of chloramphenicol and cycloheximide in medium).

Sample collection: Samples can be collected by scraping the infected surface by a sterile scalpel blade. Samples can be also collected by using transparent vinyl adhesive tape. The tape is applied to the affected area and then peeled off. Infected plant material can be placed directly on the medium. Hair samples should be plucked and not cut. Sputum samples can be collected in vials or with the help of cotton swabs and stored for 3 days in a refrigerator. Samples from nose, throat and vagina can be collected by a sterile swab.

Identification of fungi:

This is done on the basis of :

Colony morphology: Shape, size, colour, nutritional behaviour, oxygen requirement etc.

Cellular morphology: Unicellular/hyphal; aseptate/septate

Spore morphology: Conidiospore/Sporangiospore/Arthrospore/Chlamydospore.

Identification of yeast:

This done by studying:

Morphology: Size, thickness of walls, capsule presence/absence.

Biochemical tests Behaviour in broth and serum (**germ tube** formation).

Behaviour on cornmeal agar (**pseudohyphae** formation).

1. CULTURE OF FUNGI IN MOIST CHAMBER

Requirements: Glass slide, petri plate, alcohol, filter paper, spirit lamp, potato dextrose agar (PDA) or glucose agar, sterile loop, fungal sample (molds), brush, nail paint, blade or scalpel, forcep, cover slip, cotton blue and lactophenol.

Method:

Preparation of moist chamber: Line a sterile petri plate with sterile filter paper or sterilise the plate with the filter paper. Place a sterilised glass slide on the filter paper. Place a drop of PDA or glucose agar medium on the slide and allow it to solidify. With a sterile blade or scalpel, cut the agar drop into half. Remove one half and inoculate the cut surface of the other with the fungal sample. Place a cover slip on the agar drop and seal the edges with nail paint. Moisten the filter paper and cover the petri plate. Incubate in covered position at 25°C. Keep the filter paper moist by adding water to it every day. When fungal growth appears, remove the cover slip. With the help of a sterile forcep remove a small amount of growth. Stain it with cotton blue and mount it in lactophenol. Observe under microscope for morphology, sporangiophores, spores etc.

2. ISOLATION PURIFICATION AND IDENTIFICATION OF FUNGI

Requirements: Glass slide, petri plates, alcohol, filter paper, spirit lamp, potato dextrose agar (PDA) or glucose agar and broth, sterile loop, soil/water/infected tissue/fungal sample (molds), brush, nail paint, blade or scalpel, forcep, cover slip, culture tubes, cotton blue and lactophenol.

Method:

- Surface sterilise infected tissue/prepare soil solution and inoculate a PDA slant with it. Incubate at 25°C.
- When profuse coloured growth is seen, it indicates sporulation. With a sterile needle pick few spores from the slant and add in 10 ml of sterile water.
- Make serial dilutions of this spore suspension and use it to inoculate agar plates. Incubate at 25°C.
- Observe for growth. Initially more than one type of fungal colonies will appear, which can be distinguished by their colour.
- Repeat the procedure till all growth in the plate is identical.
- Observe a stained preparation on glass slide under microscope. Identify the fungus by using standard keys.

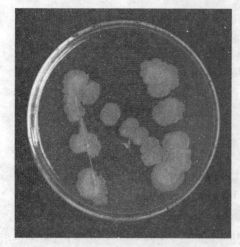

Saccharomyces colonies on Rose Bengal Agar

Fungal Colonies on PDA

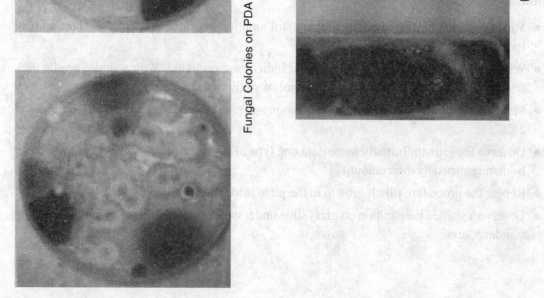

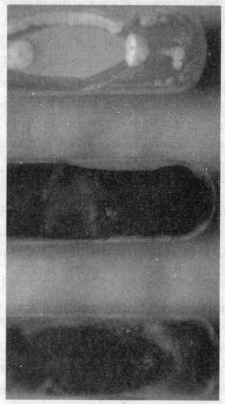

Fungal growth on Agar Slants

Fungal culture

3. STUDY OF YEAST MORPHOLOGY

Requirements: Glass slide, petri plate, alcohol, filter paper, spirit lamp, water–iodine solution, lactophenol–cotton blue solution, sterile loop, yeast sample, brush, nail paint, blade or scalpel, forcep, cover slip, *Saccharomyces cerevisiae* and *Candida albicans* cultures.

Method:

- Prepare wet mount by spreading a loopful of yeast culture in water–iodine (3:1) solution on a clean glass slide.
- Prepare wet mount by spreading a loopful of yeast culture in lactophenol–cotton blue solution on a clean glass slide.
- Observe the wet mount under high power.
- Note cell morphology, size and number of budding cells.

4. ISOLATION OF YEASTS

Requirements: Test tubes, alcohol, filter paper, spirit lamp, potato dextrose agar (PDA), sabouraud agar, cornmeal agar, glucose agar and broth, sterile loop, soil/water/infected tissue, spirit lamp and LAF bench.

Method:

- Surface sterilise infected tissue/prepare soil solution,
- Inoculate sterile potato dextrose agar (PDA) slant, and glucose broth.
- Incubate at 25°C.
- When profuse growth is seen, with a sterile needle pick a little amount of inoculum and add in 10 ml of sterile water.
- Make serial dilutions of this suspension and use it to inoculate sabouraud agar, cornmeal agar and glucose agar plates.
- Incubate at 25°C.
- Observe for growth. Initially more than one type of fungal colonies will appear, which can be distinguished by their colour.
- Repeat the procedure till all growth in the plate is identical and corresponds to yeast characters.
- Observe a stained preparation on glass slide under microscope.
- Identify the fungus by using standard keys.
- Isolate on selective media.
- Record your results.

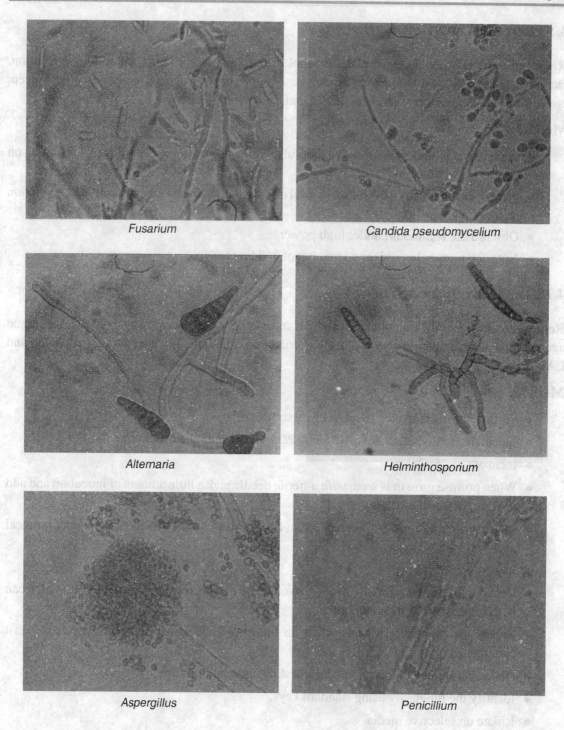

Fusarium

Candida pseudomycelium

Alternaria

Helminthosporium

Aspergillus

Penicillium

Photomicrographs of Fungal pathogens

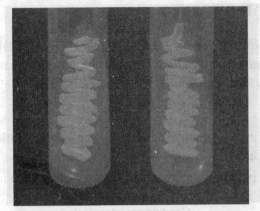

Culture of *Candida utilis*

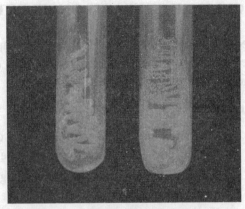

Culture of *Saccharomyces* sp.

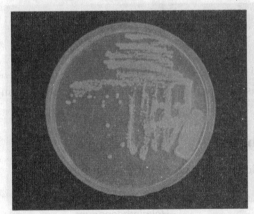

Candida utilis Colonies

Saccharomyces sp. Colonies

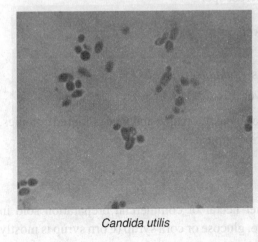

Candida utilis

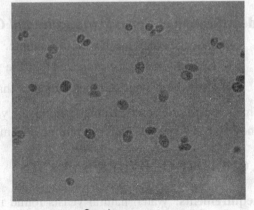

Saccharomyces sp.

Culture of *Candida* and *Saccharomyces*

5. CULTURE OF YEASTS ON SELECTIVE MEDIUM

Requirements: Test tubes, alcohol, filter paper, spirit lamp, potato dextrose agar (PDA), YEDP medium, Rose Bengal agar, *Saccharomyces cerevisiae* and *Candida albicans* cultures, dry yeast granules, spirit lamp and LAF bench.

Method:

- Prepare inoculum from dry yeast by dissolving the granules in warm water to which little sugar has been added cultures can be used directly.
- Melt YEDP medium, Rose Bengal agar and pour in respective petri dishes. Allow them to solidify.
- Take a loopful of inoculum and inoculate the agar plates.
- Incubate at 25°C.
- Observe a stained preparation on glass slide under microscope.
- Store cultures.
- Sabouraud's agar and malt peptone agar can also be used to culture yeasts.
- Mixed culture of yeasts is easy to distinguish on malt agar due to difference in colony morphology.

Identification of *Candida albicans* by Germ tube test:

- Pick the inoculum by lightly touching the colony with a needle or Pasteur pipette.
- Emulsify the inoculum with 0.5 ml serum.
- Incubate at 37°C in water bath for 2–4 h.
- Observe for germ tube formation under a microscope.

Identification of *Candida tropicalis and C. krusei*

- Inoculate malt broth with the sample.
- Growth on surface indicates that the yeast is *Candida tropicalis*.
- Growth on sides of tubes indicates that the yeast is *C. krusei*.

Characteristics of Yeasts; Human pathogenic yeasts form pasty, opaque and pale coloured colonies. They give out a sweet smell resembling the smell of ripe apples.

6. CULTURE OF BAKER'S YEASTS

Requirements: Water, natural yeasts found in flower nectar or commercial preparation sold in markets or sap running out of tree wounds or fruit juice, glucose or corn syrup (Corn syrup is mostly glucose which most microbes can use), meat broth, peptone, or any another source of amino acids.

Method:

- Prepare nutrient medium by adding 10 g of glucose (If using corn syrup, add 10 ml of corn syrup) to a litre of water.
- Inoculate tubes with a few bits of the dry baking yeast or dissolve a little dry yeast in water and add one drop of the suspension to each tube of medium. Limit the amount of yeast added so the culture is not cloudy (In case natural yeasts are being used make and add suspension of these).
- Prepare uninoculated control.
- Incubate control as well as inoculated experimental tubes.
- Incubate at different temperatures such as 4°C (refrigerator), 25°C (room temperature), 35°C and 15°C (incubator).
- Examine the tubes for growth daily or at 12 h intervals.
- Discard any tubes that get moldy.
- Record the results.

7. CULTURE OF BAKERS YEAST ON DIFFERENT CARBON AND ENERGY SOURCES

Requirements: Starch, glucose, maltose, water, natural yeasts found in flower nectar, sap running out of tree wounds or commercial preparation available as yeast granules, meat broth, peptone, or another source of amino acids.

Method:

- Dissolve 10 ml of cornsyrup in a litre of water: "Glucose Water".
- Dissolve 10 ml of table sugar per litre of water: "Sucrose Water".
- Dissolve 10 ml of starch per litre of tap water: "Starch Water".
- Dissolve 10 ml of dry milk per litre or use skim milk: "Milk".
- Sterilise the medium.
- Prepare inoculum by dissolving yeast in tap water.
- Inoculate each tube by adding one drop of yeast suspension to each culture tube.
- Maintain an uninoculated control.
- Check for gas formation by keeping an inverted Durham gas collection tube in each culture tube.
- Add an acid–base indicator to check for acid production.
- Incubate control as well as inoculated experimental tubes.
- Examine your tubes for growth at daily or 12 h intervals.
- Discard any tubes that get moldy.
- Record the results.
- Perform experiments in triplicate.

18

STUDY OF CYANOBACTERIA

About 200 species of Cyanobacteria have been identified so far. Majority are free-living, nonsymbiotic and oxygenic phototrophs. Cyanobacteria are found both in marine as well as fresh waters in and terrestrial environments where water and light are available for growth. They are an important part of plankton layer. Cyanobacteria are actually blue green algae (BGA), which were earlier grouped with algae due to the presence of the pigments and oxygenic photosynthesis. Lately they have been renamed as Cyanobacteria and transferred to Monera, the reason being that Cyanobacteria are prokaryotic organisms and their cell walls are made up of peptidoglycan or murein. They play an important role in carbon, oxygen and nitrogen cycles. They also have an evolutionary significance as it is supposed that ancient Cyanobacteria are responsible for the present day oxygenic environment.

GENERAL CHARACTERISTICS

Cyanobacteria exist in unicellular as well as filamentous forms. Their cell wall is gram-negative type although some show a distant phylogenetic relationship with gram-positive bacteria. They are usually non-motile. Some forms show gliding or rotating motility. They possess photopigments present in sac like, single membrane bound thylakoids. These pigments i.e. chlorophyll *a*, carotenoids, and accessory pigments: phycobilins, phycocyanin (blue pigment) and phycoerythrin (red pigment) absorb wavelengths of light that are missed by chlorophyll and the carotenoids. They store food as glycogen. Biochemically, cyanobacteria are similar to chloroplasts of red algae (Rhodophyta).

They mainly reproduce by vegetative methods only. They break into fragments, called **hormogonia**, which separate and develop into new colonies. Few species form spores called **akinetes,** which are functionally analogous to bacterial endospores. Akinetes are resistant to heat, freezing and drought (desiccation) and can survive in unfavourable environmental conditions.

Planktonic cyanobacteria contain gas vesicles which help them to float. Cyanobacteria produce compounds called **geosmins.** These are responsible for the smell and flavour of water. Cyanobacteria form coloured "blooms" in water bodies. The colour of cells and blooms is due to the pigments present in the mucilaginous sheath or coating present around cells. The colours of the sheaths in different species include light gold, yellow, brown, red, green, blue, violet, and blue–black. A planktonic species related to *Oscillatoria* gives rise to the redness (and hence the name) of the Red Sea. Some blooms forming species secrete poisonous substances that are toxic for animals. They can also grow in highly extreme conditions such as high temperature and salinity. They are the only oxygenic phototrophs present in many hot springs of the Yellowstone ecosystem, and in frigid lakes and oceans of Antarctica. They are absent in acidic waters where the algae are abundant.

Some cyanobacteria can fix atmospheric nitrogen. Nitrogen fixation usually occurs in **heterocysts**, which are specialised, enlarged cells, distributed along the length of a filament or at the end of a filament. Heterocysts contain very little amount of phycobilin pigments and only photosystem I. They lack the oxygen evolving photosystem II and are surrounded by a thick, special glycolipid cell wall, which regulates the rate of diffusion of O_2 into the cell. Nitrogen fixation by enzyme nitrogenase occurs in anaerobic conditions, so the organism maintains these oxygen-free compartments for N_2 fixation to occur.

Few cyanobacteria form symbiotic relationship with liverworts, ferns, cycads, flagellated protozoa, and algae. They sometimes occur as endosymbionts of the eukaryotic cells. E.g. Water fern *Azolla* and a species of *Anabaena*, lichens, corolloid roots. Planktonic cyanobacteria act as "primary producers" and are the basis of the food chain in marine environments. They are also responsible for oxygenic environment of earth.

Prochlorophytes are another group of phototrophic prokaryotes containing both chlorophyll *a* and *b*. Phycobilins are absent in these organisms. Prochlorophytes, resemble cyanobacteria (chlorophyll *a*) as well as plant chloroplasts (chlorophyll *b* instead of phycobilins). *Prochloron*, the first prochlorophyte discovered, is phenotypically very similar to certain plant chloroplasts and is the leading candidate for the type of bacterium that might have undergone endosymbiotic events that led to the development of the plant chloroplast.

ISOLATION AND CULTURE OF CYANOBACTERIA

Requirement: Chu's medium, BGA medium, and other specific media for blue green algae, pond water, flasks and growth chamber (See annexure 4 for composition).

Method:

- Inoculate sterile medium with pond water and incubate in continuous light in growth chamber.
- Examine microscopically and note cell size, presence or absence of sheath, akinetes, heterocyst etc.
- Identify accordingly as suggested by Subramanian et. al. (1999) and APHA (1985).
- Record observations.
- Cultures can be maintained at room temperature in non-direct light.

19

STUDY OF BACTERIAL DISEASES IN PLANTS

Bacterial diseases in plants can be studied by rapid screening tests, which facilitate presumptive diagnosis of bacterial diseases in plants.

1. STEM STREAMING TEST

This is a presumptive test for detecting the presence of *Pseudomonas solanacearum* in vascular tissue of wilting potato stems. The stem is cut just above the soil level and the cut surface is placed in a beaker filled with water. Positive test is indicated if threads of bacterial slime stream spontaneously out of the vascular bundles. Any other bacteria causing vascular infection in potato plants will not show this phenomenon.

2. OOZE TEST

Ooze test is a test to detect the presence of vascular pathogens.

A creamy white bacterial ooze consisting of thousands of microscopic, rod-shaped bacteria may sometimes be seen in the xylem vascular bundles of an affected stem if it is cut crosswise near the ground and squeezed. This bacterial ooze will string out forming fine, shiny threads (like a spider's web) if a knife blade or finger is pressed firmly against the cut surface, then slowly drawn away about 1/4 inch. Two cut stem ends can also be put together, squeezed, and then separated to look for shiny strands of bacteria. The sap of a healthy plant is watery and will not string. Sometimes it helps to wait several minutes after cutting to perform the test. This technique is useful in field diagnosis to separate this disease from other vascular wilts.

3. DETECTION OF POLY-ß-HYDROXYBUTYRATE GRANULES

- Prepare a smear of the ooze or the suspended tissue on a microscope slide or prepare a smear of a 48-hour culture on YPGA or SPA medium.

- Maintain a positive control by making smear of a known strain.

- Air-dry the smear and then heat fix it. Stain with Nile blue A or with Sudan black B.

a. Nile Blue Staining

Flood the smear with 1% aqueous solution of Nile blue A and incubate for 10 min at 55°C. Drain off the staining solution, wash gently in running tap water and blot dry. Cover the smear with 8% aqueous acetic acid for one minute at room temperature. Wash gently in running tap water and blot dry. Examine the stained smear with an epifluorescence microscope at 450 nm under oil immersion at a magnification of 100. Observe for bright orange fluorescence of PHB granules.

b. Sudan Black Staining

Flood the fixed smear with 0.3% Sudan black B solution in 70% ethanol and incubate for 10 min at room temperature. Wash with tap water and blot dry. Dip the smear in xylol for half a minute and blot dry. Flood the smear with 0.5% (w/v) aqueous safranin for 10 seconds at room temperature. Wash with tap water and blot dry. Apply a cover slip. Examine the stained smear with a light microscope using transmitted light under oil immersion at a magnification of 100. PHB granules stain blue–black. The cell wall stains pink.

4. ISOLATION OF THE PLANT PATHOGENIC BACTERIA

Pathogen can be isolated from plant material with typical symptoms by dilution plating of the pathogen.

- Remove ooze or sections of diseased tissue and suspend in sterile distilled water or 50 mm phosphate buffer for 5–10 min.

- Prepare serial dilutions of the suspension.

- Transfer known amount suspension and the dilutions on to a general nutrient medium such as Nutrient Agar, YPGA and SPA, and/or on selective medium.

- Spread or streak with an appropriate dilution plating technique.

- Incubate the plates for three days at 28°C.

- If the isolation test is negative, but disease symptoms are typical, then isolation must be repeated, preferably by a selective plating test.

Problems

- Culture may fail.
- Saprophytes colonising diseased tissue may outgrow or inhibit the pathogen on the isolation medium.

5. IDENTIFICATION OF PURE CULTURE

Pure culture of the isolated strain is developed by repeated serial dilution and plating (See Chapter 4 for details). Identify pure cultures by at least one of the following procedures:

- **Nutritional and enzymatic tests** (Include appropriate control strains in each test used).
- **Pathogenicity test:** This test is done for confirmation and assessment of the virulence of identified cultures. Prepare an inoculum of 10^6 cells/ml from the culture and a positive control strain. Inoculate 5–10 host plants which are preferably at the third true leaf stage or older. Incubate for up to two weeks at 22–28°C and high relative humidity with daily watering. Observe for appearance of characteristic symptoms.
- **Selective plating**
- **Enrichment test**
- **Bioassay test:** The bioassay test is used for the isolation of pathogen from host plant by selective enrichment in a host plant.
- **IF-test ELISA test:** PCR test Fluorescent *in-situ* hybridisation (FISH)
- **Protein profiling:** By polyacrylamide gel electrophoresis (PAGE)
- **Fatty acid profiling (FAP)**

Strain characterisation:

Strain characterisation is optional but is recommended for each new case using at least one of the following:

- Biovar determination on the basis of its biochemical and physiological properties.
- Additional tests differentiate biovar 2 in sub-phenotypes.
- Race determination is done on the basis of a pathogenicity test by a hypersensitivity reaction.

The culture can be further characterised by:

Genomic fingerprinting by RFLP analysis, Repetitive sequence PCR REP-, ERIC- and BOXPCR.

6. ENRICHMENT CULTURE OF *XANTHOMONAS*

Xanthomonas is a gram-negative, motile bacillus known to cause several plant diseases. It forms yellow mucoid colonies on sugar containing medium.

Requirements: SX agar (modified Starr's medium), soil/infected sample, test tubes, petri plates, spirit lamp and inoculation loop.

Method:

- Prepare sample by adding 1 g of soil to 10 ml of distilled water.
- Melt agar, pour into petri plates and allow to solidify.
- Prepare serial dilutions of the sample.
- Inoculate 0.1 ml on agar plates.
- Incubate for 72 h at 30°C.
- Observe for colonies.

Confirm by performing other identifying tests mentioned in Chapter 4.

20

STUDY OF MEDICAL MICROBIOLOGY AND IMMUNOLOGY

1. ANTIBIOTIC SENSITIVITY TEST

Gram-positive bacteria are more susceptible to common antibiotics as they lack the outer membrane of the gram-negatives. Several non-pathogenic strains change to pathogenic forms due to the development of drug resistance. *Shigella*, *E. coli* and *Salmonella* carrying drug-resistance plasmids are known to cause epidemics. Emergence of new antibiotic resistant strains in particular environments makes it increasingly important to adequately monitor resistant strains.

Sensitivity test by Kirby–Bauer Disc Method or Disc Diffusion Method

Disc diffusion is a qualitative method based on an approximation of the effect of antibiotics on bacterial growth on a solid medium. Agar plates are first inoculated with test microorganisms and then antibiotics discs are placed on the agar surface. Antibiotic-impregnated discs release antibiotics into the surrounding medium. If the organism is susceptible to the antibiotic, a zone of growth inhibition around the antibiotic disc will develop.

There is a logarithmic decrease in antibiotic concentration with the distance from the disc. Standardised regression curves have been developed that correlate inhibition zone size to the minimum inhibitory concentration of the antibiotic.

Requirements: Culture sample, sterile loops, antibiotic discs, agar plates and bunsen burner.

Method:

1. Inoculate the agar plate by spread plate method with culture sample to grow a uniform lawn of bacteria. Alternatively spread the inoculum with a sterile swab.
2. Place discs of different antibiotics onto the agar surface. Press down with another bacteriological loop to secure.
3. Observe the growth pattern after at least 24–48 h of growth (standard read time is 48 h) by measuring the size of inhibition zone.
4. Compare with standard values.
5. Report result as a degree of sensitivity to different antibiotics as +,++,+++,++++.

Observation and Results:

Name of organism	Antibiotic 1	Antibiotic 2	Antibiotic 3	Antibiotic 4

Determination of Minimum Inhibitory Concentration (MIC) and Minimum Bactericidal Concentration (MBC)

The MIC test measures the lowest concentration of an antibiotic that inhibits the visible growth of test microorganisms. The MBC is the concentration of antibiotic included in the first tube from which colonies calculated as less than 99.9% growth is recovered. Once an MIC has been performed, an MBC can subsequently be determined. The MBC is set up with subcultures made from each MIC well that appears visually clear.

Requirements: Cumlture sample, sterile loops, antibiotic discs, agar plates and bunsen burner.

Method:

1. Inoculate the agar plate culture with sample made from each MIC well that appears visually clear.
2. Observe the growth pattern after at least 24–48 h of growth (standard read time is 48 h).
3. If no growth appears take this concentration as MBC.

2. ISOLATION OF MICRO FLORA OF SKIN

1. Swab the selected skin surface. Rotate the swab to ensure that enough inoculum has been obtained.
2. Touch the swab to the agar surface in a single spot and transfer the bacteria to the agar culture dish.

3. Discard the swab.

4. Incubate the plate at 37°C for 24–48 h.

5. Note the types of colonies that develop.

6. Perform Gram staining to identify bacteria.

7. Develop pure cultures.

8. Observe presence of dermatophytic fungi microscopically.

3. ISOLATION OF ORAL MICRO FLORA

Throat Swab Sample

- Gently pass dry sterile swab over the tongue and into the posterior pharynx. The mucosa behind the vulva and between the tonsils should then be gently swabbed with a back-and-forth motion.

- Streak carefully onto a blood agar plate (BAP), pulling a three-part pattern of inoculation to obtain isolated colonies and incubate for 24–48 h in the presence of CO_2.

Tooth Swab Sample

- Swab the tooth surface as if brushing your teeth. Swab at the gum line and in the folds surrounding the molars at the back of the mouth.

- Streak carefully onto BAP.

- Incubate for 24–48 h in the presence of CO_2.

- Observe plates for evidence of haemolysis.

4. STUDY OF SPUTUM OR URINE SPECIMENS FOR TUBERCULOSIS

First early morning samples of sputum/urine are recommended as bacteria are present in more concentrated forms. Direct examination of slides stained with acid-fast stain is performed to identify parasites.

5. STUDY OF BACTERIA IN STOOL SAMPLE

Use antibiotic-containing and pathogen-selective media to limit the growth of **normal intestinal flora** (such as SS Agar for *Salmonella* and *Shigella*). Direct examination of concentrated samples is performed to identify parasites.

6. IDENTIFICATION OF STAPHYLOCOCCI BY COAGULASE TEST

Coagulase positive staphylococci release an extra cellular substance called procoagulase which combines with an activator present in the normal plasma to form a clotting agent coagulase. This coagulase causes blood to clot, which is a positive test for the presence of staphylococci.

Requirements: Plasma, test tubes, pipettes, cell sample, 24-hour yeast broth culture of staphylococci or agar plate containing staphylococci colonies.

Method:

- Add 0.5 ml of plasma in two test tubes.
- To one tube add enough amount of unknown sample culture to make the solution turbid.
- In the second tube add a loopful of staphylococcal culture. This tube will serve as positive control.
- Incubate at 37°C for a few hours.
- Examine for formation of blood clots.
- Note the time taken. If clots are formed after three hours, sample is coagulase negative.

7. PHAGE TYPING OF COAGULASE POSITIVE STAPHYLOCOCCAL STRAINS

Bacteriophages are not only species-specific but are also strain-specific. Thus, in this test strain-specific phages are used to identify and type the staphylococci.

Requirements: Sterile petri plates, glass marker, tryptone or yeast extract agar, pipettes, 24-h yeast broth culture of coagulase positive staphylococci or agar plate containing staphylococci colonies and homologous phage solution.

Method:

- Demarcate petri plate into equal squares with a marker.
- Add molten and tempered tryptone or yeast extract agar.
- Allow the agar to solidify and then inoculate it with staphylococcal culture by spread plate technique.
- Pipette a drop of different phage preparations in each square separately.
- Incubate at 37°C for a few days.
- Examine for plaque formation.
- Record the phage type, which causes cell lysis.

8. IDENTIFICATION OF STAPHYLOCOCCI BY DNASE TEST

The test is based on the ability of pathogenic staphylococci to produce DNAse enzyme. This enzyme hydrolyses the DNA present in the medium, which is indicated by the development of a clear zone around the colony when the plate is flooded with 1 N HCl.

Requirements: Sterile petri plates, glass marker, DNA agar, pipettes, 24-h yeast broth culture of coagulase positive staphylococci or agar plate containing staphylococci colonies, homologous phage solution and 1 N HCl.

Method:

- Inoculate sections of tryptose agar medium containing DNA with inoculum from test colonies.
- Inoculate controls of known *Staphylococcus aureus* and *S. epidermidis* as positive and negative controls.
- Incubate the plate at 37°C for 18–24 h.
- Flood the plate with 1 N HCl that precipitates DNA and turns the medium cloudy.
- Presence of a zone of clearing around the area of growth indicates positive test.

9. IDENTIFICATION OF GRAM-POSITIVE STREPTOCOCCI

Requirements: BAP plates, sterile cotton swabs in sterile saline, inoculating loop, incubator, spirit lamp and LAF bench.

Method:

- Inoculate BAP (Blood Agar) and CA (Chocolate Agar: heated blood agar) with oropharyngeal swab samples (Nutrient agar plates do not support the growth of many of the streptococci).
- Stab the agar after streaking to increase haemolysis on BAP (increases activity of strepto-lysin O, which is oxygen labile, stabbing the agar with the inoculating loop after the streak is another technique that facilitates streptococcal growth).
- Incubate for 48 h at 37°C in a CO_2 environment.
- Observe colony morphology, Gram-staining and haemolysis.
- Apply latex agglutination, and ELISA to confirm Lancefield grouping ("C" substance recognised by the Lancefield grouping sera is a polysaccharide, and this permits the arrangement of the streptococci into a number of antigenic groups identified as Lancefield group A, B, C, D etc.).

Hints:

- *Streptococcus viridans* and *S. pneumoniae* will produce α-haemolysis.
- Gram-staining will reveal purple-staining cocci that are smaller than the staphylococci.
- *S. pneumoniae* is lancet-shaped and typically occurs in pairs.
- Most members of this group also require low oxygen tension. This means that they will grow best in the presence of 5% CO_2.
- Streptococci grown in broth culture will appear as long chains of cocci upon Gram-staining.
- Older cultures may appear Gram-negative due to autolysis.
- The catalase test distinguishes Staphylococci from Streptococci because Streptococci do not produce catalase.

Presumptive tests for group A Streptococci

1. Bacitracin susceptibility test: Paper disks of low concentration (0.02–0.04 units) of bacitracin are seeded on blood–agar medium inoculated with Streptococci. The diameter of a zone of inhibition around the disc indicates sensitivity (>10 mm = sensitive). Group A, C and G Streptococci are sensitive to bacitracin.

2. PYR test: It is based on detection of the presence of enzyme L-pyrroglutamyl aminopeptidase. It is a more reliable test than bacitracin sensitivity test.

- Put a drop of freshly prepared solution of L-pyrrolidonyl-alpha-naphthylamide on a piece of filter paper.
- Rub test colony with it.
- Development of red colour within 5 min suggests that the organism is *Streptococcus pyogenes*.

Presumptive tests for group B Streptococci

The CAMP Test: This is based on detection of the CAMP factor, an extra cellular substance produced by Group B Streptococci, which enhances the haemolytic activity of Staphylococcal-haemolysin. Therefore, when the two meet, there will be an increased haemolysis.

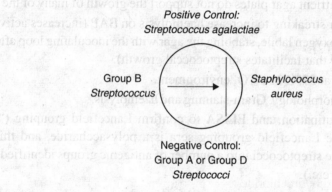

Procedure:

- A single streak of *Streptococcus* is made across the medium.
- Perpendicular to this, is a strain of *Staphylococcus aureus* that is known to produce β-haemolysin is also streaked.
- The two lines should not touch one another.
- Incubate aerobically at room temperature. In an anaerobic environment many Group A Streptococci tests positive.
- Any Bacitracin-negative, CAMP-positive, bile-esculin-negative *Streptococcus* can be reported as: Group B *Streptococcus*, presumptive by CAMP.

Positive control: *S. agalactiae*.

Presumptive tests for group D Streptococci

Lancefield Group D Streptococci are divided into two groups: (1) Enterococci, and (2) Nonenterococci.

Salt Tolerance Test for Enterococci: This is based on the ability of the Enterococci to grow in 6.5% of NaCl. It separates Enterococci from the "Group-D Streptococci and non-Enterococci".

Loopful of culture is inoculated into 6.5% of NaCl broth and incubated at 37°C for 18 h (overnight). If growth occurs it is an indication of positive test for *Enterococcus faecalis*.

Bile-Esculin Hydrolysis Test: This is used to differentiate group D Streptococci from other "non-group D Streptococci" and is based on the ability of an organism to hydrolyse the glycoside esculin to esculatin and glucose in the presence of bile (10–40%). The organism is inoculated into the bile esculin medium (stab and streak on slant) and slants are incubated at 37°C for 24 h.

Presence of a black to dark brown colour on the slant (half or more of the medium is blackened) is a **Positive Test** for *E. faecalis*. Absence of blackening of the medium or blackening of less than half the tube after 72 h of incubation is a **Negative Test** for *Streptococcus agalactiae*.

10. CULTURE AND IDENTIFICATION OF FASTIDIOUS BACTERIA

1. Neisseriae: are Gram-negative cocci, which are **oxidase positive and catalase positive**.

Requirements: BAP and chocolate agar plates, oxidase and catalase reagents, nasal culture (sterile swab needed).

Method:

- Prepare a Gram-stain slide. Observe staining reaction and cell morphology.
- Streak the nasal swab on each of BAP and chocolate agar plates.
- Incubate in CO_2 incubator at 37°C for 24 h.
- Observe colony characteristics.
- Pick a colony and touch to the surface of slide or filter paper impregnated with tetramethyl-*p*-phenylenediamine dihydrochloride.
- Observe the colour change (The reaction colour will change from pink to maroon to dark purple).
- Read the test results within 10 s. Some organisms may show slight positive reactions after this period and such results are NOT considered definitive.
- Add two drops of 3% hydrogen peroxide reagent to the colony.
- Observe for immediate bubbling (gas liberation) and record the result.

2. Haemophilus influenza
Haemophilus contains a number of species of fastidious, Gram-negative coccobacilli, which grow on specially enriched culture media such as chocolate agar and microaerophilic conditions of incubation. *H. haemolyticus* can be differentiated from *H. influenzae* by the use of blood agar (*H. haemolyticus* causes haemolysis).

Requirements: BAP plates, chocolate agar plates, blood agar plates, spirit lamp, inoculation tools and incubator.

Method:

- Inoculate blood agar and chocolate agar plates.
- Incubate anaerobically at 37°C overnight or longer.
- Observe growth in media, haemolysis, and morphology. (Growth only on the chocolate agar is characteristic of the *Haemophilus* group of organisms).

11. ANTIGEN–ANTIBODY REACTIONS

a. Identification of Blood groups and Rh typing by an Antigen–Antibody complex Formation

Requirements: Glass slide, glass marker, antisera, 70% alcohol, sterile needle, wooden sticks and spirit lamp.

Procedure:

1. Make three sections on a clean glass slide with a marker pencil and label them as A, B and O.
2. Place a drop of the appropriate antiserum on the sections of the slide so labelled.
3. Rub a piece of cotton soaked in 70% alcohol on the tip of third finger.
4. Press the end and prick the finger with a sterile, disposable needle.
5. Place a drop of blood on each section of the slide.
6. Mix blood and antisera with separate pieces of a wooden stick and record blood group and Rh type according to the presence or absence of clumping of the red blood cells.

b. Identification of an Antigen–Antibody complex by Gel Diffusion Analysis (Immunodiffusion) or Ouchterlony Plate Method

Gel diffusion combines diffusion with precipitation and facilitates the separation of different antigen–antibody reactions. Clarified agar gel is used as a matrix. Antigen and antibody are filled in separate wells. These diffuse through the gel towards each other and when **equivalence points** are reached, precipitation results. Single **line of precipitation** indicates a single antigen in combination with its specific antibody. When two antigens are present in a system, each behaves independently of the other. If more then one bands of precipitation are formed, it indicates many antigen–antibody combinations. The size and time taken for these lines to develop depends on the antigen–antibody titre and therefore is also a measure of these.

Requirements: Ouchterlony agar plates, cork borer, rabbit horse antiserum and antigens.

Procedure:

- With the help of size 2 cork borer make five holes, one in the centre and the rest four on the four corners surrounding the central hole in ouchterlony agar plates.

- Fill the central hole with rabbit horse antiserum and the surrounding holes with different antigens.
- Incubate the plates at 37°C for seven days.
- Check for precipitation lines every day.

c. Identification of an Antigen–Antibody Complex by Precipitin Test

This test is based on the reaction of soluble antigen with its antibody to form a visible precipitate. There are several methods for performing this test. It can be performed on a slide or in a test tube. Precipitate is formed at the interface where the two solutions meet.

Requirements: Rabbit antiserum containing antibodies, horse serum (antigen), test tubes, pipette and normal saline.

Procedure:

- Transfer 0.1 ml of antigen into a test tube containing 0.9 ml of saline.
- From this tube, take 0.1 ml and add to another test tube containing 0.9 ml of saline.
- In this manner make four ten-fold serial dilutions of the antigen.
- Take five test tubes and add antiserum into them.
- Now overlay the antiserum with antigen to a depth of about 4 mm by slowly running it down the side of the test tube. Take care that the two solutions do not mix with each other.
- Use different pipettes for transferring different antigen dilutions.
- Label the tubes and observe them for the appearance of precipitation bands at 5, 10 and 30 min.
- Record the highest dilution giving precipitation reaction.

d. Identification of an Antigen–Antibody Complex by Slide Agglutination Test

This test is based on the reaction between whole cells (antigen) and antibodies in serum resulting in clumping or agglutination of cells. The cells are brought together and linked to each other by the formation of antibody bridges.

Requirements: Serum containing antibodies, antigen, glass slides and pipette.

Procedure:

- Place a drop of serum containing antibodies on a clean glass slide.
- Add a drop of antigen to it and mix well.
- Observe under a microscope for clumping.
- Maintain a control by using known antigen and its specific antibodies.
- At times, cells are coloured to make the reaction visible.

e. Identification of an Antigen–Antibody Complex by Complement Fixation

This is a test to detect the presence of a specific antigen in blood. When an antigen reacts with an antibody in blood, fixation of complement, an unstable thermolabile group of proteins takes place. This test is based on this property of complement. Since complement–Antigen–antibody complex is not coloured, an indicator system comprised of sheep RBC and sheep haemolysin is used. If complement is fixed, there will be no lysis of RBC indicating positive reaction. If lysis of RBC takes place, then it means that complement was not fixed as antigen–antibody complex was not formed.

Requirements: Required dilutions of haemolysin, 2% suspension of sheep RBC, 2 units of guinea pig complement, horse serum (antigen), rabbit antihorse serum (antibody), barbital buffered saline, test tubes and pipette.

Procedure:

- Heat antigen and antibody solution at 56°C for 30 min to inactivate inherent complement.
- Dilute horse serum to 1/100 dilution with barbital buffered saline.
- Dilute antibody solution also accordingly.
- Refrigerate the dilute inactivated complement till ready to use.
- Mix 1.5 ml of 2% sheep RBC and 1.5 ml of haemolysin and refrigerate.
- Prepare four successive serial dilutions of 1/100 dilution of inactivated horse serum ranging from 1/1000 to 1/ 100,000,000.
- Label 10 test tubes from 1 to 10.
- Add inactivated horse serum, inactivated rabbit antihorse serum, complement and barbital buffered saline according to Table:

Tube No.	0.1 ml inactivated horse serum dilutions	inactivated rabbit antihorse serum (ml)	complement (2 Units) (ml)	barbital buffered saline (ml)
1	1/1000	0.1	0.1	–
2	1/10,000	0.1	0.1	–
3	1/100,000	0.1	0.1	–
4	1/1,000, 000	0.1	0.1	–
5	1/10,000, 000	0.1	0.1	–
6	1/100,000,000	0.1	0.1	–
7*	1/10,000	–	0.1	0.1
8*	–	0.1	0.1	0.1
9*	–	–	0.1	0.2
10*	–	–	–	0.3

Tubes 7–10 are control. Tube 7 is a test for search of anticomplementary factors in horse serum, which can hinder the reaction. Similarly tube 8 is a test for search of anticomplementary factors in rabbit horse antiserum which can hinder the reaction. Tube 9 tests whether the complement system is sufficient to lyse RBCs or not. Tube 10 shows that in the absence of complement system, RBC do not lyse even though saline is present. Thus tubes 7–9 should show RBC lysis and tube 10 should give negative result.

- Incubate the tubes immediately in water bath at 37°C for one hour.
- Add 0.2 ml on RBC and haemolysin mixture.
- Incubate again in water bath at 37°C for one hour.
- Observe for complement fixation in tubes 1–6.
- Report result as:
  1. +++: Complement fixation and no haemolysis hence antigen–antibody are homologous.
  2. –/0: Complete lysis, complement not fixed, hence antigen–antibody are not homologous.
  3. Tubes showing partly lysed RBC.
- Express titer of antigen as reciprocal of dilution one lower than dilution giving complete lysis.

f. Identification of an Antigen–Antibody Complex by Flourescent Antibody Technique

In order to detect an antigen–antibody complex, fluorescent dye is used as an indicator. Either antigen or antibody is tagged and the resultant complex shows fluorescence.

Requirements: Glass slides, pipettes, fluorescent microscope, tagged antibodies, 24 h culture of bacteria homologous to tagged antibody and unknown culture.

Procedure:

- Prepare smears of both cultures on separate slides.
- Heat fix the smear.
- Spread a drop of tagged antibody solution on the smear.
- Place the slides in moist chamber at room temperature for 30 min.
- Wash slide with 1% phosphate buffer of pH 7.2 and then immerse in the same buffer for 10 min.
- Remove and blot dry.
- Spread glycerol adjusted to pH 7.2 on the slide.
- Observe under fluorescence microscope with UV light for antigen–antibody reaction, which will show fluorescence.

Reagents used:

Oxidase reagent: Dissolve 1.0 g of *p*-aminodimethyaniline oxalate in 100 ml of distilled water by gentle heating. Refrigerate and always store in a brown bottle.

Barbital buffered saline: Make stock solution by dissolving 2.875 g of barbital, 1.083 g of barbital sodium, 0.083 g of calcium chloride, 0.238 g of magnesium chloride, 42.50 g of NaCl in 250 ml of hot distilled water. Cool and dilute to 1000 ml. Refrigerate. Before use dilute one part with four parts of distilled water.

Saline: 9.0 g of NaCl dissolved in one litre of distilled water.

70% alcohol: 70 ml of alcohol in 30 ml of distilled water.

Blood Agar: 10.0 g of tryptone, 5.0 g of NaCl, 15 g of agar, 500 ml of meat infusion or 5.0 g of yeast extract and 1000 ml of distilled water. Dissolve tryptone and NaCl in meat infusion. Heat for a minimum of 20 min. Add agar and water and sterilise.

Acid-fast stain: See Chapter 4.

Tryptone or yeast extract medium: 10.0 g of tryptone, yeast extract 5.0 g, 5.0 g of K_2HPO_4, 1.0 g of glucose and 1000 ml of distilled water. Add 1.5% of agar at the time of addition of glucose to make medium.

Ouchterlony agar: Dissolve 17 g of ion agar #2 in 1980 ml of borate saline by heating. Add 10 ml of Merthiolate and pour into petriplates.

DNA agar: 20.0 g of tryptose, 2.0 g of DNA, 5.0 g of NaCl, 15 g of agar and 1000 ml of distilled water.

1% phosphate buffer of pH 7.2.

Glycerol adjusted to pH 7.2.

Bile-esculin medium.

1 N HCl.

6.5% NaCl broth.

3% hydrogen peroxide reagent.

Chocolate agar plates (See Annexure IV for composition of chocolate agar).

21

GOOD LABORATORY PRACTICES

When an experiment or set of experiments are conducted in which a test item is examined *in vivo* or *in vitro* in order to collect or generate data on its properties and/or its safety then it constitutes a non-clinical health and environmental safety study. A quality control system, which is related with the managerial process and the conditions under which non-clinical health and environmental safety studies are planned, performed, monitored, recorded, archived and reported is called **Good Laboratory Practice (GLP).** Thus it can be said that these are a set of rules and regulations, which are required to be followed strictly in order to ensure safe and successful conduction of all the studies being done in the laboratory.

One of the most common hazards of working in a microbial laboratory is the danger of infection by pathogens. Since microbes are ubiquitous, chances of contamination as well as infection are very high. Therefore, it is absolutely essential to keep the laboratory clean. Apart from infections, cuts, burns electric shocks, explosion etc. are also a possibility. If, proper conditions are maintained and all rules and regulations followed, chances of mishap can be greatly reduced. Maintenance of safe and aseptic conditions is an essential part of working in a microbial laboratory.

All personnel working in the lab should be properly trained in the necessary practices and techniques, they should be aware of the potential hazards and conversant with ways and means of safe handling of bio-hazardous material as well as any crisis. To prevent injury and infection following rules and regulations should be strictly followed:

A. RULES FOR MAINTAINING ASEPTIC CONDITIONS

1. Consider all living material potentially hazardous.

2. Maintain sterile conditions. Wipe working area with disinfectant.

3. Never place contaminated and used tools such as inoculation needles, forceps, glass slides etc. on the working table.

4. Never store used instruments without sterilizing.

5. Sterilize loops, needles by dipping in alcohol and then heating in the flame. Pipettes, forceps etc. should be sterilized either by autoclaving or in the oven.

6. Always wash hands with spirit before working.

7. Long hair should be tied back to keep it from coming into contact with lab chemicals or flames.

8. Appropriate and protective clothing that covers the arms, legs, torso and feet must be worn in the laboratory.

9. Wear a sterilized cotton apron, eye goggles and masks while working with microbes. The apron must be resistant to liquid penetration to protect clothing from contamination.

10. Aprons and gloves should be worn when working with chemicals, radio isotopes, etc., while working in the laboratory area, when handling infected animals and when there is possibility of skin contact with biohazards.

11. Apron must be either disposable or it must be capable of withstanding sterilization.

12. It should be comfortable, closed type, antistatic and durable.

13. Never work with bare hands. Always wear disposable or sterilized rubber gloves.

14. When working with hazardous materials, make sure that the glove overlaps the lower sleeve and the cuff of the laboratory garment.

15. Wear a long sleeved glove or disposable arm-shield for protection of the garment.

16. Double gloving and tight fitting cuffs on laboratory clothing or sleeve protectors are useful.

17. Double gloving can provide added protection in case of a spill. Hands will remain be protected after the contaminated outer gloves are removed.

18. Latex or vinyl gloves should be worn when working with bio hazardous substances. Gloves should be periodically checked for leaks and should be changed frequently. Outer surface of gloves should not touch the skin while removing them. Hands should be washed immediately after removing gloves.

19. Gloves must be disposed of when contaminated, removed when work with infectious materials is completed and never worn outside the laboratory.

20. Disposable gloves must not be washed or reused.

21. Gloves should be not be worn outside the laboratory area. Other non contaminated items should be handled only after removing the gloves. Do not reuse latex gloves after disinfection.

Decontamination of gloves prior to disposal is a must. Rubber household gloves which can be decontaminated and reused, should be worn when doing chores like cleaning and decontamination. Always check gloves for punctures, tears or other evidence of deterioration before using.

22. When working with sharp edges use of stainless steel mesh gloves is recommended as they protect against injury.

23. Do not bring shoes in the laboratory.

24. Keep separate rubber slippers or cotton shoes, for wearing inside the lab.

25. Always dispose cultures, samples, used/ left over or contaminated medium after autoclaving.

26. Never transport cultures without proper packaging.

27. When using glass jars for blenders cover it with a polypropylene jar. This will prevent spraying of glass and contents in case of accidental breakage.

28. Molded surgical masks or respirators, full face shields or any other effective, facial barrier protection should be used wherever there are chances of splattering of biohazardous substances and their coming in contact with mucous membranes of the mouth, nose and eyes.

29. Adequate eye protection (minimum: chemical splash goggles) must be worn at all times in the laboratory. Plastic wrap around safety glasses that fit over regular glasses should be used for protecting the eyes. Safety goggles with a plastic cushion seal provide better protection.

30. Aerosols must be minimized. Procedures which produce aerosols can be classified into:
    - **Most Severe:** Shaking and blending with high speed mixers, forceful ejection of fluid from pipette or syringe.
    - **Severe:** Opening lyophilized culture.
    - **Less severe:** Grinding tissue with mortar and pestle, decanting supernate fluid after centrifugation, releasing vacuum on freeze-dryer, releasing bubbles from a pipette, spilling liquid on a hard surface, opening tube within which air pressure may be different from that of room*.
    - **Least severe:** Re-suspending packed cells, inserting hot loop in culture, withdrawing culture sample from vaccine bottle.

31. Place a towel moistened with disinfectant over the top of the blender during use.

32. Never open the blender immediately. Wait for at least one minute. This will settle the aerosol.

33. Disinfect and decontaminate all instruments immediately after use.

34. Handling of cultures should be minimized and vapor traps should be used wherever possible.

35. If possible open ampoules containing liquid or lyophilized culture material in a biological safety cabinet or a hood.

36. To store in biohazrdous material in liquid nitrogen use polypropylene cryovials. Wear gloves when opening ampoules or cryovials.

37. To open a sealed-glass ampoule, wrap it in disinfectant soaked disposable towel, nick the neck of the ampoule with a file, hold the ampoule upright and snap it open at the nick.

38. Add liquid very slowly into the contents of the ampoule in order to avoid aerosolization of the dried material.

39. Contents of the ampoule should be mixed without bubbling and emptied into a fresh container.

40. Never discard the disposable towel and ampoule top and bottom without autoclaving or incinerate it.

41. If possible, use disposable plastic loops and needles for culture work.

42. Continuous flame gas burners produce turbulence and should not be used in biological safety cabinets. They disturb the protective airflow pattern of the cabinet. Continuous heat produced by the flame can also damage the HEPA filter.

43. Allow the blower in the bio-safety cabinet to run during cleanup.

44. Wipe the walls, work surface and any equipment in the cabinet with a disinfectant-soaked cloth.

45. Contaminated disposable materials should be put in appropriate bio-hazardous waste container(s) and autoclaved before discarding as waste.

46. Expose non-autoclavable material to disinfectant and allow 20 minutes contact time before removing from the biological safety cabinet.

47. Remove protective clothing used during cleanup and place in a biohazard bag for autoclaving. If disposable, treat as medical waste.

48. Run bio-safety cabinet 10 minutes after cleanup before resuming work or turning cabinet off.

49. Decontaminate all items within the spill the area.

50. When applying disinfectant to any area, allow minimum 20 minutes contact time before starting work to ensure germicidal action of disinfectant.

51. Wipe equipment with 1:10 bleach, followed by water, then 70% ethanol or isopropanol.

52. Use disposable needle/ syringe if possible.

53. Work in a biological safety cabinet whenever possible.

54. Fill the syringe carefully to minimize air bubbles.

55. Expel air, liquid and bubbles from the syringe vertically into a cotton plug moistened with disinfectant.

56. Do not use a syringe to mix infectious fluid by aspiration.

57. Use a separate pan of disinfectant for reusable syringes.

58. Check the sterility of biological safety cabinet periodically.

59. Always leave the biological safety cabinet running whenever the cabinet is in use.

60. Use proper disinfectants to avoid damage to the interior.

61. Activities such as application of cosmetics, eating or placing any thing in mouth or any other contact with mucous membrane, smoking, storage of food, preparation of food, use of laboratory glassware for keeping or eating food should not be allowed in the laboratory.

62. Food should be stored outside the laboratory in special refrigerators earmarked for this purpose only.

63. Laboratory clothing should not be worn outside the laboratory and it should be decontaminated before laundering or disposal.

64. Traffic in laboratories must be restricted. Biohazard sign should be displayed when use of biohazards is in progress.

65. High risk entry laboratories should have a double set of doors which remain closed at all times and the Biohazard sign should be permanently affixed to the door.

66. Proper scrubbing of hands with a disinfectant is necessary every time biohazards or animals are handled, after removing gloves and before leaving the laboratory area.

67. All working space should be kept tidy, cleaned regularly and decontaminated once a day.

68. Wet mopping is a better way for cleaning as sweeping or use of ordinary vacuum creates aerosols.

69. Laboratory should be kept pest free by ensuring sanitary conditions, proper storage, collection and disposal of solid wastes and sealing of cracks and crevices.

B. RULES FOR MAINTAINING SAFETY

1. Maintain the workplace free of physical hazards. Never clutter the desk tops with books, overcoats etc. Place these things in their designated place.

2. After washing hands with spirit, wait till the hands dry completely before lighting the burner.

3. Tie long hair and cover head with a cotton sterilized cap to prevent contamination as well as accidental burning.

4. Do not smoke or eat in the lab.

5. Never pipette out reagents, solutions or culture with mouth. Always use mechanical devices such as a rubber bulb or micropipettes.

6. Do not lick labels. Use self-sticking labels or glue to stick labels.

7. Check the magnehelic gauge of the biosafety cabinet regularly for any indication of problem, which normally runs at a relatively fixed value. On observing any deviation stop using the cabinet till the cause of the deviation has been identified and fixed.

8. Do not disrupt the protective airflow pattern of the biological safety cabinet. Rapid arm movement, traffic behind you, and open laboratory doors may disrupt the airflow pattern and reduce the effectiveness of the cabinet.

9. Minimize the accumulation of materials in the biological safety cabinet to reduce turbulence and to ensure proper airflow.

10. Use appropriate precautions to avoid UV-related injuries.

11. Always carry cultures in sturdy racks or trays without overloading them.

12. Immediately cover any breakage or spill with paper towels and then saturate the area with disinfectant such as 1:10 dilution of household bleach. Clean after 15-20 minutes.

13. Maintain electrical safety especially when using extension cords. Avoid overloading of electrical circuits and creation of electrical hazards in wet areas.

14. Grounding of equipment should be proper. Properly secure all compressed gas cylinders

15. Gloves must be worn when working with bio-hazardous and/or toxic materials and physically hazardous agents.

16. Wearing of contact lenses is inappropriate in the laboratory setting

17. Temperature-resistant gloves must be worn when handling hot material, dry ice or materials being removed from cryogenic storage devices.

18. Delicate work requiring a high degree of precision dictates the use of thin walled gloves.

19. Blenders, vacuum pump exhausts, lyophilizers, culture ampoules, loop sterilizers and bunsen burners produce considerable aerosol. Make sure they are leak proof. Test leakage with sterile saline, soap solution or dye solution prior to use.

20. If using glass jars for blenders cover it with a polypropylene jar to prevent spraying of glass and contents in case of breakage.

21. Care should be taken that an open flame is ignited only when no flammable solvents are in the vicinity.

22. Volatile, flammable solvents such as ether, acetone and methanol should be handled with extreme precaution. They should never be evaporated on a hot plate in an open system. An efficient condenser system must be used.

23. All mishaps must be reported immediately to the supervisor and to all those who may be affected e.g. in case of fire.

24. Eyes must be washed immediately if some chemical is accidently splashed into them.

25. The location and operation of fire extinguishers, safety showers, eye wash stations, fire alarm boxes, exit doors, telephones etc. should be known to everyone.

26. Pipetting should never be done by mouth suction. Inhalation and tasting of any chemical also should not be done.

27. Glass tube should never be inserted into rubber stoppers forcibly. Both the stopper and the tubing should be lubricated properly (glycerol or soapy water) before insertion. Glass tube should be wrapped with a towel while inserting to prevent breakage due to pressure. Hands should be adequately protected.

28. All used glassware should be washed immediately after finishing the experiment.

C. RULES FOR EFFICIENT WORKING

1. Plan out the work carefully and carry it out meticulously

2. Use clean dry slides. Mark them with glass marker or labels. Write the name of the organism on the slide.

3. Label on reagents, cultures etc. should carry the name of reagent/culture, name or initials of person responsible, date of making/subculture, expiry time/date of next subculture, name of medium, dilution of sample etc.

4. Label containing complete information should be on the cover of the petridishes and the name of the test organism should be written on the lower surface (as petriplates are inverted for incubation) and should not hinder observation.

5. Incubate petridishes in inverted position to prevent drops of condensed water on the medium. Deposition of water on the medium acts as a vehicle for spread of microbes and formation of confluent colonies instead of individual colonies.

6. Record all observations in proper and meticulous previously formed tables.

7. Never take observations on loose paper. Keep a separate notebook or laboratory manual for this purpose.

8. Make camera lucida drawings or if possible take microphotographs for documentation.

9. Before leaving the lab, place everything in its proper place, do not leave the working table cluttered or dirty, check if all instruments have been shut down except those which need to remain working.

10. Access to exits, sinks, eyewashes, emergency showers and fire extinguishers must not be blocked.

11. Maintain a users log for all equipments by noting date, time, operator's name, contact information, lab, room number, phone number, was the material/culture bio-hazardous, temperature, pressure (in case of autoclaves), and length of time the equipment was used.

12. Maintenance records of equipment should also be made.

13. Before performing any experiment following preparation is necessary.

    - Student should read the theory as well as any suggested additional reading regarding the experiment.
    - A checklist of chemicals and equipments required should be made before starting the experiment.
    - A brief outline of the experiment, calculations required for preparing reagents, observation tables, charts and other pre-experimental exercises should be written down.

14. All data should be recorded meticulously and date wise in a sturdy laboratory notebook. Observations should never be taken on pieces of paper as they can get lost.

15. Smallest detail about the experiment such as room temperature, humidity levels, color changes,

endothermic or exothermic changes, physical state changes, boiling points, melting points, freezing points, etc. should be noted.

16. The safety do's and don'ts of the laboratory should be strictly followed for your own as well as for your colleagues safety.

17. Samples or containers of chemicals or biological samples should be carried in double containers. If the containers are struck or dropped or if the inside one breaks the outside container prevents spillage.

18. To avoid conflicting needs for space and equipment all the lab partners should plan their work and hold weekly planning meetings for this purpose.

19. All containers should be labeled. Label should have the following information: name of the chemical, date of preparation, expiry date or date of disposal, person responsible for the container and hazard warning so that people completely unfamiliar with laboratory's operations can identify the contents of the containers and take proper precautions..

Important: Following substances are hazardous:

Phenol - can cause severe burns

Acrylamide - potential neurotoxin

Ethidium bromide - carcinogen

Always wear gloves when using these chemicals and never mouth-pipette them. If you accidentally splash any of these chemicals on your skin, **immediately** rinse the area thoroughly with water.

Ultraviolet Light: Exposure can cause acute eye irritation to damage to retina. Damage may not be noticed until 30 min to 24 hours after exposure. Never look at UV light.

Always protect eyes when using UV lamps.

D. RECORDING PRACTICAL WORK

All the work done should be recorded meticulously in the workbook under following heads:
- Date and title of the experiment
- Aim or objective of the experiment
- Materials required
- Details of Methodology/ Procedures. (If standard procedures are used give reference).
- Results and observations
- Conclusions or Comments on results and observations.
- Suggestions for further experimentation.
- Diagrams/drawings/photomicrographs

ANNEXURE I

Some Common Bacteria and their Importance

Acetobacter aceti: effects formation of acetic acid from alcohol.

Actinomycetes: source of valuable antibiotics including streptomycin, erythromycin, and the tetracyclines.

Agrobacterium rhizhogenes: causes hairy root disease.

A. tumefaciens: causes crown gall disease of plants. It's Ti plasmid is used as a vector in genetic engineering studies.

Azorhizobium: forms nitrogen fixing nodules in the stem of *Sesbania*.

Bacillus anthracis: causes anthrax in animals.

B. brevis: produces antibiotic thyrothricin.

B. licheniformis: produces antibiotic bacitracin.

B. megaterium: causes curling of tea and tobacco leaves.

B. polymyxa: produces antibiotic polymixin.

B. ramosus, B. vulgaris, B. mycoides: are ammonifying bacteria. Degrade proteins into aminoacids and then ammonia.

B. subtilis: common soil bacterium. Produces antibiotic subtilin.

B. thuringensis: organism, its toxin, and even the gene (also plasmid-encoded) for the toxin are used as a biocontrol agent against a variety of insect pests.

Bacterium cartovorus: causes soft rot of mango.

Beijerinckia: aerobic, non-symbiotic nitrogen fixing bacteria.

Borrelia burgdorferi: causes Lyme disease.

Bordetella pertussis: causes whooping cough.

Campylobacter jejuni: causes gastrointestinal upsets.

Cellulomonas: cellulolytic bacterium.

Chlamydia psittaci: causes psittacosis (also known as ornithosis).

C. trachomatis: causes trachoma.

Clostridium tetani: causes tetanus.

C. acetobutylicum: produces riboflavin, Vit. B_{12}.

C. botulinum: causes acute food poisoning.

C. butyrium: causes retting of fibres.

C. pasteurinum: anaerobic, non-symbiotic nitrogen fixing bacteria.

C. perfringens: causes gangrene.

Cornyebacterium diphtheriae: causes diphtheria.

C. tritici: causes Tundu disease of wheat.

Cytophaga: cellulolytic bacterium.

Desulphovibrio desulphuricans: effects corrosion of metallic pipes.

Desulphovibrio: anaerobic, non-symbiotic nitrogen fixing bacteria.

Erwinia atroseptica: causes black rot of potato.

E. ceratovora: causes soft rot of vegetables.

E. amylovora: causes fruit blights.

Escherichia coli: intestinal symbiont. Checks growth of putrifying bacteria and produces Vitamin B and K. Mutant strains cause gastric infection.

Francisella tularensis: causes tularemia, disease of small mammals.

Frankia: forms nitrogen fixing nodules in the stem of nonlegumes: e.g. *Casuarina, Alnus, Rubus*.

Haemophilus influenzae: causes bacterial meningitis and middle ear infections in children and pneumonia in adults.

Helicobacter pylori: causes stomach ulcers.

Klebsiella: aerobic, non-symbiotic nitrogen fixing bacteria.

Lactobacillus: several species are used to convert milk into cheese, butter, and yogurt.

L. bulgaricus: used to produce yogurt and cheese.

Leptospira ictero-heamorrhagiae : causes jaundice.

Leuconostoc: imparts flavour to curd.

Methylobacterium: converts methane to methanol (found in rotten bananas).

Methylococcus: converts methane to methanol (found in coconut).

Methylophilus: produces single cell protein from methane or methanol-rich medium.

Methylotropous: produces single cell protein from methane or methanol-rich medium.

Micrococcus aureus: causes food poisoning.

Mycobacterium leprae: causes leprosy.

M. tuberculosis: causes tuberculosis (TB).

Mycoderma: forms vinegar from sugar syrup.

Mycoplasma genitalium: causes infection of testes.

M. pneumoniae: causes atypical pneumonia.

Neisseria gonorrhoeae: causes gonorrhoea, sexually-transmitted diseases (STDs).

N. meningitidis: causes meningococcal meningitis.

Nitrobacter: nitrifying bacterium.

Nitrosomonas: nitrifying bacterium, oxidises NH_3 produced from proteins to (NO_2^-).

Pasteurella pestis: causes plague.

Phytobacterium solanacearum: causes wilt of tobacco.

Propionibacterium: used in flavouring of swiss cheese.

Pseudomonas aeruginosa: present in soil and water, resistant to most antibiotics, can cause serious illness in humans with defective immune systems, serious burns and cystic fibrosis.

P. denitrificans: denitrifying bacteria.

P. solanacearum: causes wilt of solanaceae.

Rhizobium: fixes nitrogen (N_2), in the air into compounds that can be used by living things.

Rickettsia prowazekii: causes typhus fever transmitted to humans by lice and Rocky Mountain spotted fever, transmitted by ticks.

Salmonella enterica var typhimurium: confined to the intestine, it is a frequent cause of human gastrointestinal upsets, also known *as S. typhimurium*.

Salmonella enterica var typhi: causes typhoid fever, also known as *S. typhi*.

S. enterides: causes food poisoning.

Shigella dysenteriae: causes dysentery.

Staphylococcus aureus: can cause acne, food poisoning, trachoma. Some strains of *S. aureus* secrete a toxin and can cause life-threatening toxic shock syndrome.

Streptococcus pyogenes: causes strep throat.

S. thermophilus: used to produce yogurt and cheese.

S. cremoris: forms curd from milk.

S. lactis: forms curd from milk.

S. pneumoniae: causes pneumonia.

Streptomyces: source of antibiotics like streptomycin, chloromycetin, aureomycin, terramycin, neomycin.

Thiobacillus denitrificans: denitrifying bacteria.

Treponema pallidum: causes syphilis.

Vibrio cholerae: causes cholera.

Xanthomonas citri: causes citrus canker.

X. juglandes: causes blight of walnut.

X. malvacearum: causes angular leaf spot of cotton.

X. oryzae: causes blight of rice.

X. pruni: causes bacterial spot of peach.

X. solanacearum: causes ring spot disease of potato.

X. campestris: causes soft rot of solanaceae.

Yersinia pestis: causes bubonic plague. Transmitted to humans by the bite of an infected flea.

ANNEXURE II

Metric Measurements and Conversions

1. Length

1 Angstorm (Å) = 0.1 nanometre, 0.0001 micrometre

1 nanometre (nm) = 10 Å, 0.001 micrometre

1 micrometre (μm) = 1000 nanometre, 0.001 millimetre

1 millimetre (mm) = 1,000 micrometre, 0.1 centimetre, 0.0394 inch

1 centimetre (cm) = 10 millimetre, 0.394 inch

1 metre (m) = 100 centimetres, 39.37 inches, 3.28 feet, 10^2 cm, 10^3 mm, 10^6 μm, 10^9 nm, 10^{10} Å

1 kilometre = 1000 metres, 3,280 feet, 0.62 mile

1 inch = 2.5 cm

1 metre =39 inches, about a yard

1g = about 1/30 of an ounce

1 litre = about 1.06 quarts

To convert basic unit $10°=1$

into deci multiply by $10^{-1}(0.1)$

into centi multiply by $10^{-2}(0.01)$

into milli multiply by $10^{-3}(0.001)$

into micro multiply by $10^{-6}(0.000001)$

into nano multiply by $10^{-9}(0.000000001)$

into pico multiply by $10^{-12}(0.000000000001)$

2. Mass

1 picogram (pg) = 0.000000000001grams (10^{-12})

1 nanogram (ng) = 0.000000001 grams (10^{-9})

1 microgram (μg) = 0.000001 grams (10^{-6})

1 milli grams (mg) = 0.001 grams (10^{-3})

1 gram (g) = 1000 milligrams, 10^3 mg, 10^6 μg, 10^9 ng, 10^{12} pg

1 kilograms (kg) = 1000 grams, 2.2 pounds

3. Volume

1 microlitre (μl) = 1 cubic millimetre (mm^3), 0.000001 litre

1 millilitre (ml) = 1 cubiccentimetre (cm^3, cc) 1 (mg), 0.001 litre, 0.061 cubic inch,

0.03 fluid ounce (fl.oz)

1 litre = 1000 millilitres, 1.06 quarts, 10^3 ml, 10^6 μl.

4. Notations

10^{-15} = quadrillionth, prefix; femto (f); 0.000000000000001

10^{-12} = trillionth, prefix; pico (p); 0.000000000001

10^{-9} = billionth, prefix; nano (n); 0.000000001

10^{-6} = millionth, prefix; micro (m); 0.000001

10^{-3} = thousandth, prefix; milli (m); 0.001

10^{-2} = hundredth, prefix; centi (c); 0.01

10^{-1} = tenth, prefix; deci (d); 0.1

10^2 = hundred, prefix; hecto (h); 100

10^3 = thousand, prefix; kilo (k); 1,000

10^6 = million, prefix; mega (M); 1,000,000

10^9 = billion, prefix; giga (G); 1,000,000,000

5. Temperature

Fahrenheit to Celsius: °C = 5/9 (°F– °C)

Celsius to Fahrenheit: °F = 9/5 (°C) + 32

Kelvin to Celsius: K= °C +273.2

32°F = 0°C

98.6°F = 37°C; normal human body temperature.

212°F = 100°C; boiling point of water.

250°F = 121°C; temperature for sterilisation in autoclave.

338°F = 170°C; temperature for sterilisation in hot air oven.

pH scale

S. No	pH value	H⁺ ion concentration	OH⁻ ion concentration	pH of common substances and solutions
1	0	10^0	10^{-14}	HCL
2	1	10^{-1}	10^{-13}	
3	2	10^{-2}	10^{-12}	Stomach acid (1–3), lemon juice (2.3)
4	3	10^{-3}	10^{-11}	Vinegar, wine, soft drinks, beer, orange juice, acid rain
5	4	10^{-4}	10^{-10}	Tomatoes, grapes, bananas (4.6)
6	5	10^{-5}	10^{-9}	Black offee, bread, rain water
7	6	10^{-6}	10^{-8}	Urine (5–7), milk (6.6), saliva (6.2–7.4)
8	7 (Neutral)	10^{-7}	10^{-7}	Pure water, blood (7.3–7.5)
9	8	10^{-8}	10^{0-6}	Egg white (8.0), sea water (7.8–8.3)
10	9	10^{-9}	10^{-5}	Baking soda, phosphate based detergents, antacids
11	10	10^{-10}	10^{-4}	Soap solutions, milk of magnesia
12	11	10^{-11}	10^{-3}	Household ammonia (10.15–11.9), non phosphate detergents
13	12	10^{-12}	10^{-2}	Washing soda
14	13	10^{-13}	10^{-1}	Oven cleaner
15	14	10^{-14}	10^{-0}	NaOH

pH range and colour change of some commonly used dyes

Indicator	pH range	Change in colour from acidic to alkaline medium
1. Cresol red (pK_1)	0.2- 1.8	red to yellow
2. Thymol blue (pK_1)	1.2- 2.8	red to yellow
3. Bromophenol blue	3.0- 4.6	yellow to blue
4. Bromocresol green	3.8- 5.4	yellow to blue
5. Methyl red	4.4- 6.0	red to yellow
6. Chlorophenol red	4.8- 6.4	yellow to red
7. Bromocresol purple	5.2- 6.8	yellow to purple
8. Bromothymol blue	6.0- 7.6	yellow to blue
9. Phenol red	6.8- 8.4	yellow to red
10. Cresol red (pK_2)	7.2- 8.8	yellow to red
11. Thymol blue (pK_2)	8.0- 9.6	yellow to blue
12. Phenolphthalein	8.3- 10.0	colourless to red
13. Tolyl red	10.0- 11.6	red to yellow
14. Parazo orange	11.0- 12.6	yellow to orange
15. Acyl blue	12.0- 13.6	red to blue

PREPARATION OF SOLUTIONS

Molar Solutions

A molar solution is one in which 1 litre of solution contains the number of grams equal to its molecular weight.

$$\text{molarity} = \frac{\text{wt. of solute in g/l of solution}}{\text{mol. wt. of solute}}$$

E.g. To make one litre of 0.1M solution of NaOH.

Mol. wt. of NaOH= 40

Amount of NaOH /l of solution = mol. wt. of NaOH $\times$ molarity $= 40 \times 0.1 = 4$ g

   Weigh 4g of NaOH and dissolve it in little amount of solvent, then make up the final volume to 1000ml.

Percent Solutions

Percentage (w/v) = weight (g) in 100 ml of solution.

Percentage (v/v) = volume (ml) in 100 ml of solution.

   E.g. To make a 0.7% solution of sucrose, take 0.7g of sucrose and bring up volume to 100 ml with distilled water.

Concentrated Stock Solutions

Many enzyme buffers are prepared as concentrated solutions, e.g. 5x or 10x (five or ten times the concentration of the working solution) and are then diluted such that the final concentration of the buffer in the reaction is 1x.

E.g. For 25 µl of buffer, take 2.5 µl of a 10x buffer and add water to make final volume of 25 µl.

PREPARATION OF WORKING SOLUTIONS FROM CONCENTRATED STOCK SOLUTIONS

$$N_1 V_1 = N_2 V_2$$

N_1 = concentration of solution to be prepared

V_1 = volume of solution to be prepared

N_2 = concentration of stock solution

V_2 = volume of stock solution

$$V_2 = \frac{N_1 V_1}{N_2}$$

To prepare 150 ml of solution containing 10mM Tris from a stock of 75mM Tris

$$\frac{10 \times 150}{75} = 20 \text{ ml}$$

20 ml of 75 mM Tris is added to water to get a final volume of 150 ml solution containing 10mM.

Solution of known concentration is called Standard solution. E.g. Standard solution used to prepare reference curve contains accurately measured concentration of solute. **Saturated solution** contains excess amounts of solute.

Steps in Solution Preparation

- Refer to the laboratory manual for any specific instructions on preparation of the particular solution and the bottle label for any specific precautions in handling the chemical.
- Weigh out the desired amount of chemical(s). Use an analytical balance if the amount is less than 0.1 g.
- Place chemical(s) into appropriate size beaker with a stir bar.
- Add less than the required amount of water. Prepare all solutions with double-distilled water.

- When the chemical is dissolved, transfer to a graduated cylinder and add the required amount of distilled water to achieve the final volume. An exception is in preparing solutions containing agar or agarose. Weigh the agar or agarose directly into the final vessel.

- If the solution needs to be at a specific pH, first calibrate the pH metre with fresh buffer solutions and follow instructions for using a pH metre.

- Autoclave, if possible, at 121°C for 20 min. Some solutions cannot be autoclaved, for example, SDS. These should be filter sterilised through a 0.22 μm filter. Media for bacterial cultures must be autoclaved the same day it is prepared, preferably within an hour or two. Store at room temperature and check for contamination prior to use by holding the bottle at eye level and gently swirling it.

- Solid media for bacterial plates can be prepared in advance, autoclaved, and stored in a bottle. When needed, the agar can be melted in a microwave, any additional components, e.g. antibiotics, can be added and the plates can then be poured.

- Concentrated solutions, e.g. 1M Tris-HCl pH = 8.0, 5M NaCl, can be used to make working stocks by adding autoclaved double-distilled water in a sterile vessel to the appropriate amount of the concentrated solution.

BUFFERS

Citrate Buffer

Stock solution A: 0.1 mol/litre solution of citric acid (19.12g in one litre).

Stock solution B: 0.1 mol/litre solution of sodium citrate (29.41g of $C_6H_5O_7Na_3 \cdot 2H_2O$ in one litre).

Take X ml of A and Y ml of B, make upto 100 ml.

pH	Amount of A (X ml)	Amount of B (Y ml)	Amount of Water (ml)
3.0	46.5	3.5	50
3.2	43.7	6.3	50
3.4	40.0	10.0	50
3.6	37.0	13.0	50
3.8	35.0	15.0	50
4.0	33.0	17.0	50
4.2	31.5	18.5	50
4.4	28.0	22.0	50
4.6	25.5	24.5	50
4.8	23.0	27.0	50
5.0	20.5	29.5	50
5.2	18.0	32.0	50
5.4	16.0	34.0	50
5.6	13.7	36.3	50
5.8	11.8	38.2	50
6.0	9.5	40.5	50
6.2	7.2	42.8	50

Acetate Buffer

Stock Solution A: 0.2 mol/litre solution of acetic acid (11.55g in one litre).

Stock Solution B: 0.2 mol/litre solution of sodium acetate.
(16.4 g of $C_2H_3O_2Na$ or 27.2g of $C_2H_3O_2Na \cdot 3H_2O$ in one litre)

Take X ml of A and Y ml of B, make upto 100 ml.

pH	Amount of A (X ml)	Amount of B (Y ml)	Amount of Water (ml)
3.6	46.3	3.7	50
3.8	44.0	6.0	50
4.0	41.0	9.0	50
4.2	36.8	13.2	50
4.4	30.5	19.5	50
4.6	25.5	24.5	50
4.8	20.0	30.0	50
5.0	14.8	35.2	50
5.2	10.5	39.5	50
5.4	8.8	41.2	50
5.6	4.8	45.2	50

Phosphate Buffer

Stock solution A: 0.2 mol/litre solution of monobasic sodium phosphate (31.2g of $NaH_2PO_4 \cdot 2H_2O$ in one litre).

Stock solution B: 0.2 mol/litre solution of dibasic sodium phosphate (28.39g of Na_2HPO_4 or 71.7 g of $Na_2HPO_4 \cdot 12H_2O$ in one litre).

Take X ml of A and Y ml of B, make upto 200 ml.

pH	Amount of A (X ml)	Amount of B (Y ml)	Amount of Water (ml)
5.8	92.0	8.0	100
6.0	87.7	12.3	100
6.2	81.5	18.5	100
6.4	73.5	26.5	100
6.6	62.5	37.5	100
6.8	51.0	49.0	100
7.0	39.0	61.0	100
7.2	28.0	72.0	100
7.4	19.0	81.0	100
7.8	8.5	91.5	100
8.0	5.3	94.7	100

Barbitone (Veronal) Buffer

Stock Solution A: 0.2 mol/litre solution of sodium barbitone (sodium diethyl barbiturate 36.9g in one litre).

Stock solution B: 0.2 mol/litre HCl.

Take 50 ml of A and X ml of B, make upto 200 ml.

pH	Amount of A (X ml)	Amount of B (Y ml)	Amount of Water (ml)
6.8	50.0	45.0	105.0
7.0	50.0	43.0	107.0
7.2	50.0	39.0	111.0
7.4	50.0	32.5	117.5
7.6	50.0	27.5	122.5
7.8	50.0	22.5	127.5
8.0	50.0	17.5	132.5
8.2	50.0	12.7	137.3
8.4	50.0	9.0	141.0
8.6	50.0	6.0	144.0
8.8	50.0	4.0	146.0
9.0	50.0	2.5	147.5
9.2	50.0	1.5	148.5

Tris (Hydroxymethyl) Aminomethane HCl (Tris HCl) Buffer

Stock Solution A: 0.2 mol/litre solution of Tris (hydroxymethyl) aminomethane (24.2g in one litre).

Stock solution B: 0.2 mol/litre of HCl.

Take 50 ml of A and X ml of B, make upto 200 ml.

pH	Amount of A (X ml)	Amount of B (Y ml)	Amount of Water (ml)
7.2	50.0	44.2	105.8
7.4	50.0	41.4	108.6
7.6	50.0	38.4	111.6
7.8	50.0	32.5	117.5
8.0	50.0	26.8	123.2
8.2	50.0	21.9	148.1
8.4	50.0	16.5	133.5
8.6	50.0	12.2	137.8
8.8	50.0	8.1	141.9
9.0	50.0	5.0	145.0

TE Buffer

100 ml of TE buffer (10 mM Tris, 1 mM EDTA), Mix 1 ml of a 1 M Tris solution and 0.2 ml of 0.5 M EDTA and 98.8 ml sterile water.

Saline: 0.85% NaCl in water.

PREPARATION OF MOLAL SOLUTION

$$\text{Molarity} = \frac{\text{Wt. of solute in g/kg of solution}}{\text{mol. wt. of solute}}$$

To prepare 1 molal solution dissolve 1 mole of solute in 1000ml of water

$$\text{Normality} = \frac{\text{Amount of substance in g/l of solution}}{\text{Eq. wt. of substance}}$$

To prepare 0.1 N solution of Na_2CO_3 dissolve 5.3 g (eq. wt. of Na_2CO_3 = 5.3) in final volume of 1 litre of solution.

PPM (PARTS PER MILLION SOLUTION)

Gram of solute per million grams of solution or one g of solute/million ml of solution

$$\text{ppm} = \frac{\text{Mass of the component}}{\text{Total mass of solution}} \times 10^6$$

or

$$\text{ppm} = \frac{\text{g or ml of solute or substance}}{\text{g or ml of solution}} \times 10^6$$

therefore 1 ppm =1mg NaCl/l or 1mg NaCl /1000 ml or 1 mg NaCl/ml

PREPARING SOLUTIONS FROM CONCENTRATED ACIDS

$$V_1 = \frac{\text{Eq. wt. of acid} \times V_2 \times \text{normality} \times 100}{10000 \times \text{sp. gravity} \times \text{purity (\%)}}$$

sp. Gravity of HCl =1.84, purity = 96%, mol. wt. = 98 therefore for making 750 ml of 1N sulphuric acid from concentrated acid

$$V_1 = \frac{49 \text{ (Eq. wt. of acid)} \times 750 \text{ } (V_2) \times \text{normality} \times 100}{1000 \times 1.84 \text{ (sp. gravity)} \times 96 \text{ (\% purity)}} = 20.8 \text{ ml}$$

20.8 ml conc. acid is diluted with water to make final solution 750 ml to get 1 N acid solution.

ANNEXURE III

Composition of Reagents

S. No	Reagent	Composition Chemical	Quantity	Analysis
1.	Acetate Buffer Solution	A. Sodium acetate solution Sodium acetate Distilled water B. Acetic acid solution glacial acetic acid Distilled water Mix A and B in equal proportion	68 g 250 ml 25 ml 175 ml	Acetate buffer
2.	Ammonium Buffer solution	EDTA $MgSO_4 \cdot 7 H_2O$ NH_4Cl Conc. NH_4OH Mix well & dilute to 250 ml with Distilled water	1.179 g 0.780 g 16.9 g 143 ml	Ammonium buffer
3.	Calcium Chloride Solution	$CaCl_2$ Distilled water	27.5 g 1000 ml	Biological oxygen demand
4.	CAS Reagent	A. Chrome Azure S Distilled water B. Fe III Soln. (27 mg $FeCl_3$ 83.3 ml conc. HCl and 100 ml Distilled water). C. *HDTMA Distilled water *Hexadecyl Trimethyl Ammonium Bromide. Mix A & B, Add HDTMA slowly while stirring to get a dark blue solution.	60.5 mg 50 ml 10 ml 72.0 mg 40 ml	Detection of siderophore producing ability

S. No	Reagent	Composition		Analysis
		Chemical	Quantity	
5.	Crystal Violet – Ammonium Oxalate	A. Crystal violet	2 g	
		95% Ethanol	20 ml	
		B. Ammonium oxalate	0.8 g	
		Distilled water	80 ml	Gram's stain
		Mix A and B		
6.	Ferric Chloride Solution	$FeCl_3.6H_2O$	0.25 g	
		Distilled water	1000 ml	Biological oxygen demand
7.	Ferroin Indicator	1,10 –phenathroline monohydrate	1.485 g	
		$FeSO_4.7H_2O$	0.695 g	Chemical oxygen demand
		Distilled water	100 ml	
8.	0.25M Ferrous Ammonium Sulphate	$Fe(NH_4)_2(SO_4)_2 \cdot 6H_2O$	98 g	
		Conc. H_2SO_4	20 ml	Chemical oxygen demand
		Dilute with Distilled water	1000 ml	
9.	Gram's Iodine	KI	2 g	
		Iodine	1 g	Gram's stain
		Distilled water	300 ml	
10.	50% Hydrochloric Acid	Conc. HCl	50 ml	Iron
		Distilled water	50 ml	
11.	3% Hydrogen Peroxide	Hydrogen peroxide	3 ml	Catalase test
		Distilled water	97 ml	
12.	1% Iodine solution	Iodine	1 ml	Amylase test
		Distilled water	99 ml	
13.	10% KI solution	KI	10 g	Dissolved organic matter & Dissolved inorganic matter
		Distilled water	90 ml	
14.	Lugol's solution	Iodine	1 g	Preservation of algae in sample
		KI	2 g	
		Distilled water	300 ml	
15.	Magnesium sulphate solution	$MgSO_4 \cdot 7H_2O$	22.5 g	Biological oxygen demand
		Distilled water	1000 ml	

S. No	Reagent	Composition		Analysis
		Chemical	Quantity	
16.	5% Malachite Green	Malachite Green Distilled water	5 g 95 ml	Endospore Stain
17.	0.05% Methyl Orange	Methyl orange Distilled water	0.05 g 99.9 ml	Alkalinity
18.	α-Napthol Solution	α-Napthol Absolute alcohol	5 g 100 ml	V P test
19.	0.5% Phenolphthalein Indicator	Phenolphthalein 50% Ethanol	0.5 g 100 ml	Alkalinity
20.	Phosphate buffer Solution	KH_2PO_4 K_2HPO_4 $Na_2HPO_4 \cdot 7H_2O$ NH_4Cl Distilled water	8.2 g 21.75 g 33.4 g 1.7 g 1000 ml	Biological oxygen demand
21.	0.0417M Potassium Dichromate	$K_2Cr_2O_7$ Distilled water	12.259 g 1000 ml	Chemical oxygen demand
22.	40% Potassium Hydroxide Reagent	KOH Creatine Distilled water	40 g 0.3 g 100 ml	V P test
23.	N/80 Potassium Permanganate	$KMnO_4$ Distilled water	0.3950 g 1000 ml	Dissolved organic Matter & Dissolved inorganic matter
24.	0.2N Potassium Permanganate Solution	$KMnO_4$ Distilled water	600 mg 100 ml	Iron
25.	Safranin	Safranin Distilled water	1 g 100 ml	Gram's Stain
26.	1% Starch Solution	Starch Powder Distilled water	1 g 99 ml	Dissolved organic matter & Dissolved inorganic matter
27.	2% Starch Solution	Starch Hot Distilled water	2 g 98 ml	Dissolved oxygen

S. No	Reagent	Composition		Analysis
		Chemical	Quantity	
28.	N/10 Sodium Thiosulphate	$Na_2S_2O_3 \cdot 5H_2O$	24.83 g	
		Distilled water	1000 ml	Dissolved oxygen
29.	N/40 Sodium Thiosulphate	$N/10\ Na_2S_2O_3$	25 ml	
		Distilled water	75 ml	Dissolved oxygen
30.	N/80 Sodium Thiosulphate	$N/10\ Na_2S_2O_3$	12.5 ml	Dissolved organic
			87.5 ml	matter & Dissolved inorganic matter
31.	Sulphuric Acid Reagent	Ag_2SO_4	5.5 g	Chemical oxygen
		Conc. H_2SO_4	272 ml	demand
32.	25% Sulphuric Acid Reagent	Conc. H_2SO_4	25 ml	
		Distilled water	75 ml	Dissolved organic matter & Dissolved inorganic matter
33.	0.02N H_2SO_4	1N H_2SO_4	20 ml	
		Distilled water	980 ml	Alkalinity
34.	Sulphuric Acid Solution	Conc. H_2SO_4	500 ml	Nitrate
		Distilled water	75 ml	
35.	Winkler A; Manganous Sulphate Solution	$MnSO_4.H_2O$	364 g	Dissolved oxygen
		Distilled water	1000 ml	
36.	Winkler B; Alkali –iodide-azide Reagent	KOH	700 g	Dissolved oxygen
		KI	150 g	
		Distilled water	1000 ml	
		NaN_3	10 g in 40 ml DW	

ANNEXURE IV

COMPOSITION OF VARIOUS MEDIA USED FOR CULTURE OF MICROBES

A. Media Used for Bacterial Culture

S. No	Name of nutrient medium	Type	Composition of Medium		pH
			Chemical	Quantity (g/1)	
1.	Blood Agar	Selective medium enrichment medium	Tryptone	10	7.3
			Beef extract	3.0	
			NaCl	5.0	
			Agar	15	
			Distilled water	1000ml	
			Blood	10%	
			Before adding blood, the prepared medium is clear and yellowish-brown, then blood colored and not haemolytic.		
2.	Brilliant Green Lactose Agar (Bile Broth)	Confirmed test for detection of coliforms	Peptone	10	7.2±
			Lactose	10	0.2 at
			Ox bile, dried	20	25°C
			Brilliant Green	0.0133	
			Agar	15	
			Distilled water	1000 ml	

S. No	Name of nutrient medium	Type	Composition of Medium		pH
			Chemical	Quantity (g/1)	
3.	Bromothymol Blue Agar	Selective medium for *E.Coli*	Peptone	10.0	7
			Oxoid Yeast extract	5.0	
			NaCl	5.0	
			Agar	20	
			Distilled water	1000ml	
			Autoclave and add :		
			1ml 5% Maranil solution		
			2ml 50%Sodium thiosulphate		
			10ml 1% Bromothymol blue		
			25ml 33% Lactose solution		
			1.2ml 33% Glucose solution		
4.	Cetrimide Agar	Selective medium for *Pseudomonas aeruginosa*	Peptone	20	7
			$MgCl_2$	1.4	
			Potassium sulphate	10	
			Cetrimide (Cetrimethyl diammonium bromide)	0.3	
			Agar	13.6	
			Distilled water	1000ml	
5.	Choclate Agar (Heated blood agar)	Medium for *Haemophilus influenzae* and other fastidious bacteria such as *Neisseria* and *Pneumococcus*	Proteose Peptone	20	7
			Dextrose	0.5	
			NaCl	5.0	
			Agar	15	
			Disodium Phosphate	5.0	
			Distilled water	1000ml	
			Melt desired amount of agar. Cool in water bath at 75°C. Add 10% sterile blood and allow medium to remain at 75°C mixing agar and blood by gentle stirring till blood becomes chocolate coloured. (about 10 min.)		
6.	Capryllate Thallus (CT) Agar	Medium for isolation of *Serratia*			7.2
			H_3PO_4	1.96	
			$FeSO_4 \cdot 4H_2O$	0.0556	
			$MnSO_4 \cdot 4H_2O$	0.0223	
			$CuSO_4 \cdot 5H_2O$	0.025	
			$CO(NO_3)_2 \cdot 6H_2O$	0.03	
			H_3BO_3	0.0062	
			Distilled water	1000ml	

S. No	Name of nutrient medium	Type	Composition of Medium		pH
			Chemical	**Quantity (g/1)**	
			Solution A:		
			$CaCl_2 \cdot 2H_2O$	0.0147	
			$MgSO_4$	0.123	
			KH_2PO_4	0.0680	
			K_2HPO_4	2.610	
			* Trace element solution	10ml	
			Caprylic acid	1.1ml	
			Thallous sulphate	0.25	
			Distilled water	500ml	
			Yeast extract (5% wt/vol)	2ml	
			Adjust pH to 7.2, autoclave.		
			Solution B:		
			NaCl	7.0	
			$(NH_4)_2SO_4$	1.0	
			Agar	15	
			Distilled water	500ml	
			Adjust pH to 7.2. autoclave, Mix A & B		
			Trace element solution (store at 4^0C)		
7.	Deca Broth	Medium for bacteriophage isolation	Peptone	100	7.6
			Yeast extract	50	
			NaCl	25	
			K_2HPO_4	80	
8.	Deoxycholate Citrate Agar	Enrichment medium for *Proteus*	Peptone	10	7.0
			Lactose	10	
			Sodium deoxycholate	1.0	
			NaCl	5.0	
			K_2HPO_4	2.0	
			Ferric citrate	1.0	
			Sodium citrate	1.0	
			Neutral red	0.03	
			Agar	15	
			Distilled water	1000ml	
9.	DNAse Test Agar	Test for DNAse activity	Tryptose	20	7.3
			DNA	2.0	
			NaCl	5.0	
			Agar	15	

S. No	Name of nutrient medium	Type	Composition of Medium		pH
			Chemical	Quantity (g/1)	
10.	Egg Yolk Agar	Lipase activity	Egg yolk	100 ml (5%)	
			Plate count agar	900ml	
11.	Endo Agar	Confirmed test for coliforms	Peptone	10	
			Lactose	10	
			K_2HPO_4	3.5	
			Na_2SO_3	5.0g	
			Basic Fuchsin	0.5g	
			Agar	15	
			Distilled water	1000ml	
12.	Enrichment Medium		Yeast extract	6.0	7.0
			Proteose peptone	40.0	
			K_2HPO4	3.0	
			NaCl	5.0	
13.	Eosin Methylene Blue (EMB) Agar	Selective media for coliforms, Confirmed test for presence of coliforms	Peptone	10	7.2
			Lactose	5.0	
			K_2HPO_4	2.0	
			Sucrose	5.0	
			Eosin Y	0.4	
			Methylene Blue	0.065	
			Agar	15.0g	
			Distilled water	1000ml	
14.	Gelatin Agar	Test for proteolytic activity	Nutrient Agar	1000 ml	7
			KH_2PO_4	0.5	
			K_2HPO_4	1.5	
			Gelatin	4.0	
			Glucose	0.05	
15.	Gluconate Broth	Gluconate test	Peptone	1.5	7
			Yeast extract	1.0	
			K_2HPO_4	1.0	
			Potassium gluconate	40.0	
			or Sodium gluconate	37.25	
			Distilled water	1000ml	
16.	Glucose Nitrate Salt Agar	Medium for isolation of Staphylococci	Glucose	10	
			$NaNO_3$	10	
			KH_2PO_4	10	
			KCl	10	
			$MgSO_4.7H2O$	10	
			Agar	15	
			Distilled water	1000ml	

S. No	Name of nutrient medium	Type	Composition of Medium		pH
			Chemical	Quantity (g/1)	
17.	Glucose Peptone Agar	Culture of *Rhizobium*	Glucose	10	7.2
			Peptone	20	
			NaCl	5.0	
			Agar	15	
			Distilled water	1000ml	
			1.6 % Bromocresol purple in 95% Ethanol	1.0 ml.	
18.	Glucose Phosphate Broth	Indole reaction	Glucose	5.0	7.0-
			Proteose peptone	5.0	7.2
			K_2HPO_4	5.0	
			Distilled Water	1000 ml.	
19.	Hofer's Alkaline Medium	Culture of *Rhizobium*	YEMA	1000ml	7
			Bromothymol blue stock solution	5ml	
			(Stock solution ; 5 gm BTB in 100 ml ethanol)		
20.	KF Agar	Medium for feacal *Streptococcus* test	Proteose Peptone	10.0	7.2
			Yeast extract	10.0	
			NaCl	5.0	
			Sodium Glycerophosphate	10	
			Maltose	20.0	
			Lactose	1.0	
			Sodium azide	0.4	
			Bromocresol purple	0.015	
			Sodium carbonate	0.636	
			Distilled water	1000ml	
			Autoclave and add 1% unsterilized solution of TTC (Tetrazolium chloride) at room temperature.		
21.	Knight and Munaier's Medium	Selective medium for actinomycetes	Dextrose	1.0	7
			KH_2PO_4	0.1	
			$NaNO_3$	0.1	
			KCl	0.1	
			$AgSO_4 \cdot 7H_2O$	0.1	
			Agar	15	
			Distilled Water	1000 ml.	
22.	Lactose Broth	Presumptive test for standard analysis of water	Beef extract	6.0	6.9
			Peptone	10.0	
			Lactose	10.0	
			Distilled water	1000ml	

S. No	Name of nutrient medium	Type	Composition of Medium		pH
			Chemical	**Quantity (g/1)**	
23.	MacConkey Agar	General Purpose medium , selective medium for Enterobacteria, *Proteus*	Peptone NaCl Sodium taurocholate Distilled water Agar Dissolve all in DW. Adjust pH to 8 with 40% NaOH. Add agar Dissolve agar by steaming, take two egg whites, and dissolve them in 50 ml water,. Add this to 1 litre of medium and steam for 2 hrs. Adjust pH to 7.4 with NaOH solution. Before use add 100 ml of 10% sterile aqueous lactose and 3.5 ml of 1% neutral red solution (dissolved in 50% alcohol). Omission of NaCl prevents spreading of *Proteus*.	20.0 5.0 5.0 1000 ml 20.0	7.4
24.	Meat Infusion Broth	General purpose, enrichment medium, selective medium	Lean meat Peptone NaCl Distilled water	500 10.0 5.0 1000ml	7.4
25.	M- Endo Broth	Medium for total coliform test by membrane filter technique	Yeast extract Peptone Lactose Sodium deoxycholate K_2HPO_4 KH_2PO_4 NaCl Sodium lauryl sulfate Na_2SO_3 Basic Fuchsin DW Heat to boiling, do not autoclave, cool and dispense in sterile absorbent pads	1.5g 5.0 12.5 0.1 4.37 5.0 5.0 0.05 2.1 1.05 1000ml	7.2 +/− 2 at 25°C
26.	M-FC Broth	Medium for feacal coliform test	Tryptose Proteose Peptone Yeast extract NaCl Bile salts Lactose Aniline blue Add 10ml of 1% rosonilic acid in 0.2 N NaOH. Continue heating for 1 min. Cool to room temp, do not autoclave.	10 5.0 3.0 5.0 1.5 12.5 0.1	7.4

S. No	Name of nutrient medium	Type	Composition of Medium Chemical	Quantity (g/1)	pH
27.	Milk Agar	Test for proteolytic activity	Equal volumes of double strength nutrient agar and reconstituted skimmed milk		7.2
			or		
			Nutrient agar	87.5 ml	
			Reconstituted skimmed milk	12.5 ml	
28.	Nitrate Broth	Test for enzyme activity	Beef Extract	3.0	7.0
			Peptone	5.0	
			NaCl	5.0	
			$NaNO_3$	5.0	
			DW	1000 ml	
29.	Novobiocin-Penicillin Cycloheximide Agar	Medium for *Pseudomonas* culture	Novobiocin	45 µg	
			Penicillin G	44.9 mg	
			Cycloheximide	75 mg	
			Mix these in 3ml ethanol, dilute with 50ml sterile DW and then add to one litre sterile *Pseudomonas* F Agar		
30.	Nutrient Agar	General purpose medium for bacteria	Peptone	5.0	6.8 – 7.2
			NaCl	3.0	
			Beef Extract	3.0	
			Agar agar	15.0	
			DW	1000 ml	
31.	Nutrient Broth	General purpose medium for bacteria	Peptone	5.0	7.0
			NaCl	3.0	
			Beef Extract	3.0	
			DW	1000 ml	
32.	Oxidation Fermentation (OF) Medium. Heat to dissolve and add bromothymol blue	Carbohydrate fermentation	Peptone	2.0	7.2
			NaCl	1.5	
			K_2HPO_4	1.0	
			Agar	3.0	
			Distilled water	1000 ml	

S. No	Name of nutrient medium	Type	Composition of Medium		pH
			Chemical	Quantity (g/1)	
33.	Peptone Broth (Peptone Water)	To prepare bacterial suspension	4% aqueous solution of peptone		7.2
34.	Peptone Water Broth	Test for indole activity	NaCl	1.4	7.2
			$NaHCO_3$	0.02	
			KCl	0.04	
			$CaCl_2$	0.04	
			KH_2PO_4	0.24	
			$Na_2HPO_4 \cdot 2H_2O$	0.88	
			Peptone	0.8	
			Distilled water	1000 ml	
			Steam the ingredients at 100^0C for 20min., filter through double fold Whatman no.1 paper. Adjust pH and autoclave.		
35.	Phenyl Pyruvic Acid (PPA) Slope Medium	Medium for identification of *Proteus*	DL-phenylalanine	2.0	
			Meat extract	3.0	
			Na_2HPO_4	1.0	
			NaCl	5.0	
			Agar	1.2	
			Distilled water	100ml	
36.	Plate Count Agar	Lipase activity, Standard plate count	Tryptone	5.0	7.2
			Yeast extract	2.5	
			Glucose	1.0	
			Agar	15	
			Distilled water	1000ml	
37.	Pikovskaya Medium	Medium for detection of phosphate solubilizing ability	Glucose	10.0	
			$Ca_3(PO_4)$	5.0	
			$(NH_4)_2SO_4$	0.5	
			NaCl	0.2	
			$MgSO_4 \cdot 7H_2O$	0.1	
			KCl	0.2	
			Yeast extract	0.5	
			$MnSO_4 \cdot H_2O$	0.002	
			$FeSO_4 \cdot 7H_2O$	0.002	
			Agar	15.0	
			Distilled water	1000 ml	

S. No	Name of nutrient medium	Type	Composition of Medium — Chemical	Quantity (g/1)	pH
38.	*Pseudomonas* F Agar	Selective medium for fluorescein pigment producing *Pseudomonas*	Peptone/casein enzymic hydrolysate	10.0	7+/−
			Proteose Peptone	10.0	
			Dipotassium phosphate	1.5	
			Magnesium sulphate	1.5	
			Agar	15	
			Distilled water	1000ml	
39.	*Pseudomonas* P Agar	Selective medium for pyocin pigment producing *Pseudomonas*	Peptone	20.0	7
			$MgCl_2$	1.4	
			Potassium sulphate	10.0	
			Agar	15.0	
			Distilled water	1000ml	
40.	Reid's Modified Wilson & Blair's Medium	Medium for culture of *Vibrio*	Add 1% NaCl and mannose to Wilson & Blair's Medium		9.2
41.	Robertson's Cooked Meat Medium	It permits the growth of strict anaerobes and preserves delicate organisms.	Nutrient broth and pieces of fat free minced cooked meat of ox heart.		7.2
42.	Semisolid Medium	Medium for studying motility	0.2-0.5% agar in 1 litre medium.		7
43.	Simmon's Citrate Agar	Test for utilization of specific substrate	$MgSO_4$	0.2	6.9
			$(NH_4)_2HPO_4$	1.0	
			K_2HPO_4	1.0	
			Sodium citrate	2.0	
			NaCl	5.0	
			Bromothymol blue	0.08	
			Agar	15g	
			DW	1000 ml	
44.	Soil Extract Agar *	Medium for culture bacteria of soil fungi, actinomycetes	Tap water	900ml	7.0 – 7.2
			Soil extract stock	100ml	
			Agar	15	
			K_2HPO_4	0.5	
			Glucose	1.0	

S. No	Name of nutrient medium	Type	Composition of Medium		pH
			Chemical	**Quantity (g/1)**	
45.	Starch Agar	Test for enzyme activity	Beef Extract	3.0	
			Peptone	5.0	
			NaCl	5.0	
			Agar	15.0	
			10% Starch solution	200 ml	
			Distilled water	1000 ml	
			Steam starch separately for 1 hour and then add to molten agar.		
46.	SL-Medium (medium should be clear light straw colored firm gel)	Medium for isolation of *Lactobacilli*	Yeast extract	5.0	
			Trypticase	10.0	
			KH_2PO_4	6.0	
			Diammonium citrate	2.0	
			$MgSO_4 \cdot 7H2O$	0.58	
			$MnSO_4 \cdot 4H2O$	0.28	
			Glucose	10	
			Arabinose	5.0	
			Sucrose	5.0	
			Tween 80	1.0 ml	
			Sodium acetate ($5H_2O$)	2.5	
			Agar	15	
			Distilled water	1000 ml	
			Dissolve and heat agar by steaming in 500ml water. Dissolve all other ingredients except sodium acetate in molten agar and steam for 5 minutes. Dissolve Sodium acetate in 15ml DW and add 10%v/v of 99.5%aqueous acetic acid to adjust pH to 5.4. Make up the vol. to 20ml. Add this buffer to hot medium in sterile screw capped bottles. Avoid repeated melting and cooling.		
47.	SS Agar	Medium for culture of *Salmonella*	Beef extract	5.0	7
			Lactose	10.0	
			Bile salt mixture	8.5	
			Sodium citrate	8.5	
			Sodium thiosulphate	8.5	
			Ferric citrate	1.0	
			Agar	13.5	
			Distilled water	1000 ml	

S. No	Name of nutrient medium	Type	Composition of Medium		pH
			Chemical	Quantity (g/1)	
48.	SX Agar based on Starr's Medium.	Medium for culture of *Xanthomonas* (*Phytomonas*)	Beef extract	1.0	6.8
			Soluble potato starch	10.0	
			Ammonium chloride	5.0	
			Potassium diphosphate	2.0	
			Methyl violet (1% in 20% ethanol)	1ml	
			Methyl green (1 % solution)	2ml	
			Cycloheximide	250mg	
			Agar	15	
			Distilled water	1000ml	
49.	Tetrathionate Reductase Broth	Medium for species identification of *Serratia*	Potassium tetrathoinate	5.0	7.4
			Bromothymol blue (0.2 % aqueous solution)	25ml	
			Peptone water	1000ml	
50.	Thioglycollate Broth	Anaerobic medium	Yeast extract	5.0	7.1
			Casein hydrolysate		
			Pancreatic digest	15	
			L-cystine	5.0	
			Agar	0.75	
			NaCl	2.5	
			Sodium thioglycollate	0.5	
			Glucose	5.0	
			Resazurin sodium solution (freshly prepared)	1 ml	
			Distilled water	1000ml	

Dissolve all ingredients except thioglycollate and resazurin by steaming at 100ºC. Add thioglycollate and adjust pH to 7.3. If precipitate forms, heat without boiling, filter hot through moist filter paper. Add resazurin, mix and autoclave. Cool at once at 25ºC and store in dark at 20ºC to 30ºC.

S. No	Name of nutrient medium	Type	Chemical	Quantity (g/1)	pH
51.	Thiosulphate Citrate Bile Salt Sucrose (TCBS) Agar	Medium for *Vibrio*	Yeast extract	5.0	8.6
			Peptone	10.0	
			Sucrose	20.0	
			Sodium thiosulphate	10.0	
			Sodium citrate	3.0	
			Ox-gall	5.0	
			NaCl	10.0	
			Thymol blue	0.04	
			Bromothymol blue	0.04	
			Ferric citrate	1.0	
			Agar	15.0	
			Distilled water	1000ml	

S. No	Name of nutrient medium	Type	Composition of Medium		pH
			Chemical	**Quantity (g/1)**	
52.	Toluidine Blue Agar (TDA)	Medium for isolation of Staphylococci	DNA	0.3	
			0.01M CaCl$_2$	1.0ml	
			NaCl	1.0	
			Agar in tris buffer pH 9	10.0	
			Distilled water	1000ml	
			Boil the mixture to dissolve DNA and agar, add 3ml of 0.1M toulidine blue		
53.	Triple Sugar Iron (TSI) Agar	Production of hydrogen sulphide and differentiation of coliforms	Peptone	20.0g	7.4
			Proteose peptone	5.0	
			NaCl	5.5	
			Lactose	10.0	
			Sucrose	10.0	
			Dextrose	1.0	
			Ferrous ammonium sulphate	0.2	
			Sodium Thiosulphate	0.3	
			Phenol red	0.024	
			Beef extract	3.0	
			Yeast extract	3.0	
			Agar	12.0	
			Distilled water	1000ml	
54.	Urea Agar	Urease activity	Peptone	1.0	6.8-
			Glucose	1.0	6.9
			NaCl	5.0	
			K$_2$HPO$_4$	5.0	
			Phenol Red	6.0 ml of	
			Agar	20	
			Distilled water	1000ml	
			Prepare agar base and autoclave. Prepare 40% urea solution separately and sterlize by a bacteriological filter. Add urea solution to agar base to make final concentration of 20% urea. Dispense in tubes and autoclave.		

S. No	Name of nutrient medium	Type	Composition of Medium		pH
			Chemical	Quantity (g/1)	
55.	Wilson & Blair's Medium	Medium for culture of *Salmonella*	**Solution A: Bismuth sulphite** Bismuth ammonium citrate Distilled water Mix the two and then add anhydrous sodium sulphite Make volume upto 1000ml with DW Add $Na_2HPO_4 \cdot 12H_2O$ **Solution B: Iron citrate brilliant green mixture** 1% solution Ferric citrate with 1% brilliant green make upto 25 ml with DW Add 70ml of (A) and 4ml of (B) to 1000ml sterile nutrient agar.	30.0 100ml 100 21.0	
56.	Yeast Extract Broth	Yeast medium for sampling of air by membrane filter	Yeast extract Malt extract Peptone Dextrose Agar Distilled Water	3.0 3.0 5.0 10.0 20.0 1000ml	4.5
57.	Yeast Extract Mannitol Agar (YEMA)	Medium for culture of *Rhizobium*	K_2HPO_4 $MgSO4 \cdot 7H_2O$ NaCl Mannitol Yeast extract powder Agar Distilled water	0.5 0.2 0.1 10 0.4 15 1000ml	7
58.	Yeast Extract Mannitol Agar with 1% Congo Red (CRYEMA)	Culture of *Rhizobium* culture of *Rhizobium*	YEMA 1.0% congo red	1.01 2.5 ml.	7
59.	Yeast Medium	Medium for sampling of air by membrane filter	Yeast extract Malt extract Peptone Dextrose Agar Distilled water	3.0 3.0 5.0 10 20 1000	4.5

B. Composition of Media Used for Fungal Culture

Name	Type	Composition in g/l		pH
1. Czapek's Dox Agar	General purpose medium	Sucrose	30.0	7.3
		NaNO₃	2.0	
		K₂HPO₄	1.0	
		MgSO₄	0.5	
		KCl	0.5	
		FeSO₄	0.01	
		Agar	20.0	
		DW	1000 ml	

Dissolve all salts except phosphate in 500 ml DW, add sucrose to it. Dissolve phosphate separately in 500ml DW and add to the rest of the solution.

2. Corn meal Agar	Selective medium for culture of yeasts	Corn meal	20	
		Peptone	20	
		Agar	15	
		Dextrose	20	
		DW	1000 ml	

Before addition of agar boil corn meal in a little quantity of water to make a smooth cream. Add water and filter through cheese cloth.

3. Malt Extract Agar	Medium for culture of wood destroying fungi	Malt extract	25	
		Agar	15	
		DW	1000 ml.	

4. Potato Dextrose Agar	General purpose medium	Potato	250	6.0-
		Agar	15	6.5
		Dextrose	20	
		DW	1000 ml.	

Peel the potatoes, cut them into small pieces and boil 250 ml water. Mash the boiled potatoes to form a thick paste. Strain through cheese cloth. Add dextrose and agar and water to make upto a litre. Autoclave and use.

5. Richard's Solution	Selective medium for *Aspergillus*	KNO₃	10.0	6.6-
		KH₂PO₄	5.0	7.2
		MgSO₄	2.5	
		FeCl₃	0.02	
		Sucrose	50.0	
		DW	1000 ml	

	Name	Type	Composition in g/l		pH
6.	Rose Bengal Agar	Selective medium for yeasts	Dextrose	10.0	4.0
			Peptone	5.0	
			KH_2PO_4	1.0	
			$MgSO_47H_2O$	0.5	
			Streptomycin	0.03	
			Agar	20.	
			Rose Bengal	0.035 .	
			DW	1000 ml.	

The antibiotic should be filter sterilized separately and added aseptically to the sterilized medium. Aureomycin (33-2000mg) can be used in place of streptomycin.

	Name	Type	Composition in g/l		pH
7.	Sabouraud Agar	General purpose medium	Glucose	20	5.6
			Peptone	10	
			Agar	20.	
			DW	1000 ml.	

Medium can be supplemented with 10 mg /ml Aureomycin or 16 mg /ml chloramphenicol or 400 mg /ml cycloheximide or 5 mg /ml gentamycin to prevent bacterial growth. The antibiotic should be filter sterilized separately and added aseptically to the sterilized medium.

	Name	Type	Composition in g/l		pH
8.	YEDP Agar	General purpose medium for fungi	Dextrose	20	
			Peptone	10	
			Yeast extract	3	
			Agar	15 .	
			DW	1000 ml.	

	Name	Type	Composition in g/l		pH
9.	Waksman's Agar	Medium for isolating soil fungi	Glucose	10	4.0
			Peptone	5	
			K_2HPO_4	1.0	
			$MgSO_47H_2O$	0.5	
			Agar	20	
			DW	1000 ml.	

C. Composition of Media Used for Algal Culture

Name	Type	Composition in g/l		pH
1. BG–11	Selective medium for blue green algae	$NaNO_3$	1.5g	
		K_2HPO_4	0.04 g.	
		$MgSO_4 \cdot 7H_2O$	0.075 g	
		$CaCl_2 \cdot 2H_2O$	0.036g	
		Ferric Ammonium Citrate	0.006g	
		Na_2CO_3	0.02g	
		EDTA	0.001g	
		Trace metal mix	1.0ml	
		Citric acid	0.005gm	
		DW	1000 ml	

Trace metal mix : $CO(NO_3)_2\ 6H_2O$; 1.0ml

Add KNO_3 1 .5 g/l for algae other than Blue green algae.

A_5 Micronutrient Solution		H_3BO_3	2.86g	
		$MgCl_2 \cdot 4H_2O$	1.81g	
		$ZnSO_4 \cdot 4H_2O$	0.222g	
		$Mo\ O_3$	0.0177g	
		$CuSO_4 \cdot 5H_2O$	0.079g	
		DW	1000 ml	
2. Chu's Medium	Selective medium for cyanobacteria	Calcium nitrate	0.04	
		Magnesium sulphate	0.025	
		K_2HPO_4	0.005	
		Sodium carbonate-	0.020	
		Sodium silicate	0.025	
		Ferric chloride	0.008	
		DW	1000 ml	

Name	Type	Composition in g/l		pH
3. Fe – EDTA Solution 1 ml/l		EDTA	26.19	
		KOH (1N)	268 ml	
		$FeSO_4 \cdot 7H_2O$	24.9	
4. Fogg's Nitrogen Free Medium	Selective medium for cyanobacteria	KH_2PO_4	0.2	7.5 –
		$MgSO_4 \cdot 7H_2O$	0.2	8.0
		$CaCl_2 \cdot 2H_2O$	0.1	
		Fe – EDTA	1ml	
		A_5 solution	1ml	

Trouble Shooting

S No.	Problem	Causes
1.	Abnormal color of medium	Deteriorated dehydrated medium, Improperly washed glassware, Impure water.
2.	Atypical precipitate	Deteriorated dehydrated medium, Improperly washed glassware, Impure water, Incorrect weighing, Incomplete mixing, Overheating.
3.	Incomplete solubility	Incomplete mixing, Inadequate heating, Inadequate convection in a very small flask.
4.	Caremelization or dark colour of mediium	Deteriorated dehydrated medium, Incorrect weighing, Incomplete mixing, Overheating.
5.	Toxicity	Improperly washed glass ware, Impure water, Burning or scorching.
6.	Medium remains liquid or loses gelation property	Incorrect weighing, Incomplete mixing, Overheating, Too much dilution due to large inoculum, Repeated remelting.
7.	Loss of property such as selectivity, differential nature or nutritive value	Deteriorated dehydrated medium, Impure water, Incorrect weighing, Incomplete mixing, Overheating, Too much dilution due to large inoculum, Repeated remelting.
8.	Contamination	Improper sterilization, Improper culture technique, Absence of sterile conditions.
9.	Incorrect pH	Deteriorated dehydrated medium, Improperly washed glass ware, Impure water, Incorrect weighing, Incomplete mixing, Overheating, Repeated remelting, Storage at high temperature, Hydrolysis of ingredients, pH determined at wrong temperature.

ANNEXURE V

List of Institutes offering M.D/ M. Sc/ B. Sc / Research Facilities in Microbiology and Biotechnology

1. Agharkar Research Institute Animal Sciences Division Agarkar Road, Pune 411004, Maharashtra

2. Aligarh Muslim University, Aligarh, Uttar Pradesh, 202002

3. All India Institute of Medical Sciences, Ansari Nagar, New Delhi 110029.

4. Amity University Noida./ Jaipur/ Lucknow

5. Assam University, Silchar, Silchar, Assam.

6. Banaras Hindu University, Varanasi, Uttar Pradesh, 221005.

7. Banasthalli Vidhyapeeth Rajasthan 304022.

8. Barakatullah Vishwavidyalaya, Faculty of Science & Faculty of Life Sciences, Hoshangabad Road, Bhopal, Madhya Pradesh, 462026

9. Bhabha Atomic Research Centre, Trombey, Mumbai

10. Bharathiar University, Coimbatore, Coimbatore, Tamil Nadu, 641046

11. Bharathidasan University, Tiruchirappalli, Palkalaiperur, Tiruchirappalli , Tamil Nadu, 620024

12. Bhavnagar University, Gauyrishanker Lake Road, Bhavnagar, Gujarat, 364002.

13. Birla Institute of Scientific Research Statue Circle, Jaipur-302001 (Rajasthan) Phone ; 91-141-2385283,5108163 fax: 91-141-2385121, website:www.bisrjaipur.com. Email: research@bisrjaipur.com.

14. Calicut University Kozhikode, Kerela 673635.

15. Central Drug Research Institute (CDRI), Lucknow

16. Central Food Technological Research Institute (CFTRI), Mysore, 570013

17. Central Institute of Medicinal And Aromatic Plants Lucknow, 226015, Uttar Pradesh

18. Central Research Institute (CRI), Kasauli, Himachal Pradesh.

19. Centre for Biotechnology, Pondicherry University Pondicherry 605014.

20. Centre for Cellular and Molecular Biology, Hyderabad.

21. Centre for Plant Molecular Biology, Tamil Nadu Agricultural University Coimbtore 641003.

22. Chaudhary Charan Singh Haryana Agricultural University, Faculty of Basic Sciences & Humanities, Hisar, Haryana, 125004.

23. Chhatrapati Shahu Ji Maharaj University, Kalyanpur, Kanpur, Uttar Pradesh, 208024.

24. Christian Medical College, Bagayam, Vellore, Tamil Nadu, 632001.

25. Consortium India Ltd. G-6 (3rd Floor), NDSE Part 1New Delhi.

26. Council for Scientific and Industrial Research, New Delhi.

27. Defence Research and Development Organisation.

28. Department of Biosciences & Biotechnology, Banasthali Vidyapeeth Banasthali –304022 (Rajasthan), Phone 91-1438-228302, Fax: 91-1438-228365 email: ban_bioinfo@yahoo.com" website: http://www.banasthali.ac.in

29. Department of Biosciences, M S University, Baroda, Gujrat.

30. Department of Biotechnology, Devi Ahilya Vishwavidhalaya, Indore, MP.

31. Department of Biotechnology, Guru Nanak Dev University Amritsar.143005

32. Department of Biotechnology, Jadavpur University, Calcutta.700032.

33. Department of Life Sciences, Jawahar Lal Nehru University (JNU), Delhi.

34. Department of Marine Sciences Goa University Goa.403005.

35. Department Of Microbiology National Institute of Immunology New Delhi 110067.

36. Department of Microbiology, All India Institute of Medical Sciences (AIIMS), Ansari Nagar, NewDelhi.

37. Department of Microbiology, C.U.Shah Medical College, Surendernagar.

38. Department of Microbiology, CBSH, G.B.Pant University of Agriculture and Technology, Pantnagar

39. Department of Microbiology, Choudhary Charan Singh Haryana Agricultural University. (CCSHAU).

40. Department of Microbiology, Christ College, Rajkot.

41. Department of Microbiology, Delhi University (DU), Delhi.

42. Department of Microbiology, Kakatiya University, Warangal.

43. Department of Microbiology, M.J. College, Jalgaon

44. Department of Microbiology, Maharaja Sayajirao University Baroda.

45. Department of Microbiology, Mithibai College, Mumbai.

46. Department of Microbiology, Panjab University, Chandigarh.

47. Department of Microbiology, Postgraduate Institute of Medical Education and Research (PGIMER), Chandigarh.

48. Department of Microbiology, Pune University, Pune.

49. Department of Microbiology, Shivaji University, Kolhapur.

50. Department of Microbiology, Shri M&N Virani Science College, Rajkot.

51. Department of Microbiology, University of Delhi South Campus, New Delhi

52. Department of microbiology, University of Pune, Pune, Maharashtra

53. Department of Microbiology, University School of Sciences, Gujarat University, Ahmedabad

54. Department Of Molecular Biology Institute Of Microbial Technology P.B. No. 1304, Sector-39-A Chandigarh, 160014.

55. Dept. of Microbiology, Gulbarga University, Gulbarga-585106

56. Dept. of Biotechnology, Panjab University, Chandigarh

57. Dept. of Microbiology, C.B.H. & H., C.C.S.H.A.U.,Hisar-125004 Haryana

58. Dept. of Microbiology, C.B.S. & H., P.A.U., Ludhiana-141 001

59. Dept. of Microbiology, MDS University, Ajmer (Raj.)

60. Dept. of Microbiology, Panjab University, Chandigarh

61. Deptt. Of Bioscience, Saurashtra University, Rajkot, Gujarat- 360005

62. Directorate Of Research (Veterinary and Animal Sciences) Rajasthan Agriculture University, Bikaner-334006 Telephone: 0151-2545359 Fax 0151-2250336.Email:directorvas@yahoo.co.uk

63. Division Of Microbiology Central Drug Research Institute Chattar Manzil Palace, Lucknow

64. Dr. Babasaheb Ambedkar Marathwada University, Aurangabad-431004, Maharashtra

65. Enterovirus Research Centre Haffkine Institute Compound, Parel, Bombay

66. Gauhati University, Faculty of Science, PO Gopinath Bordoloi Nagar, Guwahati, Assam,781014

67. Goa University, Faculty of Science, Sub PO Goa University, Taleigao Plateau, Goa, 403203.

68. Govind Ballabh Pant University of Agriculture & Technology, Pantnagar, Dist. Nainital, Uttranchal, 263145.

69. Gujarat University, Faculty of Science, Navrangpura, Ahmedabad, Gujarat, 380009.

70. Gulbarga University, Faculty of Science & Technology, Jnana Ganga, Gulbarga, Karnataka, 585106

71. Guru Nanak Dev University, Faculty of Science, Amritsar, Punjab, 143005.

72. Gurukulakangri Vishwavidyalaya, P O Gurukulakangri, Hardwar, Uttar Pradesh, 249404.

73. Haryana Agricultre University Hissar, Haryana.

74. Himachal Pradesh Krishi Vishwavidyalaya, Faculty of Basic Sciences, Palampur, Distt. Kangra, Himachal Pradesh, 176062.

75. Himachal Pradesh University, Faculty of Science, Summer Hill, Shimla, Himachal Pradesh, 171005.

76. IIT Delhi/Kharagpur/Mumbai/ Kanpur.

77. Indian Agricultural Research Institute, New Delhi.

78. Indian Council for Medical Research, New Delhi.

79. Indian Institute of Science, Bangalore.

80. Indian Institute of Science, Banglore,560012.

81. Indira Gandhi Centre for Human Ecology, Environmental & Population studies, University of Rajasthan, Jaipur. Email: heeps@uniraj.ernet.in

82. Institute for Microbial Technology (IMTECH) Chandigarh .

83. Institute of Engineering & Technology, Biotechnology college, MIA Alwar (Rajasthan).

84. Institute of Microbial Technology (IMT), Post Box No 1304, Sector 39-A Chandigarh , 160036.

85. International College for Girls Gurukul Marg, SFS, Mansarovar, Jaipur (Rajasthan)Phone: 91-141-2395494, website: http://www.icfia.org. Email: icg@icfia.org

86. ITRC, Industrial Technology Research Center, Lucknow, P.B. No 173, Uttar Pradesh, 226001.

87. Jai Narayan Vyas University, Bhagat Ki Kohti Pali Road Jodhpur-342001, (Rajasthan). Phone 91-291-2649733, 2432947 Fax no. 91-291-2649733. Web site: http://www.jnvu.org.

88. Jaipur Engineering College & Research Centre, Shri Ram Ki Nangal, Via Vatika, Tonk Road, Jaipur (Rajasthan)

89. Jawaharlal Nehru Technological University, Masab Tank, Mahaveer Marg, Hyderabad Andhra Pradesh.

90. Jawaharlal Nehru University, New Delhi. 110067

91. Jiwaji University, Faculty of Sciences & Faculty of Life Sciences, Vidya Vihar, Gwalior, Madhya Pradesh, 474011.

92. Kakatiya University, Vidyaranyapuri, Warangal, Andhra Pradesh, 506009.

93. Kurukshetra University, Faculty of Science, Kurukshetra, Haryana 136119.

94. Lalit Narayan Mithila University, Faculty of Science, Kameshwaranagar, Darbhanga, Bihar, 846008.

95. M L Sukhadia University, Udaipur, (Rajasthan) 313001 Tel:+91-294-2413035 Fax 91-294-2413150. E-mail: adm@phynt.mlsu.ac.in

96. M N Institute of Applied Sciences, Bikaner (Rajasthan)

97. Maharishi Dayanand University , Ajmer, (Rajasthan)

98. Maharshi Dayanand University, Faculty of Science, Rohtak, Haryana 124001.

99. Mahatama Gandhi Instt. of Applied Science Jaipur Shri Ram Ki Nangal, Via Vatika, Tonk Road, Jaipur (Rajasthan) Phone:0141-2770274, Fax-0141-2770803, Email:mgijaipur@hotmail.com

100. Mahatama Jyoti Roa Phoole, Shikshan sanasthan,Ram nagar road,New Sanganer Road Sodala Phone : 0141-2294680, 2295101, Fax: 91-141-2292146.

101. Mahatma Gandhi University, Faculty of Applied Science, Priyadarshini Hills PO, Kottayam, Kerala, 686560.

102. Mangalore University, Faculty of Science, University Campus, Mangalagangothri, Karnataka, 574199.

103. Manipal Academy of Higher Education, Manipal, Karnataka.

104. Manonmaniam Sundaranar University, University Campus, Abishekapatti, Tirunelveli, Tamil Nadu, 627018.

105. Medical Research Society Mumbai, Maharashtra, 400004.

106. Microbial Type Culture Collection (MTCC) Institute of Microbial Technology (IMT), Post Box No 1304, Sector 39-A, Chandigarh, 160036.

107. Microbiology Deptt., MVM Science& Home Science College, Rajkot.

108. Microbiology Division, Indian Agricultural Research Institute (IARI), Pusa, New Delhi.

109. Modi Institute of Management & Technology Dadabari Etx., Kota

110. Mody College of Arts, Science & Commerce, Lakshmangarh, Sikar, (Rajasthan). Phone: 91-01573-225001-225012 (12 lines), Fax 01573-225044, Email:modycollege@yahoo.com, Website:www.modyinst.com.

111. Molecular Microbiology, Biotechnology and Biochemistry, Microbiology Unit, College of Science & Technology, Andhra University, Visakhapatnam-530 003, Andhra Pradesh, INDIA.

112. Nagpur University, Ravindra Nath Tagore Marg, Nagpur, Maharashtra, 440001.

113. National Centre for Biological Sciences Tata Institute of Fundamental Research, Homi Bhabha Road, Mumbai, Maharashtra, 400005.

114. National Centre for Cell Science (NCCS), Ganesh Khind, Pune.

115. National Centre for Plant Genome Research (NCPGR), New Delhi.

116. National College for Girls, Vijay Nagar, Alwar (Rajasthan).

117. National Dairy Research Institute (NDRI), Karnal.

118. National Environmental Engineering Research Institute, Nagpur.

119. National Facility for Tissues and Cell Culture, Pune.

120. National Institute of Immunology (NII), New Delhi.

121. National Institute Of Virology 20-A, Dr. Ambedkar Road P.B. No.11, Pune 411001 Maharashtra.

122. North Maharashtra University, Faculty of Science, Post Box No. 80, Umavinagar, Jalgaon, Maharashtra, 425002.

123. Orissa University of Agriculture & Technology, Bhubaneswar, Orissa, 751003.

124. Osmania University, Hyderabad, 500007, Andhra Pradesh.

125. Padmashree College of Medical Laboratory Technology, India

126. Plant Biotechnology Center Rajasthan Agriculture University, Bikaner 334006, (Rajasthan), Phone:91-151-2250689, Fax:91-151-250576

127. Pondicherry University, R.V Nagar, Kalapet, Pondicherry, 605014.

128. Postgraduate Institute of Medical Education and Research, Chandigarh.

129. Pt. Ravishankar Shukla University, Faculty of Life Science, Raipur, Madhya Pradesh, 492010.

130. Punjab Agricutural University, Ludhiana Punjab.

131. Punjab University, Sector-14, Chandigarh 160014.

132. Rajendra Agricultural University Pusa, Samisthipur, Bihar.

133. Rani Durgavati Vishwavidyalaya, Faculty of Science & Life Sciences, Saraswati Vihar, Pachpedi, Jabalpur, Madhya Pradesh, 482001

134. Sai College of Medical Science & Technology, Opp Chaubepur Police Station, Chaubepur, Kanpur, Uttar Pradesh, 209203.

135. Sardar Patel University, Vallabh Vidyanagar Gujarat, 388120.

136. Saurashtra University, Faculty of Science, University Road, Rajkot, Gujarat, 360005.

137. School of Biological Sciences Madurai kamraj University Madurai 625021.

138. School of Biotechnology, Banaras Hindu University Varanasi 221005.

139. School of Life Sciences, Central University Hyderabad.

140. Seedling Academy of Design Technology and Management. Khorebariyon, Jagatpura, Jaipur (Rajasthan) phone:0141-2753377,2754399. Email: seedlingacademy@hotmail, Website: www.seedlingeducation.com.

141. Seth Moti Lal (P.G) College, Centre for Biology Reverence, Vigyan Bhawan Jhunjhunu 333001 (Rajasthan).

142. Shivaji University, Vidyanagar, Kolhapur, Maharashtra-416004.

143. Shobhasaria College of Engineering NH-11 Gokulpura, Sikar-332001, Sikar Phone: 91-1572-2 2 2 650 –653 Fax: 91-1572-222654.

144. South Gujarat University, Faculty of Science, Post Box No. 49, Udhna Magdalla Road, Surat, Gujarat, 395007.

145. Sri Krishandevaraya University, Sri Venkateswarapuram, Anantapur, Andhra Pradesh, 515003.

146. Sri Sathya Sai College for Women, Jawahar Nagar Bye Pass, Jaipur, (Rajasthan).Offers graduation courses in Biotechnology.

147. Stani Memorial College, IRRM campus, Mansarovar, Jaipur-302020.

148. Swami Ramanand Teerth Marathwada University, Faculty of Science, Dnyanteerth, Gautami Nagar, Vishnupuri, Nanded Maharashtra, 431603

149. Tamil Nadu Dr MGR Medical University, Chennai, Tamil Nadu.

150. The Maharaja Sayajirao University of Baroda, Fatehgani, Vadodara, Gujarat, 390002.

151. University of Calcutta, Senate House 87, College Street, Kolkata, West Bengal, 700073.

152. University of Calicut, Faculty of Science, PO Malappuram Dt. Kozhikode, Kerala, 673635.

153. University of Health Sciences, Vijayawada, Andhra Pradesh.

154. University of Hyderabad, Hyderabad 500046.

155. University of Kerala, Faculty of Science, University PO, Thiruvananthapuram , Kerala, 695034.

156. University of Madras, Faculty of Science, Centenary Building, Chepauk, Triplicane P O, Chennai Tamil Nadu, 600005.

157. University of Mumbai, M G Road, Fort, Mumbai Maharashtra, 400032.

158. University of Mysore, Crawford Hall, Mysore, Karnataka 570005

159. University of Pune, Faculty of Science, Ganeshkhind, Pune Maharashtra, 411007.

160. University of Rajasthan Jaipur, Jawahar Lal Nehru Marg, Jaipur 302004. Rajasthan, Phone No. 0141-2706813,2710995 Fax: 2711799 Web Site: "http://www.uniraj.ernet.in/" www.uniraj.ernet.in, Email: "mailto:info@uniraj.ernet.in".

161. USB College of Pharmacy and Biotechnology. Abu Road, Rajasthan.

162. Utkal University, PO Vani Vihar, Bhubaneswar, Orissa, 751004.

ANNEXURE VI

LIST OF SOME IMPORTANT JOURNALS

International Journals

1. Annals of Clinical Microbiology and Antimicrobials
2. Annual Review of Microbiology
3. Antimicrobial Agents and Chemotherapy
4. Applied and Environmental Microbiology
5. Applied Biochemistry and Microbiology
6. Applied Microbiology and Biotechnology
7. Archives of Microbiology
8. Archives of Virology
9. Canadian Journal of Microbiology
10. Clinical Microbiology and Infection
11. Clinical Microbiology Newsletter
12. Clinical Microbiology Reviews
13. Critical Reviews in Microbiology
14. Current Microbiology
15. Environment Microbiology
16. European Journal of Clinical Microbiology and Infectious Diseases
17. FEMS Immunology and Medical Microbiology
18. FEMS Microbiology and Ecology
19. FEMS Microbiology Letters
20. FEMS Microbiology Reviews

21. Food Microbiology

22. Infection and Immunity

23. International Journal of Antimicrobial Agents

24. International Journal of Food Microbiology

25. International Journal of Systematic and Evolutionary Microbiology

26. International Microbiology

27. Journal of Antimicrobial Chemotherapy

28. Journal of Applied Microbiology

29. Journal of Bacteriology

30. Journal of Basic Microbiology

31. Journal of Clinical Microbiology

32. Journal of Clinical Virology

33. Journal of Eukaryotic Microbiology

34. Journal of Molecular Microbiology and Biotechnology

35. Microbial Ecology

Indian Journals

1. Indian Journal of Microbiology , Division of Microbiology, Indian Agricultural Research Institute (IARI), Pusa, New Delhi - 110 012

2. Indian Journal of Medical Microbiology , L.V. Eye Institute, L.V. Prasad Marg, Banjara Hills, Hyderabad - 500 034

3. Current Science, Indian Academy of Sciences, C. V. Raman Avenue, P.O. Box No. 8001, Bangalore - 560 080

4. Indian Journal of Comparative Microbiology, Immunology and Infectious Diseases Room no. 338, Immunology Section, Modular Laboratories Buildings, IVRI, Izatnagar -243 122

5. Indian Journal of Pharmacology, Department of Pharmacology, JIPMER, Pondicherry - 605 006

6. Indian Journal of Pharmaceutical Sciences, Kalina, Santacruz East, Mumbai - 400 098

7. Indian Journal of Virology, CCS Haryana Agricultural University, Hissar, Haryana

8. Journal of Biosciences, Indian Academy of Sciences, C. V. Raman Avenue, P.O. Box No. 8005, Bangalore - 560 080

9. Journal of Genetics, Indian Academy of Sciences, C. V. Raman Avenue, P.O. Box No. 8005, Bangalore - 560 080

10. Phytomedica, Indian Herbs House, Sharda Nagar, Saharanpur - 247 001 (U.P.)

ANNEXURE VII

IMPORTANT MICROBIOLOGY BOOKS

1. *An Introduction to Microbiology.* Tauro P., Kapoor K. K and Yadav K.S. 1986. Wiley Eastern Limited.

2. *Bailey & Scott's Diagnostic Microbiology.* Forbes B. A., Sahm D. F., Weissfeld A. S and Baron E.J. 1998. 11th ed. The C.V. Mosby Co., St. Louis.

3. *Bergey's Manual of Determinative Bacteriology.* Buchanan R.E and N.E. Gibbons (Eds). 1974. 8th ed. The Williams and Wilkins Co., Baltimore.

4. *Bergey's Manual of Determinative Bacteriology.* Holt J. G. (Ed.) 1994. 9th ed. Williams and Wilkins Company Baltimore.

5. *Bergey's Manual of Systematic Bacteriology.* Holt J.G. (Ed.) Volume 1, 1982. Gram-negative bacteria of medical or industrial importance. Volume 2, 1986. Gram-positive bacteria of medical or industrial importance. Volume 3, 1988. Other Gram-negative bacteria, cyanobacteria, Archaea. Volume 4, 1988. Other Gram- positive bacteria.

6. *Biology 201: Microbiology Laboratory Manual.* Johnson T.R., Case C.L., Cappuccino J.G., Sherman N., Schurman V and Savage J. 1999. Addison-Wesley.

7. *Biology of Microorganisms.* Madigan M.T., Martinko J. M and Parker. J. 2000. 9th ed. Brock, Prentice-Hall, Inc., Englewood Cliffs, New Jersey.

8. *Diagnostic Molecular Microbiology: Principles and Applications.* Persing D. H., Smith T.F., Tenover F.C and White T.J. 1993. ASM Press. NY.

9. *Difco Manual.* Anonymous.1984. Dehydrated Culture Media and Reagents for Microbiology. Difco Laboratories, Detroit.

10. *Disinfection, Sterilization and Preservation.* Block S.C. 1992. 4th ed. Lea and Febiger. Baltimore.

11. *Five Kingdoms: An Illustrated Guide to the Phyla of Life on Earth.* Margulis L., and K.V Schwartz.1988. W.H Freeman. San Francisco.

12. *Fundamentals of Microbiology*. Edward A. I.1994. 5th ed.The Benjamin/Cummings Publishing Company. Inc. Redwood City CA.

13. *General Microbiology*. Atlas R.M. 1997. 2nd Ed. Wm. C. Brown publishers. Dubuque

14. *General Microbiology*. Boyd R. 1988. Times Mirror / Mosby College Publishing, St. Louis.

15. *General Microbiology*. Schlegel H. G. 1995. 7th ed. Cambridge Low price Editions. Cambridge University Press. U K.

16. *Hand Book of Microbiology*. Bisen P.S and Verma K. 1994 CBS Publishers and Distributors, New Delhi.

17. *Handbook of Microbiological Media*. Atlas R.M. 1993. CRC Press Florida.

18. *Microbiology*. Tortora G.J., Berdell R. F and Case C.L. 1989. Benjamin-Cummings Publishing Company.

19. *International Code of Nomenclature of Bacteria*. Sneath P. H. A. 1992. American Society of Microbiology. Washington. DC.

20. *Introduction to Microbiology*. Ingraham J. L and Ingraham C. A. 2000.2nd ed. Brooks/ Cole USA.

21. *Laboratory Excercises in Microbiology*. Prescott L. M and Harley J. P. 1993. 2nd ed. Wm. C. Brown Publishers.

22. *Laboratory Experiments in Microbiology*. Johnson T.R and Case C.L. 2003. Addison – Wesley.

23. *Laboratory Guide for Identification of Plant Pathogenic Bacteria*. Schaad N.W., Jones J.B and Jones W.C. 2001. 3rd ed. American Phyto-pathological Society.

24. *Laboratory Manual and Workbook in Micro-biology*. Morello, J. A., Mizer H. Li and Wilson M. Li., 1991. Wm. C. Brown Pub.

25. *Laboratory Textbook and Experiments in Microbiology*. Bartholomew J.1977. Kendall / Hunt Publishing Co., Dubuque.

26. *Light and Electron Microscopy*. Slayter E. M and Slayter H. S. 1992 Cambridge University Press NY.

27. *Manual of Basic Techniques for a Health Laboratory*. WHO, 1980.

28. *Manual of Clinical Microbiology*. Murray P. R., Brown E. J., Faller M. A., Tenover F.C and Yolken R. H. 1995. 6th ed. ASM press. Washington DC.

29. *Manual of Methods for General Bacteriology*. Gerhardt P. (ed.) 1994. ASM Press Washington DC.

30. *Medical Microbiology.* Boyd, R. F. and Marr Little J. 1980. Brown and Co., Boston.

31. *Methods in Microbiology.* Desai J.D and Desai A. J. 1995 Emkay Publications Delhi.

32. *Microbes in Action: Laboratory Manual of Microbiology.* Seeley H. W and Vandenmark P. J.1975 2nd ed. WH Freeman & Co. Pvt. Ltd.

33. *Microbial Applications.* Benson H. J. 1990. 5th ed. Complete and short versions Wm. C. Brown Publishers. Dubuque.

34. *Microbiological Applications: A Laboratory Manual in General Microbiology.* Benson, H. J. 1990. Wm. C. Brown Publishers, Dubuque.

35. *Microbiology A Laboratory Handbook.* Cappucino J. G and Sherman N. 2004. 6th ed. Pearson Education Pvt. Ltd. Singapore.

36. *Microbiology Laboratory Excercises.* Barnett M. 1990 2nd ed. Short and complete versions. Wm. C. Brown Publishers.

37. *Microbiology, Dynamics & Diversity.* Perry J. J. and Staley J. T. 1997. Saunders College Publishing. New York.

38. *Microbiology.* Pelczar M. J., Chan E. C. S and Kreig N. R: 1986. McGraw Hill book Company Singapore.

39. *Microbiology.* Prescott L. M., Harley J. P and Klein D. A. 2002 5th ed. McGraw-Hill; New York.

40. *Microbiology. Principles and Applications.* Black, J.G. 1996. 3rd ed. Prentice Hall. Upper Saddle River, New Jersey.

41. *Microbiology.* Sharma P.D. 1997 2nd ed. Rastogi Publications Meerut. India.

42. *Microbiology.* Tortora G.J., Berdell R. F and Case C L. 1989. Benjamin-Cummings Publishing Company.

43. *Microbiology: An Introduction* Tortora G., Berdell R. F and Case C.L. 2003 Addison Wesley.

44. *Microbiology: An Introduction.* Tortora, G.J., Funke, B. R., Case, C.L. 1995 5th ed. The Benjamin/Cummings Publishing, Co., Inc., Redwood City, CA.

45. *Modern Bacterial Taxonomy.* Priest F.G. 1993 Chapman and Hall. London.

46. *Molecular methods for microbial Identification and Typing.* Toener K. J and Cockayne A. 1993. Chapman and Hall London.

47. *Numerical Taxonomy: The principles and Practice of Numerical Classification.* Sneath, P. H. A., and R. R. Sokal. 1973. W.H Freeman. San Francisco.

48.   *Practical Medical Microbiology.* Mackie and McCartney 1999 14th ed. J.G Collee, Fraser A. G., Marmion B. P., Churchill A. S. (Eds.). Livingstone. NY.

49.   *Principles of Biochemical Tests in Diagnostic Microbiology.* Blazevic D. J and Ederer G. M. 1975. John Wiley and Sons, New York.

50.   *Principles of Microbiology.* Smith A. L. 1984. 10th ed. Times Mirror/Mosby College Publishing, St. Louis. Missouri.

51.   *Staining Procedures.* Clark G (ed.) 1980. 4th ed. Williams and Wilkins Baltimore.

52.   *The Nature of Life.* Postlethwait J.H. and Hopson J.L. 1995. 3rd ed. McGraw-Hill, Inc. New York.

53.   *The Prokaryotes: A Handbook of Bacteria: Ecophysiology, Isolation, Identification, Applications.* Balows A., H. Truper M.D., Harder W. and Schleifer K.H.1991. 2nd ed. Springer–Verlag.

54.   *Yeasts: Characteristics and Identification.* Payne R.W. 1991. Cambridge University Press.

55.   *Viruses.* Levine A. J. 1992. Scientific American Library. NY.

INDEX

A

Abbe-type 46
absolute alcohol 104, 109, 116, 119, 136, 138, 158, 183, 184, 208, 210, 218
absorbance 59, 60, 61, 293
Acetobacter 29, 31, 32, 343
acetone 113, 115, 116, 152, 178, 204, 340
achromatic-aplanatic 46
Achromatium 9, 33
Achromatium oxaliferium 9
acid fast staining 116, 117, 137, 208
Acidaminococcus 32
acidity test 296
Acidophiles or acid-loving 227
Acinetobacter 32, 197, 267
Acinetobacteria 32
acridine orange 50, 120
acridine orange solution 120
Actinomyces 11, 27, 32, 198, 245
Actinomycetes 3, 11, 26, 27, 28, 136, 271, 343, 365
Actinomycetous bacteria 245
Actinoplanes 32
additives 2, 151, 227, 277
aero tolerant 228
aerobic 19, 23, 27, 28, 152, 180, 185, 236, 237, 253, 266, 281
aerobic archaebacteria 32
Aeromonas 32
agar agar 142, 150, 367
agar block 104, 106, 107, 118, 119, 120, 132
agar slant 159, 160, 185, 211, 243, 257
Agostino bassi 5
Agrobacterium 32, 236, 277, 343

Ahring's microbe 2
air curtain 90
air displacement 83
air shower 90
airfuge 79
akinetes 317, 318
Albert solution 121
Albert stain I 121
Albert stain II 121
Alcaligenes 32, 198, 262, 267
Alcaligenes faecalis 198, 262
alcohol test 295
alcohols 103, 240
Alexander Flemming 7
Alician blue reagent 123
Alkaliphiles 227
alkylating agents 103, 240
alpha (α) proteobacteria 33
alpha hemolytic 181, 219
Alysiella 33
amino acids 2, 151, 15, 66, 186, 304, 314
ammonifying bacteria 271, 272, 343
Amoebacter 33
amylase test 147, 188, 385
Amylobacter 31
Anabaena 33, 318
anaerobes 2, 9, 26, 34, 36, 103, 152, 196, 228, 253
anaerobic 19, 26, 30, 95, 96, 146, 152, 180, 213, 229, 253, 318, 328, 344, 371
anaerobic archaebacteria 32
anaerobic culture jar 95
anaerobic spirochaetes 33
anaerobiosis 171, 172